METHODS IN CELL BIOLOGY

VOLUME VI

Contributors to This Volume

RAYMOND M. BAKER

FRANK S. BARNES

T. EREMENKO

GARY J. GRIMES

NELDA B. HOLMGREN

MASAKATSU HORIKAWA

HAJIM KATSUTA

H. A. LUBS

PETER P. LUDOVICI

W. H. MCKENZIE

S. MERRICK

S. R. PATIL

JEAN CLAUDE SCHAER

RICHARD SCHINDLER

FUMIO SUZUKI

TOSHIKO TAKAOKA

LARRY H. THOMPSON

ROBERT A. TOBEY

P. VOLPE

WAYNE WRAY

J. ZALTA

J-P. ZALTA

Methods in
Cell Biology

Edited by

DAVID M. PRESCOTT

DEPARTMENT OF MOLECULAR, CELLULAR AND
DEVELOPMENTAL BIOLOGY
UNIVERSITY OF COLORADO
BOULDER, COLORADO

VOLUME VI

1973

ACADEMIC PRESS • New York and London

A Subsidiary of Harcourt Brace Jovanovich, Publishers

ACADEMIC PRESS, INC.
111 Fifth Avenue, New York, New York 10003

United Kingdom Edition published by
ACADEMIC PRESS, INC. (LONDON) LTD.
24/28 Oval Road, London NW1

LIBRARY OF CONGRESS CATALOG CARD NUMBER: 64-14220

PRINTED IN THE UNITED STATES OF AMERICA

CONTENTS

4. A Method for Measuring Cell Cycle Phases in Suspension Cultures
P. Volpe and T. Eremenko

⋋ 5. A Replica Plating Method of Cultured Mammalian Cells
Fumio Suzuki and Masakatsu Horikawa

6. Cell Culture Contaminants
Peter P. Ludovici and Nelda B. Holmgren

7. Isolation of Mutants of Cultured Mammalian Cells
Larry H. Thompson and Raymond M. Baker

LIST OF CONTRIBUTORS

Numbers in parentheses indicate the pages on which the authors' contributions begin.

RAYMOND M. BAKER, The Ontario Cancer Institute and Department of Medical Biophysics, University of Toronto, Toronto, Canada (209)

FRANK S. BARNES, Department of Electrical Engineering, University of Colorado, Boulder, Colorado (325)

T. EREMENKO, Cell Biology Laboratory of the International Institute of Genetics and Biophysics, Naples, Italy (113)

GARY J. GRIMES, Department of Electrical Engineering, University of Colorado, Boulder, Colorado (325)

NELDA B. HOLMGREN, W. Alton Jones Cell Science Center, Lake Placid, New York (143)

MASAKATSU HORIKAWA, Department of Radiation Biology, Faculty of Pharmaceutical Sciences, Kanazawa University, Kanazawa, Japan (127)

HAJIM KATSUTA, Department of Cancer Cell Research, Institute of Medical Science, University of Tokyo, Takanawa, Tokyo, Japan (1)

H. A. LUBS, Department of Pediatrics, University of Colorado Medical Center, Denver, Colorado (345)

PETER P. LUDOVICI, Department of Microbiology and Medical Technology, University of Arizona, Tucson, Arizona (143)

W. H. McKENZIE, Department of Pediatrics, University of Colorado Medical Center, Denver, Colorado (345)

S. MERRICK, Department of Pediatrics, University of Colorado Medical Center, Denver, Colorado (345)

S. R. PATIL, Department of Pediatrics, University of Colorado Medical Center, Denver, Colorado (345)

JEAN CLAUDE SCHAER, Department of Pathology, University of Bern, Bern, Switzerland (43)

RICHARD SCHINDLER, Department of Pathology, University of Bern, Bern, Switzerland (43)

FUMIO SUZUKI, Department of Radiation Biology, Faculty of Pharmaceutical Sciences, Kanazawa University, Kanazawa, Japan (127)

TOSHIKO TAKAOKA, Department of Cancer Cell Research, Institute of Medical Science, University of Tokyo, Takanawa, Tokyo, Japan (1)

LARRY H. THOMPSON, The Ontario Cancer Institute and Department of Medical Biophysics, University of Toronto, Toronto, Canada (209)

ROBERT A. TOBEY, Biomedical Research Group, Los Alamos Scientific Laboratory, University of California, Los Alamos, New Mexico (67)

P. VOLPE, Cell Biology Laboratory of the International Institute of Genetics and Biophysics, Naples, Italy (113)

WAYNE WRAY, Department of Cell Biology, Baylor College of Medicine, Houston, Texas (283, 307)

J. ZALTA, Laboratoire de Chimie Biologique, Université Paul Sabatier, Toulouse, France (317)

J-P. ZALTA, Laboratoire de Chimie Biologique, Université Paul Sabatier, Toulouse, France (317)

PREFACE

In the ten years since the inception of the multivolume series *Methods in Cell Physiology*, research on the cell has expanded and added major new directions. In contemporary research, analyses of cell structure and function commonly require polytechnic approaches involving methodologies of biochemistry, genetics, cytology, biophysics, as well as physiology. The range of techniques and methods in cell research has expanded steadily, and now the title *Methods in Cell Physiology* no longer seems adequate or accurate. For this reason the series of volumes known as *Methods in Cell Physiology* will now continue under the title *Methods in Cell Biology*.

Volume VI of this series continues to present techniques and methods in cell research that have not been published or have been published in sources that are not readily available. Much of the information on experimental techniques in modern cell biology is scattered in a fragmentary fashion throughout the research literature. In addition, the general practice of condensing to the most abbreviated form materials and methods sections of journal articles has led to descriptions that are frequently inadequate guides to techniques. The aim of this volume is to bring together into one compilation complete and detailed treatment of a number of widely useful techniques which have not been published in full detail elsewhere in the literature.

In the absence of firsthand personal instruction, researchers are often reluctant to adopt new techniques. This hesitancy probably stems chiefly from the fact that descriptions in the literature do not contain sufficient detail concerning methodology; in addition, the information given may not be sufficient to estimate the difficulties or practicality of the technique or to judge whether the method can actually provide a suitable solution to the problem under consideration. The presentations in this volume are designed to overcome these drawbacks. They are comprehensive to the extent that they may serve not only as a practical introduction to experimental procedures but also to provide, to some extent, an evaluation of the limitations, potentialities, and current applications of the methods. Only those theoretical considerations needed for proper use of the method are included.

Finally, special emphasis has been placed on inclusion of much reference material in order to guide readers to early and current pertinent literature.

DAVID M. PRESCOTT

CONTENTS OF PREVIOUS VOLUMES

Volume I

Volume II

Volume III

Volume IV

Volume V

Chapter 1

Cultivation of Cells in Protein- and Lipid-Free Synthetic Media

HAJIM KATSUTA AND TOSHIKO TAKAOKA

*Department of Cancer Cell Research, Institute of Medical Science,
University of Tokyo, Takanawa, Tokyo, Japan*

I. Significance of Culturing Cells in Synthetic Media

Protein-free, chemically defined synthetic media are of great use in analyzing the chemical nature of cultured cells and of the metabolites excreted from them as well as in examining the effect of certain substances on cells, especially in cases where the substances may first interact with serum proteins in the medium. A number of studies have been carried out along these lines, as will be discussed later. However, a very limited number of kinds of cell lines have been grown indefinitely in protein-free media.

1

II. Composition of Chemically Defined Synthetic Media

Many mixtures of synthetic media have been reported. Table I lists the names of mixtures and authors of synthetic media, so far as known to us, which have been designed for the cultivation of mammalian cells. Different kinds of cells naturally have different nutritional requirements. In addition, our knowledge of cell metabolism is very limited. These are among the reasons why descriptions of so many mixtures have been published and not every kind of cell has as yet been serially grown in such media, especially in the primary culture.

Some of these mixtures were designed theoretically, all others empirically. Our work on synthetic media stemmed from our question of whether high molecular weight substances in media might be essential for cells as a nutritional source. By the use of primary culture of rat ascites hepatoma AH-130 cells, we estimated the consumption of serum proteins in the medium by the cells following cultivation. However, little decrease was detected in the amount of proteins. The addition of ^{131}I-labeled serum proteins also revealed little incorporation of the proteins into cells. These findings showed that proteins presumably do not serve cells as nutritional substances. We tried to replace serum proteins in the medium with other high molecular weight substances, especially with so-called plasma expanders, which were developed during the Second World War; alginic acid, dextran, and polyvinylpyrrolidone (PVP). All of them were more or less effective. However, the highest efficiency of substitution was obtained with 0.1% PVP (K-90, average molecular weight 700,000; Badische Anilin- und Soda-Fabrik, West Germany), which replaced approximately 99.5% by volume of serum proteins added to the medium in the optimal concentration (Katsuta et al., 1959a). This result was introduced into the cultivation of L-929 cells in protein-free media. When transferred to medium consisting of 0.05% PVP, 0.4% lactalbumin hydrolyzate (NBCo, U.S.A.), and 0.08% yeast extract (Difco Lab., U.S.A.) with no supplement of serum, the proliferation of L-929 cells was readily initiated and has continued up to the present (Katsuta et al., 1959b). This subline was designated as L·P1. In parallel to this cultivation, other trials were made with other mixtures of protein-free media consisting of (1) PVP and lactalbumin hydrolyzate and (2) lactalbumin hydrolyzate alone. Continuous cell growth in these media was eventually obtained; the sublines grown in (1) and (2) were designated as L·P2 and L·P4, respectively (Katsuta et al., 1961).

Since L-929 cells were demonstrated, as above, to be capable of grow-

TABLE I
LIST OF SYNTHETIC MEDIA AND THEIR AUTHORS

References	Medium
Baker, 1936	For fibroblasts, epithelial cells, and monocytes
Biggers and Lucy, 1960	BL 1
Biggers et al., 1961	BGJ a and b
Dubreuil and Pavilanis, 1958	SN-3
Eagle, 1955	Basal medium
Eagle, 1959	Minimum essential medium
Dulbecco and Freeman, 1959	Modified Eagle's medium
Evans et al., 1956a	NCTC 107
Evans et al., 1956b	NCTC 108
McQuilkin et al., 1957	NCTC 109
Evans et al., 1958	NCTC 110, 111, 112
Earle, 1962	NCTC 117
Sanford et al., 1963	NCTC 113, 118, 119, 120, 121, 125, 126, 127, 129, 130
Evans et al., 1964	NCTC 131, 132, 133, 135
Price et al., 1966	NCTC 138, 139, 140
Fischer et al., 1948	V-605, 612, 614
Fischer and Sartorelli, 1964	For murine leukemia cells
Garvey, 1961	For reticuloendothelial cells
Ham, 1963	F10
Ham, 1965	F12
Holmes, 1959	A
Holmes and Wolfe, 1961	A2
Katsuta and Takaoka, 1960	DM-11 $\sim$ -63
Katsuta et al., 1961	DM-114, -120
Takaoka and Katsuta, 1971	DM-120, -145
Hosokawa et al., 1971	DM-135, -144
Kelley et al., 1960	SRI-8
Leibovitz, 1963	L-15, L-15G
Ling et al., 1968	7C
Neuman and McCoy, 1956	Medium 2
McCoy et al., 1956	Medium 3
Neuman and McCoy, 1958	For Walker 256 cells
McCoy et al., 1959	5a
Neuman and Tytell, 1960	For Walker 256 cells
Tritsch and Moore, 1962	213
Hanss and Moore, 1964	RPMI 213, 1311
Iwakata and Grace, 1964	RPMI 1629
Moore et al., 1967	RPMI 1640
Moore et al., 1968	RPMI 1934
Moore and Kitamura, 1968	RPMI 1603
Mizrahi et al., 1972	RPMI 1701
Morgan et al., 1950	199
Healy et al., 1954	703
Healy et al., 1955	858
Morgan et al., 1955	M 150

(Continued)

TABLE I (*Continued*)

References	Medium
Morgan *et al.*, 1956	M 416
Parker, 1961	CMRL-1066
Healy and Parker, 1966	CMRL-1415
Paul, 1959	HERT-1
Marcus *et al.*, 1956	For human cells
Sato *et al.*, 1957	For human cells
Puck *et al.*, 1957	For human cells
Puck *et al.*, 1958	For human cells
Puck, 1961	For HeLa cells
Rappaport, 1956	SM-1
Melnick *et al.*, 1957	SM-2
Rappaport *et al.*, 1960	SM-3
Rosenberg and Kirk, 1953	S-27
Haff and Swim, 1957a	S16, S18
Haff and Swim, 1957b	73
Swim and Parker, 1958a	S 103
Swim and Parker, 1958b	705
Scherer, 1953	HeLa maintenance medium
Ginsberg *et al.*, 1955	HeLa maintenance medium
Trowell, 1955	TAPCI
Trowell, 1959	T8
Waymouth, 1955	A, B, C
Waymouth, 1956	ML 97, 94/4, 192/2
Waymouth, 1959	MB 752/1
Kitos *et al.*, 1962	MD 705/1, ML 838/1
Waymouth, 1965b	MAB 87/3
White, 1946	For chick cells
White, 1949	For chick cells
White, 1955	W_{II}

ing in medium containing lactalbumin hydrolyzate alone in buffered saline, we analyzed the amino acid composition of this material and designed the mixture DM-11 of synthetic medium (Kagawa *et al.*, 1960). The amino acid levels in DM-11 were the same as those in 0.4% lactalbumin hydrolyzate, which concentration was previously confirmed to be optimum for cell growth. As to the content of vitamins, we briefly followed that of MB 752/1 (Waymouth, 1959). We also estimated the amount of each amino acid consumed by the cells during a few days of cultivation in DM-11 and designed the mixture DM-12 following the consumption rates. After that time, we became engaged in the estimation of the optimum concentration of each amino acid until the mixture DM-63 was achieved (Katsuta and Takaoka, 1960). When we turned to the analysis of vitamin requirements, i.e., from mixtures 64 to 96, we encountered much difficulty. It took an enormous amount of time to determine the effect of vitamins on cell growth, and we gave up the investigation. The mixtures from DM-97 to DM-148 were designed for determining the requirements for amino acids and other components including inositol (Katsuta *et al.*, 1961; Takaoka and Katsuta, 1971; Hosokawa *et al.*, 1971). The subline derived from L·P1 which has been serially grown in a protein-free chemically defined synthetic medium in the absence of macromolecular substances was designated as L·P3.

Table II illustrates the composition of synthetic media which have been reported to sustain continuous and serial growth of cells with no supplement of proteins or macromolecular substances. This table was prepared by modification of that described by Waymouth (1965a). It is also useful to refer to the review by Morton (1970).

Unexpectedly, in culturing L·P4 cells, we found that there were many differences in the growth-promoting activity among the commercial lots of lactalbumin hydrolyzate. The addition, to the medium, of certain amino acids which were found to be insufficient in amount by amino acid analysis of the hydrolyzate, restored the growth of L·P4 cells (Takaoka *et al.*, 1960b).

III.   Cell Lines Serially Grown in Synthetic Media

A very limited number of cell lines have been grown serially in protein-free synthetic media. Table III gives the cell lines which have been grown continuously in our laboratory in protein- and lipid-free synthetic media. The inclusion of "P3" in the names of cell lines indicates those cultured in the synthetic media containing so-called nonessential amino acids in addition to essential amino acids, and "P5" designates those grown in the media containing no nonessential amino acids. Most of the latter lines were found to be capable of growing in Eagle's MEM alone as well. Some of the lines, e.g., JTC-16·P3, JTC-21·P3, and RSP-2·P3 have not been adapted to the media lacking nonessential amino acids.

It was interesting that most of the lines transformed by Nagisa culture, so far as examined, readily proliferated in the synthetic media. "Nagisa culture" represents the cultivation of cells for a long period without subculturing, e.g., for a few months or more, in tubes with flattened surfaces. When the tubes were incubated, being kept slanted at 5° in stationary culture, marked changes appeared in the morphology of cells scattered along the zone nearest to the air-liquid interface, named Nagisa, resulting in the production of mutant cell lines (Katsuta *et al.*, 1965; Katsuta and Takaoka, 1968). Most of these mutant cell lines were

TABLE II

COMPOSITION OF SYNTHETIC MEDIA COMMONLY USED FOR SERIAL CULTIVATION OF MAMMALIAN CELLS (mg/liter)[a]

	Eagle (1959)[b] Minimal essential medium	McQuilkin et al. (1957)[c] NCTC 109	Fischer and Sartorelli (1964)[d]	Ham (1965)[e] F12	Healy et al. (1955) 858	Katsuta et al. (1961)[f] DM-120	Neuman and Tytell (1960)[g]	Waymouth (1965b)[h] MAB 87/3
Amino acids (L-form)								
Alanine	—	31.48	—	8.91	25	400	8.91	11.2
α-Aminobutyric acid	—	5.51	—	—	—	—	—	—
Arginine	105	—	—	—	70	100	—	—
Arginine·HCl	—	31.16	15	210.70	—	—	42.14	75
Asparagine	—	—	10	13.21	—	—	—	24
Asparagine·H₂O	—	9.19	—	—	—	—	45.03	—
Aspartic acid	—	9.91	—	13.31	30	25	13.31	60
Cysteine	—	—	—	—	260	—	24.23	—
Cysteine·HCl	—	260	—	31.53	—	80	—	90
Cystine	24	10.49	20	—	20	—	—	15
Glutamic acid	—	8.26	—	14.71	75	150	14.71	150
Glutamine	292	135.73	200	146.15	100	100	219.23	350
Glycine	—	13.51	—	7.51	50	15	7.51	50
Histidine	31	—	60	—	20	30	—	—
Histidine·HCl	—	—	—	19.17	—	—	—	—
Histidine·HCl·H₂O	—	26.65	—	—	—	—	20.96	150
Hydroxy-L-proline	—	4.09	—	—	—	—	13.11	—
Isoleucine	52	18.04	75	3.94	10	150	39.35	25
Leucine	52	20.44	30	13.12	20	400	39.35	50
Lysine	58	—	—	—	70	100	—	—
Lysine·HCl	—	38.43	50	36.54	—	—	36.53	240
Methionine	15	4.44	100	4.48	15	80	14.92	50
Ornithine·HCl	—	9.41	—	—	—	—	—	—

Phenylalanine	32	16.53	60	4.96	25	80	16.52	50
Proline	—	6.13	—	34.54	40	12	11.51	50
Serine	—	10.75	15	10.51	25	80	26.27	12.8
Taurine	—	4.18	—	—	—	—	—	—
Threonine	48	18.93	40	11.91	30	100	11.91	75
Tryptophan	10	17.50	10	2.04	10	40	6.13	40
Tyrosine	36	16.44	60	5.44	40	50	18.12	40
Valine	46	25.00	70	11.72	25	85	11.72	65
Carbohydrates and related compounds								
Ascorbic acid	—	50	—	—	50	40	0.5	17.5
Sodium acetate	—	—	—	—	50	—	—	—
Sodium acetate·3H$_2$O	—	50	—	—	—	—	—	—
Ethyl alcohol	—	40	—	1 ml	16	—	600	—
Galactaric acid	—	—	—	—	—	—	21.01	—
d-Glucosamine·HCl	—	3.85	—	—	—	—	—	—
Glucose	1000	1000	1000	1801.6	1000	1000	3000	5000
Sodium glucuronate	—	—	—	—	4.2	—	—	—
Sodium glucuronate·H$_2$O	—	1.8	—	—	—	—	—	—
d-Glucuronolactone	—	1.8	—	—	—	—	—	—
Inositol	2.0	—	—	—	0.05	—	9.0	—
i-Inositol	—	0.125	1.5	—	—	—	—	—
myo-Inositol	—	—	—	18.02	—	—	—	—
myo-Inositol·2H$_2$O	—	—	—	—	—	—	—	1.0
Pyruvate	—	—	—	—	—	—	88.06	—
Sodium pyruvate	—	—	—	110.05	—	—	—	—
Nucleotide precursors								
Deoxyadenosine	—	10	—	—	10	—	—	—
Deoxycytidine	—	—	—	—	10	—	—	—
Deoxycytidine·HCl	—	10	—	—	—	—	—	—

(Continued)

TABLE II (Continued)

	Eagle (1959)[b] Minimal essential medium	McQuilkin et al. (1957)[c] NCTC 109	Fischer and Sartorelli (1964)[d]	Ham (1965)[e] F12	Healy et al. (1955) 858	Katsuta et al. (1961)[f] DM-120	Neuman and Tytell (1960)[g]	Waymouth (1965b)[h] MAB 87/3
Deoxyguanosine	—	10	—	—	10	—	—	—
Hypoxanthine	—	—	—	4.08	—	—	—	25
5-Methylcytosine	—	0.1	—	—	—	—	—	—
5-Methyldeoxycytidine	—	—	—	—	0.1	—	—	—
Thymidine	—	10	—	0.727	10	—	—	8.0
Uridine triphosphate	—	1.0	—	—	1.0	—	—	—
Metabolic cofactors and vitamins								
p-Aminobenzoic acid	—	0.125	—	—	0.05	—	—	—
Biotin	—	0.025	0.01	0.00733	0.01	0.002	0.2	0.02
Calciferol	—	0.25	—	—	0.1	—	—	—
Choline	1.0	—	—	—	0.5	—	—	—
Choline chloride	—	1.25	1.5	13.96	1.0	250	5.0	250
Cocarboxylase (TPP)	—	1.0	—	—	2.5	—	—	—
Coenzyme A	—	2.5	—	—	7.0	—	—	—
DPN	—	7.0	—	—	1.0	—	—	—
FAD	—	1.0	—	—	0.01	—	—	—
Folic acid	1.0	0.025	10	1.32	0.01	0.01	2.0	0.5
Folinic acid	—	—	—	—	—	—	0.1	—
Lipoic acid	—	—	—	0.206	—	—	—	—
Menadione	—	0.025	—	—	0.01	—	—	—
Nicotinic acid	—	0.0625	—	—	—	—	—	—
Nicotinamide	1.0	0.0625	0.5	0.0366	—	5.0	0.5	1.0
Pantothenic acid	1.0	—	—	—	—	1.0	—	—
Calcium pantothenate	—	0.025	0.5	0.258	—	—	0.2	1.0

	1	2	3	4	5	6	7	8
Pyridoxal	1.0	—	—	—	0.025	—	—	—
Pyridoxal·HCl	—	0.0625	0.5	—	—	—	0.5	—
Pyridoxine	—	—	—	—	0.025	—	—	—
Pyridoxine·HCl	—	0.0625	0.5	—	—	1.0	0.2	1.0
Riboflavin	0.1	0.025	—	0.0617	—	—	0.5	1.0
Thiamine	1.0	—	—	0.0376	—	10	—	—
Thiamine·HCl	—	0.025	1.0	0.337	—	—	—	10
α-Tocopherol phosphate	—	—	—	—	0.01	—	—	—
α-Tocopherol phosphate·Na₂	—	0.025	—	—	—	—	—	—
TPN	—	1.0	—	—	1.0	—	—	—
Vitamin A	—	0.25	—	—	0.1	—	—	—
Vitamin B₁₂	—	10.0	—	1.36	—	0.005	0.00075	0.2
Lipidic compounds and detergents								
Cholesterol	—	—	—	—	0.2	—	—	—
Linoleic acid	—	—	—	0.0841	—	—	—	—
Methyl oleate	—	—	—	—	—	—	15	—
Tween 80	—	12.5	—	—	5.0	—	15	—
Peptides, proteins, and amines								
Glutathione	—	10	—	—	10	—	0.5	15
Glutathione·Na	—	—	—	—	—	—	—	—
Insulin	—	—	—	—	—	—	1.0	8
Putrescine dihydrochloride	—	—	—	0.161	—	—	—	—
Salmine sulfate	—	—	—	—	—	—	5.0	—
pH Indicator, antibiotics								
Phenol red	—	20	5.0	1.17	20	—	2.5	—
n-Butyl p-hydroxybenzoate	—	—	—	—	0.2	—	—	—
Penicillin	—	—	250	—	—	—	—	—
Sodium penicillin G	—	—	50	—	1.0	—	—	—
Streptomycin	—	—	—	—	—	—	50	—
Streptomycin sulfate	—	—	—	—	—	—	50	—
Dihydrostreptomycin sulfate	—	—	—	—	100	—	—	—

(Continued)

TABLE II (Continued)

Salts	Eagle (1959)[b] Minimal essential medium	McQuilkin et al. (1957)[c] NCTC 109	Fischer and Sartorelli (1964)[d]	Ham (1965)[e] F12	Healy et al. (1955) 858	Katsuta et al. (1961)[f] DM-120	Neuman and Tytell (1960)[g]	Waymouth (1965b)[h] MAB 87/3
NaCl	6800	6800	8000	7599	6800	8000	6460	6000
KCl	400	400	400	223.65	400	200	400	150
CaCl₂	200	200	—	—	200	264	200	—
CaCl₂·2H₂O	—	—	91	44.11	—	—	—	120
MgSO₄	—	100	—	—	—	—	—	—
MgSO₄·7H₂O	200	—	100	—	200	100	200	100
MgCl₂·6H₂O	—	—	60	122.00	—	—	—	240
Na₂HPO₄	—	—	—	—	—	—	140	300
Na₂HPO₄·7H₂O	—	—	—	268.09	—	—	—	—
Na₂HPO₄·12H₂O	—	—	—	—	—	35	—	—
NaH₂PO₄·H₂O	150	140	69	—	140	—	—	—
NaH₂PO₄·2H₂O	—	—	—	—	—	177	—	208
KH₂PO₄	—	—	—	—	—	—	—	—

NaHCO₃	2000	2200	1125	1176	2200	1000	2200	2200	2240
CoCl₂·6H₂O	—	—	—	—	—	—	—	—	0.022
CuSO₄·5H₂O	—	—	—	0.00249	—	—	—	—	0.05
FeSO₄	—	—	—	—	—	—	—	—	0.45
FeSO₄·7H₂O	—	—	—	0.834	—	—	—	—	—
Fe(NO₃)₃	—	—	—	—	0.1	—	—	—	—
MnSO₄·H₂O	—	—	—	—	—	—	—	—	0.016
(NH₄)₆Mo₇O₂₄·4H₂O	—	—	—	—	—	—	—	—	0.025
ZnSO₄·7H₂O	—	—	—	0.863	—	—	—	—	0.03

^{a} Chemically defined components only are presented in this table. As to the other components, see below.

^{b} This synthetic mixture was originally supplemented with 5 to 10% whole or dialyzed serum.

^{c} The chemical composition of NCTC 135 is the same as that of NCTC 109 except that cysteine·HCl is eliminated in NCTC 135.

^{d} This synthetic mixture was originally supplemented with 10% horse serum.

^{e} The chemical composition of F12M is the same as that of F12 except that the concentration of $ZnSO_4 \cdot 7H_2O$ is reduced to 5×10^{-7} M.

^{f} The chemical composition of DM-145 is the same as that of DM-120 except that inositol, 2 mg/liter, is added in the former. Trace elements: Fe 0.104–0.108, Cu 0.04–0.06, Mn 0.03.

^{g} This synthetic mixture is originally supplemented with lactalbumin digested with pancreatin (EDAMIN, DR, 2000 mg/liter).

^{h} For liver cell culture, 0.2 mg/liter hydrocortisone hemisuccinate is added to the medium. The chemical composition of MB 752/1 is the same as that of MAB 87/3 except that MB 752/1 does not contain $FeSO_4$, $CuSO_4 \cdot 5H_2O$, $MnSO_4 \cdot H_2O$, $ZnSO_4 \cdot 7H_2O$, $(NH_4)_6Mo_7O_{24} \cdot 4H_2O$, $CoCl_2 \cdot 6H_2O$, insulin, serine, alanine, and asparagine.

TABLE III

CELL LINES SERIALLY GROWN IN PROTEIN- AND LIPID-FREE SYNTHETIC MEDIUM IN OUR LABORATORY

Name of subline established	Former name of line	Cells of origin	Initiation of culture	Production of mutant line by "Nagisa"	Transfer to synthetic medium	Synthetic medium
L·P3	L-929	Mouse fibroblasts	Jul. 5, 1955[b]	—	Sept. 30, 1959	DM-120
L·P5	L·P3			—	Jan. 19, 1971	DM-144
HeLa·P3	HeLa	Human uterus carcinoma	Nov. 11, 1958[b]	—	June 6, 1968	DM-145
JTC-21·P3	RLH-1·P3	Rat liver parenchymal cells	Nov. 15, 1962	Feb. 21, 1964	May 15, 1969	DM-145
JTC-22·P3[a]	RLH-2·P3	Rat liver parenchymal cells	Nov. 15, 1962	Apr. 19, 1964	May 15, 1969	DM-145
JTC-23·P3	RLH-3·P3	Rat liver parenchymal cells	Feb. 4, 1963	Nov. 16, 1964	Feb. 1, 1969	DM-120
JTC-23·P5	JTC-23·P3				Jan. 21, 1971	DM-144
JTC-24·P3[a]	RLH-4·P3	Rat liver parenchymal cells	Nov. 15, 1962	Dec. 25, 1964	May 15, 1969	DM-145
JTC-25·P3	RLH-5·P3	Rat liver parenchymal cells	Jul. 21, 1965	Dec. 25, 1966	Oct. 15, 1968	DM-120
JTC-25·P5	JTC-25·P3				Jan. 19, 1971	DM-144
JTC-20·P3	RTH-1·P3	Rat thymus reticulum cells	Apr. 17, 1965	Nov. 13, 1966	Apr. 24, 1969	DM-120
JTC-20·P5	JTC-20·P3				Nov. 17, 1969	DM-144
JTC-12·P3	JTC-12	Monkey kidney cells	Mar. 14, 1961		May 15, 1969	DM-145
JTC-12·P5	JTC-12·P3				Jan. 21, 1971	DM-144
JTC-16·P3	JTC-16	Rat ascites hepatoma cells	Mar. 8, 1964		Apr. 30, 1969	DM-145
T·P5	T	Rat liver parenchymal cells	June 30, 1963	Apr. 30, 1965	Oct. 27, 1969	DM-145
CQ-11·P5	RSC-1	Rat subcutaneous fibroblasts	Feb. 6, 1967	—	Oct. 27, 1967	DM-145
RLT-6·P5	RLT-6	Rat liver parenchymal cells	Nov. 29, 1968	—	Oct. 27, 1969	DM-145
RSP-2·P3	RSP-2	Rat spleen plasma cells	Jul. 21, 1965	—	Oct. 22, 1969	DM-145
MDCK·P3	MDCK	Dog kidney cells	Oct. 1, 1970[b]	—	Jan. 18, 1971	DM-145

[a] Lost on August 2, 1972, by an accident with a deep freezer.
[b] Date of the initiation of culture in our laboratory.

found to be capable of readily and rapidly growing in protein- and lipid-free synthetic media. There might be a connection between the change in cell properties caused by Nagisa culture and the ability to grow in synthetic media. However, the mechanism is still obscure.

The line T·P5 was derived from rat liver parenchymal cells, line RLC-5. The cells were cultivated in Nagisa culture and were further treated with 4-dimethylaminoazobenzene (DAB). The cells transformed by these treatments were found to be very resistant to DAB (Katsuta and Takaoka, 1968). They were cultured in DM-145 from October 27, 1969, but transferred to DM-144 on January 21, 1971. CQ-11·P5 has a history of being treated with 4-nitroquinoline 1-oxide (4NQO). RLT-6·P5 was derived from rat liver parenchymal cells, line RLC-10. It was transformed into malignant cells by 4NQO-treatment and later transferred to a synthetic medium (H. Katsuta and T. Takaoka, unpublished).

Some of the cell lines which have been grown in other laboratories in protein-free synthetic media are given in Table IV. There may be other cell lines serially grown in such media but the authors have no knowledge of these. It is especially valuable that three lines have been cultured from direct explants in synthetic media in the laboratory of Evans.

IV.   Initiation of Cell Growth in Synthetic Media

To achieve the continuous growth of mammalian cells in protein-free synthetic media, there are many methods of culture; this reflects the fact that many kinds of cells show different responses. These will be mentioned following our experience obtained in establishing the cell lines which are illustrated in Table III.

Briefly, we found three kinds of cell lines; (1) lines which grow readily and rapidly as soon as they are transferred to the synthetic media, (2) those which grow after some period of adaptation, and (3) those which require a very prolonged period of adaptation.

A.   Prompt Initiation of Growth without Special Treatment

When we transfer cells from their growth medium to another medium, including natural media, the cells do not always grow rapidly in the new medium. They appear to need some period of adaptation to the new environment. As described previously, however, most of the cell lines transformed by Nagisa culture so far examined, readily initiated vigorous proliferation when transferred from serum media to protein-free

TABLE IV

Some of the Cell Lines Serially Grown in Protein-Free Synthetic Media in Other Laboratories

A. National Cancer Institute (Evans et al., 1963, 1964; Sanford et al., 1963; Andresen et al., 1967; Price et al., 1966)

NCTC Cell line number	Cell origin	Species	Growth in NCTC media[a]							
			NCTC 109	NCTC 135	NCTC 117	NCTC 126	NCTC 131	NCTC 132	NCTC 133	
2071	Line L	C3H mouse	+	+	+	+	+	+	+	Adapted
3681	Fibroblast	C3H mouse embryo	+	+	+	−	−	−	+	Adapted
4247	Direct explant	C3H mouse embryo	+	+	O	O	O	O	O	Direct explant
3749	MCA-induced lymphoma	DBA/2 mouse	+	+	+	+	+	O	O	Adapted
3954	Parotid and sub-maxillary	C3H mouse	+	+	+	−	−	+	+	Adapted
4075	Parotid and sub-maxillary	C3H mouse	+	+	+	−	−	+	+	Adapted
4206	Fibroblast of peritoneal cavity sarcoma	Chinese hamster	+	+	+	−	−	−	+	Adapted
3526	Pooled kidney cells	Rhesus monkey	+	+	+	+	+	O	O	Adapted

No.	Cell type	Origin								Status
3075	Skin epithelium	Human	+	+	+	+	+	O	O	Adapted
3354	Epithysis of femur	Human fetus	+	+	+	+	+	O	O	Adapted
3952	Cervical carcinoma (HeLa)	Human	+	+	+	−	−	−	+	Adapted
4067	Liver parenchyma	C3H mouse	+	+	+	−	−	−	−	Adapted
4242	Liver parenchyma	C3H mouse	+	+	+	+	O	O	O	Adapted
4705	Minced embryo	C3H mouse	O	+	+	−	−	−	+	Direct explant
5009	Minced embryo	C3H mouse	O	O	O	O	O	O	O	Direct explant
4952	Kidney	African green monkey	O	+	+	+	O	O	O	Adapted

a Key to symbols: + = Cultures grew serially. − = Cultures did not grow serially. O = Not tested.

B. Jackson Laboratory (Waymouth, 1965b; Waymouth *et al.*, 1971; C. Waymouth, 1972, personal communication)

Cell strain	Origin	Medium	Supplement
L-929	Mouse fibroblasts	MD 705/1	—
AB 87	Mouse lung cells	MAB 87/3	—
AB 163	Mouse lung cells	MAB 87/3	—
FL 83B	Mouse liver cells	MAB 87/3	0.2 µg/ml Hydrocortisone hemisuccinate
FL 83BNH	(A subline of FL 83B)	MAB 87/3	—
FL 88	Mouse liver cells	MAB 87/3	0.2 µg/ml Hydrocortisone hemisuccinate
	Mouse liver parenchymal cells	MAB 87/3	Dexamethasone

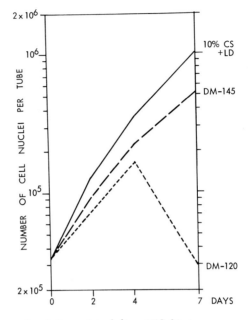

FIG. 1. The growth of the original line JTC-21 in a serum medium and in synthetic media, DM-120 and DM-145.

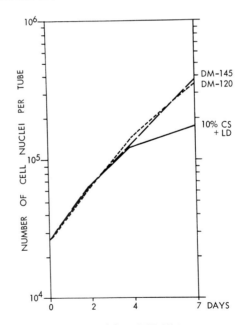

FIG. 2. The growth of the original line JTC-25 in a serum medium and in synthetic media, DM-120 and DM-145.

media. Two examples are shown in Figs. 1 and 2. Figure 1 shows growth curves of JTC-21 cells when they were subcultured into the synthetic media. In DM-120 containing no inositol, they failed to grow continuously. In DM-145 containing 2 mg/liter inositol, however, the cells readily initiated growth and continued to proliferate for a week. On the other hand, interestingly, JTC-25 cells proliferated more rapidly in DM-120 and DM-145 than in growth medium containing calf serum, in which they had been serially cultured for a long time, as shown in Fig. 2. This was also observed in the case of JTC-20.

B. Gradual Adaptation to Synthetic Media with No Supplement

When transferred to synthetic media, the lines L-929 and JTC-24 required some period of adaptation. For the adaptation of JTC-16 and MDCK, about half a year was required.

Two types of lines have been observed to belong to this group: (1) Some lines, e.g., JTC-16, showed cell proliferation, while a certain number of cells were dying in the same culture, as confirmed by cinemicrography, resulting in little net increase in cell number in the population. Net growth, however, was eventually obtained after a certain period of adaptation; (2) Other lines did not grow at all when they were subcultured to synthetic media, but they maintained cell survival for a long time and finally began to grow. The lines JTC-12 and MDCK belong to this group.

C. Very Gradual Adaptation with Decrease in Concentration of Macromolecular Substances

In some cell lines, it is necessary to decrease gradually the concentration of serum proteins in the medium in order to adapt the cells to protein-free, synthetic media. Following the decrease in serum concentration, the rate of cell growth also decreases as a matter of course. When the cells acquire an unhealthy morphology, it is necessary to increase the level of serum again. It is very important always to watch the morphological changes in cells during cultivation.

We occasionally found that the use of PVP in place of proteins contributed significantly to the adaptation of cells to protein-free media (Katsuta et al., 1959a,b). This has been confirmed also by Yasumura (1962, 1963). Such macromolecular substances may presumably play a role in balancing the colloid osmotic pressures in the medium. By such a gradual decrease in serum concentration or a replacement with PVP, some lines have eventually been adapted to synthetic media. The line HeLa was adapted, surprisingly, after 8 years of cultivation in the medium containing PVP in place of proteins (Katsuta *et al.*, 1960; Takaoka *et al.*, 1960a; Takaoka and Katsuta, 1971).

V. Biomorphological Changes Induced in Cells by Transfer to Synthetic Media

When the sublines established in protein-free media were compared in cell morphology with the original lines or sublines, they could readily be divided into three groups: (1) A group considerably changed in morphology. JTC-21·P3 (Figs. 3, 4) and JTC-24·P3 do not adhere tightly to the glass surface but grow while maintaining a spherical shape. In contrast, JTC-16·P3 has altered to close adherence to the glass surface and shows higher adhesiveness between cells (Figs. 5, 6). (2) A group changed slightly. L·P3 (Fig. 7), JTC-23·P3 (Figs. 8, 9), JTC-25·P3 (Figs. 10, 11), and JTC-20·P3 exhibit no marked morphological changes. In JTC-22·P3, cells with spherical shapes have increased slightly in number. HeLa·P3 belonged to the first group in the early stage, many cells showing spherical shape, but it has come recently to resemble the original line in cell morphology (Fig. 12). (3) A group that is scarcely changed. No evident alteration has been found in the morphology of JTC-12·P3 (Figs. 13, 14).

When observed by time-lapse cinemicrography, L·P3 showed many mitoses but few multipolar ones. Cell locomotion was scarcely detectable. Most of the cells were spread with long cytoplasmic projections before the formation of a confluent cell sheet. JTC-21·P3 exhibited many instances of cell division. Most of the cells were spherical in shape. They showed no locomotion but a high frequency of piling up. In JTC-22·P3, the cells had spherical shapes. A considerable number of cell divisions were observed including abnormal mitosis or cytoplasmic fusion of the two daughters after mitosis. No locomotion was detected. JTC-23·P3 also showed a large number of mitoses and tripolar mitoses. Active pinocytosis was observed but no locomotion occurred. In JTC-24·P3, mitosis was found with a high rate of frequency. Most of the cells extended cytoplasmic projections in the beginning of cinemicrography but became spherical later, showing a marked piling up. Locomotion was scarcely exhibited. Mitosis was also abundant in JTC-25·P3. A few slender cytoplasmic projections extended from most of the cells. No locomotion was observed. JTC-20 showed many instances of mitosis. The cells were

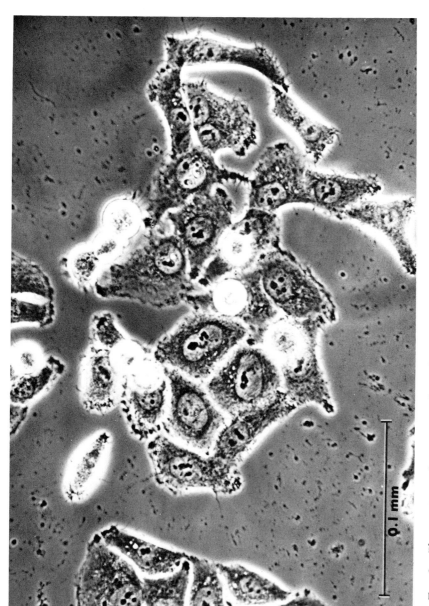

Fig. 3. Phase-contrast photomicrograph of the original line JTC-21 in a serum medium (rat liver parenchymal cells transformed by Nagisa culture). Most of the cells have adhered to the glass surface, spreading their cytoplasm. A few cells have piled up in some colonies and show spherical shapes.

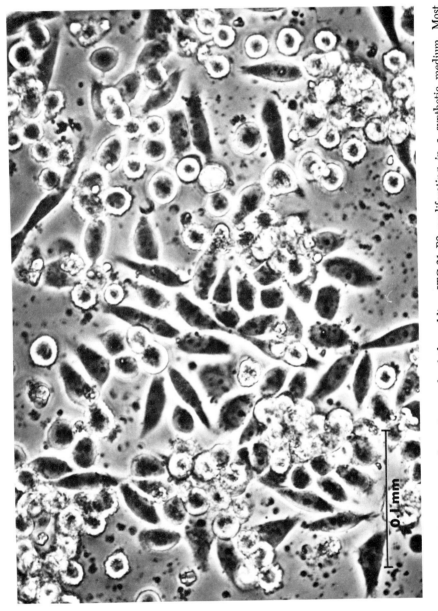

FIG. 4.   Phase-contrast photomicrograph of the subline JTC-21·P3 proliferating in a synthetic medium. Most of the cells do not adhere tightly to the glass surface and do not spread their cytoplasm.

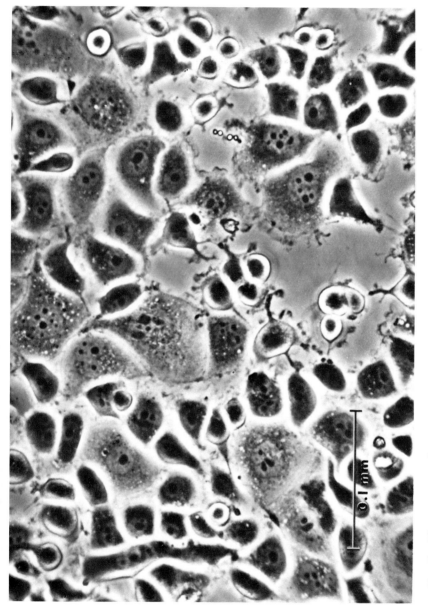

Fig. 5. Phase-contrast photomicrograph of the original line JTC-16 in a serum medium of rat ascites hepatoma AH-7974 cells. Some of them have piled up on the cell sheet and show spherical shapes. Cells in the sheet do not adhere very tightly to each other.

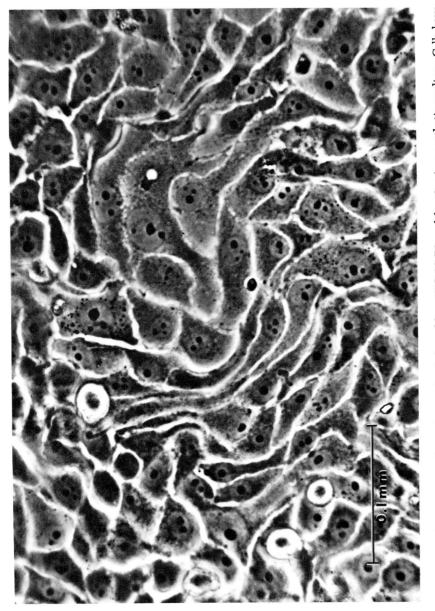

FIG. 6. Phase-contrast photomicrograph of the subline JTC-16·P3 proliferating in a synthetic medium. Cells have tightly adhered to glass surface, showing higher adhesiveness between cells than do the cells of the original line.

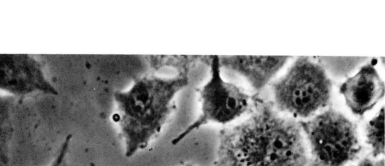

FIG. 7.   Phase-contrast photomicrograph of the subline L·P3 growing in a synthetic medium. Some of the cells have stretched cytoplasmic projections. No marked difference is observed in morphology between this subline and the original line L-929 (not shown).

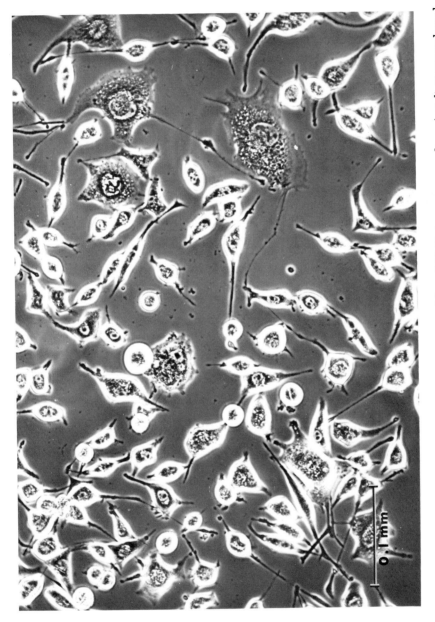

Fig. 8. Phase-contrast photomicrograph of the original line JTC-23 in a serum medium (rat liver parenchymal cells transformed by Nagisa culture). Many cells are found with stretched cytoplasmic projections, and others show spherical shapes.

0.1 mm

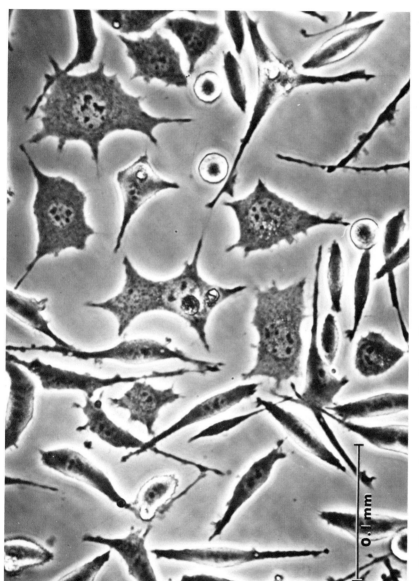

FIG. 9. Phase-contrast photomicrograph of the subline JTC-23·P3 in a synthetic medium. In general morphology, the cells resemble those of the original line, except that fewer cells show spherical shapes.

0.1 mm

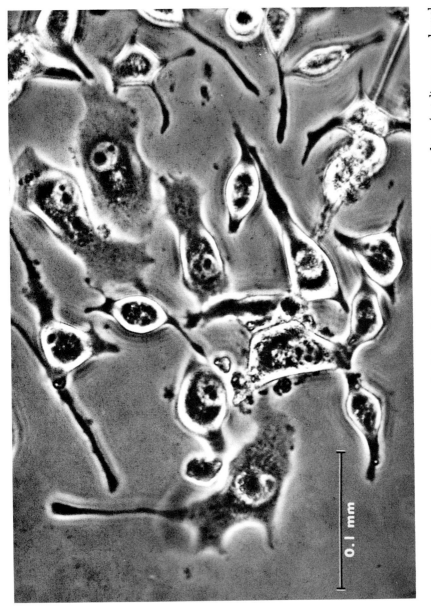

FIG. 10.   Phase-contrast photomicrograph of the original line JTC-25 in a serum medium (rat liver parenchymal cells transformed by Nagisa culture). The cells are in an early stage after subcultivation.

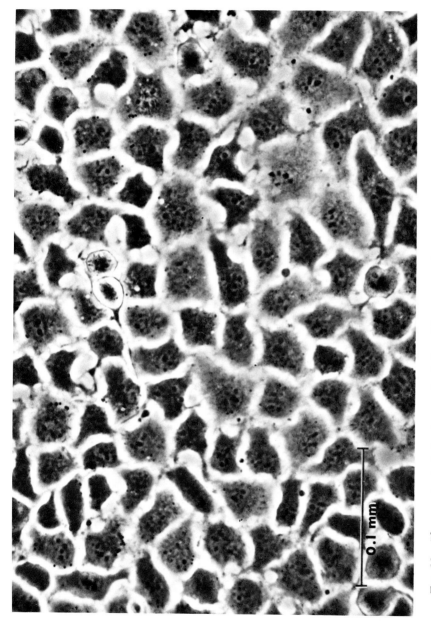

FIG. 11. Phase-contrast photomicrograph of the subline JTC-25·P3 forming a confluent cell sheet. In the early stage after subcultivation these cells closely resemble those in Fig. 10, but they have come to form a confluent sheet.

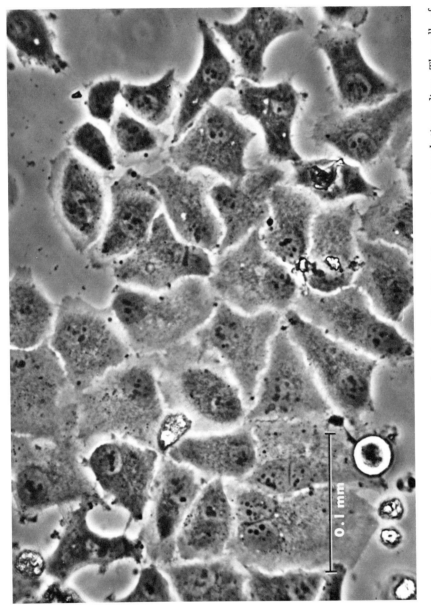

FIG. 12. Phase-contrast photomicrograph of the subline HeLa·P3 growing in a synthetic medium. The cells of this subline have recently come to resemble those of the original line HeLa, spreading cytoplasm and forming a cell sheet.

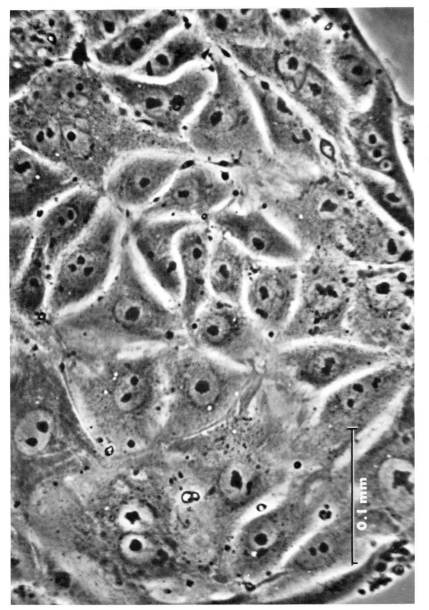

FIG. 13.   Phase-contrast photomicrograph of the original line JTC-12 in a serum medium (derived from cynomolgus monkey kidney cells). All the cells show epithelial morphology.

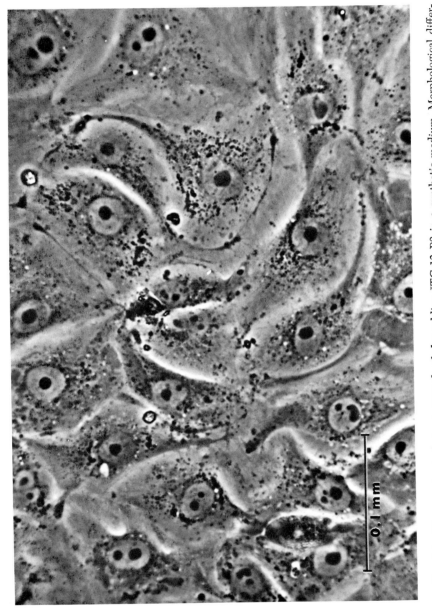

FIG. 14. Phase-contrast photomicrograph of the subline JTC-12·P3 in a synthetic medium. Morphological difference is hardly detectable between this photograph and Fig. 13.

spread slightly. Locomotion was not detected. In JTC-12·P3, a lower frequency of mitosis was detected than in the other sublines. Very strong adhesiveness was observed between cells. No locomotion was evident in isolated single cells. However, interestingly, small colonies had a considerable capacity for locomotion. JTC-16·P3 showed considerable locomotion and maintained a peculiar movement among the cells resembling a stream even after the formation of cell sheets. This would suggest that the property of the cell surface of this tumor line is much different from that of other lines. The cells of RSP-2·P3 exhibited very peculiar behavior: very rapid movement of slender cytoplasmic projections but no marked locomotion.

The sublines L-P3 (Fig. 15), JTC-16·P3 (Fig. 16), JTC-20·P3, JTC-21·P3 (Fig. 17), and JTC-25·P3 (Fig. 18) were compared in chromosome numbers with the original lines or sublines, respectively.

The modal number of chromosomes was 59 in the original line L-929 maintained in our laboratory. It changed to 64 in the subline L·P3

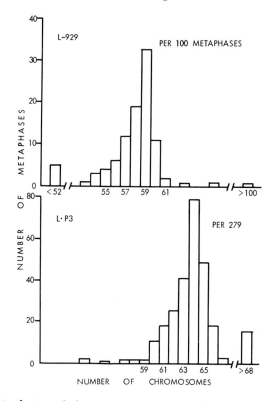

FIG. 15.   Distribution of chromosome numbers in the original line L-929 and in its subline L·P3.

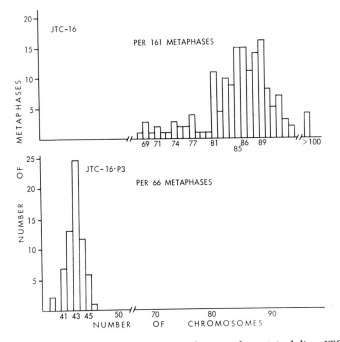

Fᴵɢ. 16.   Distribution of chromosome numbers in the original line JTC-16 and in its subline JTC-16·P3.

adapted to a synthetic medium. However, in the case of JTC-16·P3, the modal number decreased from about 89 in the original line to 43, and the range of distribution of chromosome numbers was much narrowed. Both JTC-21·P3 and JTC-25·P3 also decreased their modal numbers from 69 and 65, respectively, to 62. JTC-20·P3 exhibited no large shift, i.e., from 59 and 60 to 62.

These findings indicate that, by the growth in the protein- and lipid-free synthetic media, some shift takes place in the modal number of chromosomes. It has, however, not yet been confirmed whether such a shift was caused by selection of certain cells or by a mutation-like event.

VI.   Tumorigenicity of Cells Grown in Synthetic Media

Some of the sublines, JTC-16·P3 and JTC-25·P3, grown in synthetic media were examined for their transplantability into rats.

JTC-16·P3 was originated from a transplantable rat ascites hepatoma,

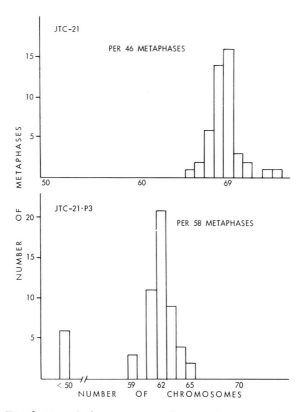

FIG. 17. Distribution of chromosome numbers in the original line JTC-21 and in its subline JTC-21·P3.

AH-7974. After about 7.5 months of cultivation in a synthetic medium, the cells were inoculated intraperitoneally with an inoculum size of 3×10^6 cells per rat into 29-day-old F12 rats of the JAR-2 line. The two injected rats died of tumors 46 days and 52 days after the transplantation. After about 1 year of growth in a synthetic medium, the cells were again inoculated intraperitoneally with the inoculum size of 5×10^6 cells into 27-day-old F14 rats of the JAR-2 line. In 20 days, the two injected rats were sacrificed to prepare specimens, and the accumulation of a large amount of ascitic fluid containing tumor cells was found in both animals. In the cases of intraperitoneal back-transplantation of the strain AH-7974 passaged through rats and the tissue culture strain JTC-16 grown in the serum medium, highly hemorrhagic ascites are obtained as usual. In the case of JTC-16·P3, however, we rarely observed bleeding into the ascitic cavity. The reason is obscure. When the tumor cells produced in animals by back-transplantation were recultured, they grew readily in the

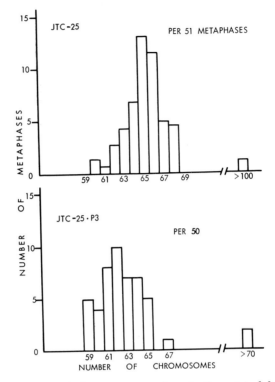

Fig. 18. Distribution of chromosome numbers in the original line JTC-25 and in its subline JTC-25·P3.

protein-free synthetic medium. This type of cell should be of great use in the investigation of tumors.

JTC-25·P3 was originated from the line JTC-25 of rat liver parenchymal cells and transformed by Nagisa culture. The back-transplantation of the original line JTC-25 has not yet resulted in tumor death of inoculated animals. To confirm whether transplantability might be changed by growth in a synthetic medium, JTC-25·P3 cells have been transplanted five times into animals. On intraperitoneal back-transplantation into each two of 6-day-old F33 and 10-day-old F37 rats of the inbred line JAR-1 with an inoculum size of 10^6 cells per rat after about 6 and 17 months of growth in the synthetic medium respectively, no tumors were produced. The rats were sacrificed after 358 days in the former and after 111 and 130 days in the latter, but no tumor formation was detected. After 18 months of culture in the synthetic medium, the cells were inoculated into both cheek pouches of two golden hamsters with an inoculum size of 3×10^6 cells per pouch. After 168 days the animals were sacrificed, but no tumor nodules were found in any pouches. After 24 months of growth in the synthetic medium, the intraperitoneal inoculation of the cells into 1-day-old F38 rats of the JAR-1 line with the inoculum size of 15×10^6 cells per rat resulted in no appearance of tumors. In 99 days the animals were sacrificed but no tumors were detected. The cells of JTC-25 were originated from liver cells of a rat of the same inbred line.

VII.   Sensitivity of Cells Grown in Synthetic Media to Light Irradiation, Chemical Carcinogen Treatment, and Deep Freezing

In some instances of treating cultured cells with chemical carcinogens, special precautions have been taken to prevent interference by visible light with the carcinogens. Earle (1943) treated cultures of mouse fibroblasts with 3-methylcholanthrene (MCA) in an operating room under deep orange light illumination. Berwald and Sachs (1965) handled materials, treated them with MCA, in a room with blacked-out windows which allowed only a small amount of diffuse electric light for visibility.

A very active carcinogen, 4NQO, has recently been employed in our laboratory to transform cells in culture (Andoh *et al.*, 1971; Katsuta and Takaoka, 1972b). However, photodynamic action on 4NQO was found when examined by Nagata *et al.* (1967) using *Paramecium caudatum*. For this reason we examined the sensitivity of several cell lines, including L·P3, to the treatment with 4NQO and the irradiation with 365-mμ light (Katsuta and Takaoka, 1972a).

Figures 19 and 20 represent the sensitivities of L-929 and L·P3 to such treatments. Two days after subcultivation cells were treated with $3.3 \times 10^6 M$ 4NQO at 37°C for 30 minutes and/or irradiated with 365-mμ

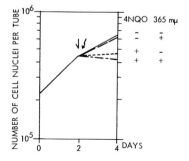

FIG. 19.   Sensitivity of the original line L-929 to 4NQO-treatment and/or to irradiation with 365-mμ light.

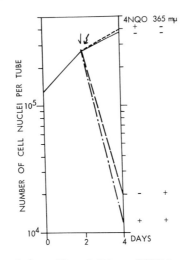

F1G. 20. Sensitivity of the subline L·P3 to 4NQO-treatment and/or to irradiation with 365-mμ light.

light at room temperature for 2 hours. The results showed marked difference in the responses of these two lines. The original line L-929 was sensitive to 4NQO but resistant to the irradiation. L·P3, in contrast, was resistant to 4NQO but very sensitive to the irradiation. With combined treatments, however, both lines were apparently inhibited. L·P3 cells were previously found to contain no essential fatty acids (Kagawa et al., 1969, 1970). To determine whether the high sensitivity of L·P3 to the irradiation might be due to the absence of essential fatty acids, 10% calf serum was added to synthetic medium DM-120 one day before, directly before, or directly after the irradiation. As shown in Fig. 21, the addition of serum before the irradiation considerably protected the cells from destruction by irradiation.

Cells grown in synthetic media are, in general, sensitive to deep freezing. The effect of the concentration of dimethyl sulfoxide (DMSO) added to DM-120 at freezing on the survival rate of JTC-25·P3 cells is shown in Table V (H. Katsuta and T. Takaoka, unpublished data). Samples of about 10^7 cells suspended in 1.5 ml DM-120 were inoculated into short test tubes. Each tube was sealed with a double rubber stopper and plastic tape and packed with several sheets of newspaper and a plastic bag. The tubes were frozen in a deep freezer (Vertis, U.S.A.) at −80°C. After 1 week of freezing the tubes were immersed in a water bath at 37°C and quickly thawed. The medium was renewed to remove DMSO. Viability counting was done within the period of 10 minutes after the addition of an equal volume of 0.04% erythrosin (Chroma, West Germany) solution. The results showed no marked difference in the survival rate between 5% and 10% DMSO.

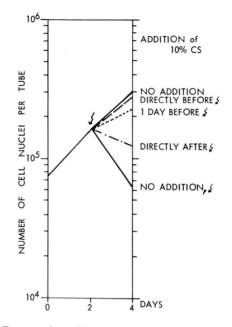

FIG. 21. The effect on the subline L·P3 of the addition of calf serum to the medium before or after the irradiation with 365-mμ light.

To determine whether cells suspended in medium or cells in mono-layer would survive at a higher rate, the experiment of Table VI (H. Katsuta and T. Takaoka, unpublished data) was carried out. JTC-20·P5 cells that had been grown in Eagle MEM alone were employed. The cell suspension was prepared by scraping of cell sheets with a rubber policeman. The medium contained 5% DMSO. Other procedures were the same as described above. The tubes were deep frozen for 1 week. Viable counting was done with erythrosin. The results, however, showed little difference between cell suspensions and cell sheets. The differences between the survival rates shown in Table V and those in Table VI stem from differences in the kind of cells examined.

TABLE V

EFFECTS OF THE CONCENTRATION OF DIMETHYL SULFOXIDE (DMSO) IN THE
SYNTHETIC MEDIUM DM-120 ON THE SURVIVAL RATE OF JTC-25·P3
CELLS AFTER DEEP FREEZING

Concentration of DMSO (%)	Viable cells (%)	
	Experiment 1	Experiment 2
0	9	3
5	82	80
10	68	83

TABLE VI

EFFECTS OF THE CELL STATE AT FREEZING ON THE SURVIVAL OF
JTC-20-P5 CELLS GROWN IN EAGLE MEM ALONE

Experiment	Cell state	Viable cells (%)
1	Suspension	54
	Cell sheet	56
2	Suspension	67
	Cell sheet	69

VIII. Factors Presumably Responsible for the Failure of Cultivation of Cells in Synthetic Media

To obtain continuous growth of cells in protein-free synthetic media, there are two important factors: (1) treatment of cultures, and (2) nutritional constitution of culture media.

The first subcultivation should not be hurried. We do not subculture cells rapidly but wait with much patience for the initiation of rapid cell growth, renewing the culture fluid about twice a week.

This is also important after the establishment of growth in synthetic media. Too frequent subculturing invites the death of cells. When L·P3 cells were subcultured every day, all the cells died out. The cells grown in synthetic media are, in general, sensitive to enzyme treatment. It is better to dislodge them from the glass surface with a rubber policeman. Some of the lines, however, are damaged by mechanical dislodging as well. In the case of JTC-16·P3, we gently shake the culture flask and transfer the cells floating in culture fluid to a new flask. Cells continue to grow in both flasks. The inoculum size of cells at subculturing should be 3 to 10 times greater than that for cells grown in serum media. Once cell sheets are formed, there will be no trouble. Frequent renewal of medium is not required, e.g., L·P3 continues its growth for 2 weeks without renewal. Decrease in pH is slower in synthetic media than in serum medium. The maximum cell population in the same volume of culture fluid is much higher in synthetic media, e.g., L-929 cells cease growth at 1×10^7 cells but L·P3 cells continue to 3×10^7 cells in the same volume (10 ml of culture fluid).

As to the nutritional constitution of media, there remains much work to do. JTC-16·P3 and JTC-21·P3 have a high inositol requirement for growth, while JTC-25·P3 does not. When the original lines were grown in medium consisting of calf serum and lactalbumin hydrolyzate, such a difference was not detected. However, when they were transferred to media consisting of dialyzed serum and synthetic media, the requirement was first found (Ohta et al., 1969; Furuya et al., 1971). In protein-free synthetic media, the difference in inositol requirement was more clearly demonstrated. In the absence of inositol, JTC-16·P3 and JTC-21·P3 underwent cell degeneration within 4 days and eventually died out. The excretion of certain metabolites from the cells into the media can also be accurately examined by the use of synthetic media. The analysis of the amino acid composition in Eagle MEM, in which JTC-25·P5 was grown, disclosed that most nonessential amino acids had appeared in the medium within 2 days of cultivation (Hosokawa et al., 1971). This kind of work will contribute to the improvement of synthetic media.

The examination of HeLa·P3 by the mixed hemadsorption method showed that this subline has maintained the blood type B, suggesting that little change may be induced in blood type antigen by the cultivation in synthetic media.

The following disadvantages, however, should also be taken into consideration for synthetic media: (1) careful treatment of cells is required; and (2) to clone cells is quite difficult. JTC-25·P3 shows less than 3% plating efficiency with an inoculum size of 10^3 cells per 6-cm dish. However, this has been improved by the use of feeder layer, i.e., routinely to almost 10% and occasionally to 30%. The use of conditioned medium is also effective.

During my recent visit to European countries and the USA, I found that plastic culture vessels are used in many laboratories in place of glass vessels. Plastic vessels are very convenient to handle and save labor. However, they are not suitable for the cultivation of cells which are infrequently subcultured. In long-term cultivation, it is very important to flame the mouth of the culture vessels immediately after the stopper is removed, as well as before a new one is inserted, every time the culture fluid is renewed. This prevents bacterial contamination. When plastic vessels are used, it is impossible to flame the mouth and, in many cases, contamination results.

It is the dream of tissue culturists to grow every kind of cells in protein-free synthetic media. This, however, has hardly yet been achieved. It has been accepted in general that this is due to the failure of the composition of medium to meet the nutritional requirements of cells. It should be possible to overcome this difficulty. However, in addition, another reason is also to be considered, i.e., most workers have given up the cultivation of cells in synthetic media and have not tried it with their cell lines. Our experience clearly indicates the necessity of such trials. We believe that many other cell lines will grow in synthetic media.

ACKNOWLEDGMENTS

We wish sincerely to thank Dr. Virginia J. Evans of National Cancer Institute and Dr. Charity Waymouth of the Jackson Laboratory for their kindness in sending us the lists of synthetic media and cell strains as well as reprints of their publications. The help of Dr. Toshiharu Matsumura of our laboratory in preparing the tables in this chapter is gratefully acknowledged.

Our published and unpublished work described herein were mainly supported by the Grants for Cancer Research from the Japanese Ministry of Education.

REFERENCES

Andoh, T., Kato, K., Takaoka, T., and Katsuta, H. (1971). *Int. J. Cancer* **7**, 455.
Andresen, W. F., Price, F. M., Jackson, J. L., Dunn, T. B., and Evans, V. J. (1967). *J. Nat. Cancer Inst.* **38**, 169.
Baker, L. E. (1936). *Science* **83**, 605.
Berwald, Y., and Sachs, L. (1965). *J. Nat. Cancer Inst.* **35**, 641.
Biggers, J. D., and Lucy, J. A. (1960). *J. Exp. Zool.* **144**, 253.
Biggers, J. D., Gwatkin, R. B. L., and Heyner, S. (1961). *Exp. Cell Res.* **25**, 41.
Dubreuil, R., and Pavilanis, V. (1958). *Can. J. Microbiol.* **4**, 543.
Dulbecco, R., and Freeman, G. (1959). *Virology* **8**, 396.
Eagle, H. (1955). *Science* **122**, 501.
Eagle, H. (1959). *Science* **130**, 432.
Earle, W. R. (1943). *J. Nat. Cancer Inst.* **4**, 165.
Earle, W. R. (1962). *In* "New Developments in Tissue Culture" (J. W. Green, ed.), p. 1. Rutgers Univ. Press, New Brunswick, New Jersey.
Evans, V. J., Bryant, J. C., Fioramonti, M. C., McQuilkin, W. T., Sanford, K. K., and Earle, W. R. (1956a). *Cancer Res.* **16**, 77.
Evans, V. J., Bryant, J. C., McQuilkin, W. T., Fioramonti, M. C., Sanford, K. K., Westfall, B. B., and Earle, W. R. (1956b). *Cancer Res.* **16**, 87.
Evans, V. J., Fioramonti, M. C., Sanford, K. K., Earle, W. R., and Westfall, B. B. (1958). *Amer. J. Hyg.* **68**, 66.
Evans, V. J., LaRock, J. F., Yosida, T. H., and Potter, M. (1963). *Exp. Cell Res.* **32**, 212.
Evans, V. J., Bryant, J. C., Kerr, H. A., and Schilling, E. L. (1964). *Exp. Cell Res.* **36**, 439.
Fischer, A., Astrup, T., Ehrensvard, G., and Oehlenschlager, V. (1948). *Proc. Soc. Exp. Biol. Med.* **67**, 40.
Fischer, G. A., and Sartorelli, A. C. (1964). *Methods Med. Res.* **10**, 247.
Furuya, M., Takaoka, T., Nagai, Y., and Katsuta, H. (1971). *Jap. J. Exp. Med.* **41**, 471.
Garvey, J. S. (1961). *Nature (London)* **191**, 972.
Ginsberg, H. S., Gold, E., and Jordan, W. S., Jr. (1955). *Proc. Soc. Exp. Biol. Med.* **89**, 66.
Haff, R. F., and Swim, H. E. (1957a). *J. Gen. Physiol.* **41**, 91.
Haff, R. F., and Swim, H. E. (1957b). *Proc. Soc. Exp. Biol. Med.* **94**, 779.
Ham, R. G. (1963). *Exp. Cell Res.* **29**, 515.
Ham, R. G. (1965). *Proc. Nat. Acad. Sci. U. S.* **53**, 288.
Hanss, J., and Moore, G. E. (1964). *Exp. Cell Res.* **34**, 243.

Healy, G. M., and Parker, R. C. (1966). *J. Cell Biol.* **30**, 531.

Healy, G. M., Fisher, D. C., and Parker, R. C. (1954). *Can. J. Biochem. Physiol.* **32**, 327.

Healy, G. M., Fisher, D. C., and Parker, R. C. (1955). *Proc. Soc. Exp. Biol. Med.* **89**, 71.

Holmes, R. (1959). *J. Biophys. Biochem. Cytol.* **6**, 535.

Holmes, R., and Wolfe, S. W. (1961). *J. Biophys. Biochem. Cytol.* **10**, 389.

Hosokawa, A., Takaoka, T., and Katsuta, H. (1971). *Jap. J. Exp. Med.* **41**, 4.

Iwakata, S., and Grace, J. T., Jr. (1964). *N. Y. State J. Med.* **64**, 2279.

Kagawa, Y., Kaneko, K., Takaoka, T., and Katsuta, H. (1960). *Jap. J. Exp. Med.* **30**, 95.

Kagawa, Y., Takaoka, T., and Katsuta, H. (1969). *J. Biochem. (Tokyo)* **65**, 799.

Kagawa, Y., Takaoka, T., and Katsuta, H. (1970). *J. Biochem. (Tokyo)* **68**, 133.

Katsuta, H., and Takaoka, T. (1960). *Jap. J. Exp. Med.* **30**, 235.

Katsuta, H., and Takaoka, T. (1968). *In* "Cancer Cells in Culture" (H. Katsuta, ed.), p. 321. Univ. of Tokyo Press, Tokyo.

Katsuta, H., and Takaoka, T. (1972a). *Jap. J. Exp. Med.* **42**, 341.

Katsuta, H., and Takaoka, T. (1972b). *J. Nat. Cancer Inst.* **49** (in press).

Katsuta, H., Takaoka, T., Hosaka, S., Hibino, M., Otsuki, I., Hattori, K., Suzuki, S., and Mitamura, K. (1959a). *Jap. J. Exp. Med.* **29**, 45.

Katsuta, H., Takaoka, T., Mitamura, K., Kawada, I., Kuwabara, H., and Kuwabara, S. (1959b). *Jap. J. Exp. Med.* **29**, 191.

Katsuta, H., Takaoka, T., Furukawa, T., and Kawana, M. (1960). *Jap. J. Exp. Med.* **30**, 147.

Katsuta, H., Takaoka, T., and Kikuchi, K. (1961). *Jap. J. Exp. Med.* **31**, 125.

Katsuta, H., Takaoka, T., Doida, Y., and Kuroki, T. (1965). *Jap. J. Exp. Med.* **35**, 513.

Kelley, G. G., Adamson, D. J., and Vail, M. H. (1960). *Amer. J. Hyg.* **72**, 275.

Kitos, P. A., Sinclair, R., and Waymouth, C. (1962). *Exp. Cell Res.* **27**, 307.

Leibovitz, A. (1963). *Amer. J. Hyg.* **78**, 173.

Ling, C. T., Gey, G. O., and Richters, V. (1968). *Exp. Cell Res.* **52**, 469.

McCoy, T. A., Maxwell, M., and Neuman, R. E. (1956). *Cancer Res.* **16**, 979.

McCoy, T. A., Maxwell, M., and Kruse, P. F., Jr. (1959). *Proc. Soc. Exp. Biol. Med.* **100**, 115.

McQuilkin, W. T., Evans, V. J., and Earle, W. R. (1957). *J. Nat. Cancer Inst.* **19**, 885.

Marcus, P. I., Cieciura, S. J., and Puck, T. T. (1956). *J. Exp. Med.* **104**, 615.

Melnick, J. L., Hsiung, G. D., Rappaport, C., Howes, D., and Reissig, M. (1957). *Tex. Rep. Biol. Med.* **15**, 496.

Mizrahi, A., Mitchen, J. R., von Heyden, H. W., Minowada, J., and Moore, G. E. (1972). *Appl. Microbiol.* **23**, 145.

Moore, G. E., and Kitamura, H. (1968). *N. Y. State J. Med.* **68**, 2054.

Moore, G. E., Gerner, R. E., and Franklin, H. A. (1967). *J. Amer. Med. Ass.* **199**, 519.

Moore, G. E., Gerner, R. E., and Minowada, J. (1968). *In* "The Proliferation and Spread of Neoplastic Cells," p. 41. Williams & Wilkins, Baltimore, Maryland.

Morgan, J. F., Morton, H. J., and Parker, R. C. (1950). *Proc. Soc. Exp. Biol. Med.* **73**, 1.

Morgan, J. F., Campbell, M. E., and Morton, H. J. (1955). *J. Nat. Cancer Inst.* **16**, 557.

Morgan, J. F., Morton, H. J., Campbell, M. E., and Guerin, L. F. (1956). *J. Nat. Cancer Inst.* **16**, 1405.

Morton, H. J. (1970). *In Vitro* **6**, 89.

Nagata, C., Fujii, K., and Epstein, S. S. (1967). *Nature (London)* **215**, 972.

Neuman, R. E., and McCoy, T. A. (1956). *Science* **124**, 124.

Neuman, R. E., and McCoy, T. A. (1958). *Proc. Soc. Exp. Biol. Med.* **98**, 303.

Neuman, R. E., and Tytell, A. A. (1960). *Proc. Soc. Exp. Biol. Med.* **104**, 252.

Ohta, S., Takaoka, T., and Katsuta, H. (1969). *Jap. J. Exp. Med.* **39**, 359.

Parker, R. C. (1961). *In* "Methods of Tissue Culture," p. 62. Harper (Hoeber), New York.

Paul, J. (1959). *J. Exp. Zool.* **152**, 475.

Price, F. M., Kerr, H. A., Andresen, W. F., Bryant, J. C., and Evans, V. J. (1966). *J. Nat. Cancer Inst.* **37**, 601.

Puck, T. T. (1961). *Harvey Lect.* **55**, 1.

Puck, T. T., Cieciura, S. J., and Fisher, H. W. (1957). *J. Exp. Med.* **106**, 145.

Puck, T. T., Cieciura, S. J., and Robinson, A. (1958). *J. Exp. Med.* **108**, 945.

Rappaport, C. (1956). *Proc. Soc. Exp. Biol. Med.* **91**, 464.

Rappaport, C., Poole, J. P., and Rappaport, H. P. (1960). *Exp. Cell Res.* **20**, 465.

Rosenberg, S., and Kirk, P. L. (1953). *J. Gen. Physiol.* **37**, 239.

Sanford, K. K., Dupree, L. T., and Covalesky, A. B. (1963). *Exp. Cell Res.* **31**, 345.

Sato, G., Fisher, H. W., and Puck, T. T. (1957). *Science* **126**, 961.

Scherer, W. F. (1953). *Amer. J. Pathol.* **29**, 113.

Swim, H. E., and Parker, R. F. (1958a). *J. Lab. Clin. Med.* **52**, 309.

Swim, H. E., and Parker, R. F. (1958b). *Can. J. Biochem. Physiol.* **36**, 861.

Takaoka, T., and Katsuta, H. (1971). *Exp. Cell Res.* **67**, 295.

Takaoka, T., Katsuta, H., Furukawa, T., and Kawana, M. (1960a). *Jap. J. Exp. Med.* **30**, 319.

Takaoka, T., Katsuta, H., Kaneko, K., Kawana, M., and Furukawa, T. (1960b). *Jap. J. Exp. Med.* **30**, 391.

Tritsch, G. L., and Moore, G. E. (1962). *Exp. Cell Res.* **28**, 360.

Trowell, O. A. (1955). *Exp. Cell Res.* **9**, 258.

Trowell, O. A. (1959). *Exp. Cell Res.* **16**, 118.

Waymouth, C. (1955). *Tex. Rep. Biol. Med.* **13**, 522.

Waymouth, C. (1956). *J. Nat. Cancer Inst.* **17**, 315.

Waymouth, C. (1959). *J. Nat. Cancer Inst.* **22**, 1003.

Waymouth, C. (1965a). *In* "Cells and Tissues in Culture" (E. N. Willmer, ed.), Vol. I, p. 99. Academic Press, New York.

Waymouth, C. (1965b). *In* "Tissue Culture" (C. V. Ramakrishnan, ed.), p. 168. Junk, The Hague.

Waymouth, C., Chen, H. W., and Wood, B. G. (1971). *In Vitro* **6**, 371.

White, P. R. (1946). *Growth* **10**, 231.

White, P. R. (1949). *J. Cell. Comp. Physiol.* **34**, 221.

White, P. R. (1955). *J. Nat. Cancer Inst.* **16**, 769.

Yasumura, Y. (1962). *13th Meet., Jap. Tissue Cult. Ass., 1962* p. 10.

Yasumura, Y. (1963). *Tissue Cult. Stud. Jap., Ann. Bibliogr. 1963* p. 33.

Chapter 2

Preparation of Synchronous Cell Cultures from Early Interphase Cells Obtained by Sucrose Gradient Centrifugation[1]

RICHARD SCHINDLER AND JEAN CLAUDE SCHAER

Department of Pathology, University of Bern, Bern, Switzerland

I. Introduction

The biochemical events underlying the normal progression of animal cells in the division cycle have become a subject of increasing interest

[1] Supported by the Swiss National Science Foundation and the Swiss Cancer League.

43

during recent years (Mueller, 1969). Because of the limited sensitivity of available methods of analysis, biochemical studies usually cannot be based on the study of individual cells. Therefore, populations of cells proliferating in synchrony are required.

Cell cultures may be synchronized by reversible inhibition of a process, such as DNA synthesis, which is specific for one phase of the cell cycle. Various antimetabolites, such as 5-fluoro-2'-deoxyuridine (Eidinoff and Rich, 1959; Schindler, 1960), amethopterin (Rueckert and Mueller, 1960; Schindler, 1960), hydroxyurea (Sinclair, 1965, 1967), and excess thymidine (Xeros, 1962; Bootsma et al., 1964; Petersen and Anderson, 1964) have been used for synchronization of mammalian cell cultures. Methods based on the effects of these inhibitors permit preparation of large numbers of synchronously proliferating cells. On the other hand, during selective inhibition of DNA synthesis, cellular metabolism is modified and no longer corresponds to that of normal, uninhibited cells. For instance, enlargement of cells due to continued synthesis of RNA and proteins is usually observed. Furthermore, application of inhibitors of DNA synthesis in concentrations commonly used may not necessarily result in complete cessation of DNA synthesis (Bostock et al., 1971).

Synchronous cultures have also been obtained by selective killing of cells in a specific phase of the division cycle, e.g., by incorporation of high activities of thymidine-^{3}H (Whitmore and Gulyas, 1966). This method, however, does not provide for immediate destruction of lethally damaged cells.

Another group of methods used for preparation of synchronous cultures is based on the separation by physical means of cells in one specific phase of the division cycle from the rest of the cell population. Mitotic cells may be separated from interphase cells attached to a solid substrate by mechanical forces (Axelrad and McCulloch, 1958). Upon further incubation, these mitotic cells progress through interphase and exhibit synchrony of DNA synthesis and of division (Terasima and Tolmach, 1963). The number of cells obtained by this technique is, however, relatively small, and application of the method is restricted to cell types that adhere to a solid substrate.

Mitchison and Vincent (1965) reported that synchronous cultures of fission yeast, budding yeast, and Escherichia coli may be prepared by centrifugation of cells in sucrose gradients and separation of slowly sedimenting, small cells from the rest of the population. Sinclair and Bishop (1965) subjected L strain cells to sucrose gradient centrifugation and obtained populations consisting primarily of early interphase cells which, upon further incubation, exhibited partial synchrony of cell division. Similar results were obtained with mouse lymphoma cells (Ayad et al., 1969) and mastocytoma cells (Warmsley and Pasternak, 1970) using Ficoll gradients. Fractionation of cell populations with respect to position in the division cycle without subsequent culture of cells has also been reported by several authors (Morris et al., 1967; MacDonald and Miller, 1970).

Further improvements in the degree of synchrony obtained by sucrose gradient centrifugation of murine mastocytoma cells have recently been described (Schindler et al., 1970; Schindler and Hürni, 1971). The present report is concerned with the description of this technique of preparing synchronous cell cultures and with an evaluation of the synchrony obtained.

II. Requirements to Be Satisfied by the Method

The following criteria are applicable to methods of preparing synchronous cell cultures which are based on a separation of cells of different size by velocity (rate) sedimentation in density gradients:

1. Separation of cells according to size should be effected with a high degree of resolution. In a homogeneous cell population, a cell immediately before entering mitosis has, in theory, twice the volume of a cell shortly after telophase. In order to achieve a good synchrony, it is essential that cells differing in volume by a factor considerably smaller than 2 can be separated from each other.

2. In order to stabilize the column of liquid in which the cells are allowed to sediment, a density gradient has to be used. No irreversible cell damage should, however, occur during sedimentation of cells in the gradient. The substance(s) used for preparation of the density gradient should, therefore, be nontoxic to the cells.

3. No swelling or shrinking of cells in the course of the separation procedure should occur. Swelling or shrinking may in itself damage the cells or may result in changes of sedimentation behavior affecting the efficiency of separation. In order to exclude changes in cell volume, all portions of the density gradient used for separation of cells should be isotonic.

4. The separation technique as well as concomitant manipulation of cells should not significantly affect the subsequent progression of cells in the division cycle. If cell cycle progression is slower after the separation procedure, and if recovery is not identical for all cells, the synchrony achieved may be unsatisfactory.

5. The time required for all manipulations used to prepare cultures of early interphase cells should be as short as possible. Otherwise, progression of cells in the division cycle may take place during the separation procedure, and the resultant changes of cell size may adversely affect the extent of separation with respect to position in the cell cycle.

III.   Cell Line and Culture Techniques

Cultures *in vitro* of a transplantable murine mast cell tumor (cell line P-815-X2) were used. This cell line was derived from the original P-815 tumor (Dunn and Potter, 1957) by a selection process *in vitro* and two consecutive cloning procedures (Schindler *et al.*, 1959; Green and Day, 1960). The general culture techniques have been previously reported (Schindler *et al.*, 1959). Cell densities (number of cells per milliliter) were determined by hemocytometer or by Coulter counts. The medium used has been described as medium I (Schaer and Schindler, 1967) and contained 10% undialyzed horse serum. The cultured cells were maintained in suspension by a suspended magnetic stirrer (Fig. 1). Under these conditions, cell number doubling times of 8.5 to 10 hours were usually observed.

Approximate cell cycle characteristics of this cell line, as determined in a number of independent experiments, are presented in Table I. The cell cycle time was derived from the periodicity of the mitotic index, labeling index, and additional parameters in partially synchronous cultures. For determining the duration of the G_2 period and half of the time required for mitosis thymidine-^{3}H was added to cultures under steady state conditions (Fig. 1), and thereafter samples were withdrawn

TABLE I

CELL CYCLE CHARACTERISTICS OF P-815-X2 MURINE MASTOCYTOMA
CELLS UNDER STEADY STATE CULTURE CONDITIONS

Period	Duration in hours[a]
G_1	1.3
DNA synthesis	5.0
G_2	1.4
Mitosis	0.3
Cell cycle time	8
Cell number doubling time	9

[a] Average values obtained in a number of separate experiments (see text).

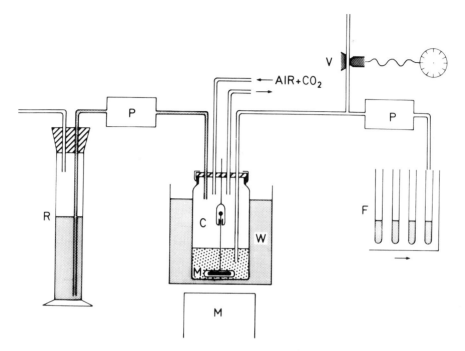

Fig. 1. Suspension culture under steady state conditions (from Schindler *et al.*, 1970). R, medium reservoir; P, peristaltic pump; C, culture flask; M, magnetic stirrer; W, water bath; V, electric valve connected to timer; F, fraction collector. The agitated cell suspension is continuously being diluted by inflow of fresh medium from a reservoir in order to compensate for the increase in cell number due to cell proliferation. At predetermined time intervals, the electric valve is closed, and the peristaltic pump (on the right side) withdraws a sample from the cell suspension. Cells in the samples are fixed by the fixing solution present in the collecting tubes of the fraction collector.

at regular time intervals for autoradiographic determination of the relative number of labeled mitoses (Schindler *et al.*, 1968; Gautschi *et al.*, 1971). The duration of mitosis was determined by the graphic method of Okada (1967) using the mitotic index of cultures under steady state conditions and the cell cycle time. Similarly, the duration of DNA synthesis was estimated from the relative number of labeled cells obtained after labeling of steady state cultures with thymidine-^{3}H during 20 minutes. The duration of G_1 represents the difference between the cell cycle time and the cumulative duration of DNA synthesis, G_2, and mitosis. It should be noted that cell number doubling times regularly exceeded cell cycle times by approximately 1 hour. Variations between cell cycle parameters observed in individual experiments may possibly be attributed to different lots of horse serum used. In addition, a selection of more rapidly multiplying cells during prolonged *in vitro* culture cannot be excluded.

In order to minimize changes in medium composition with progressive culture, synchronous cell populations were incubated under steady state conditions, as illustrated in Fig. 1. Under these conditions, the cell suspension is continuously being diluted with fresh medium which is slowly drawn into the culture flask by a peristaltic pump in order to compensate for the increase in cell number due to cell multiplication. Samples of the cell suspension are withdrawn at regular time intervals for measurements of mitotic index, cell number per milliliter, and further parameters. A constant pH in the suspension culture was maintained by equilibration with a mixture of 95% air and 5% CO_2.

IV. Preparation of Partially Synchronous Cell Populations

Spinner cultures of 300 to 1000 ml containing approximately 10^5 cells/ml of medium are incubated at 37°C. After 24 hours they are diluted with fresh medium to reduce the cell density to 10^5 cells/ml and incubated again during 24 hours. During this second incubation period of 24 hours, the cell multiplication rate should approach a level corresponding to a cell number doubling time of 8.5 to 10 hours.

Gradients are prepared by appropriate mixing of (a) medium which, instead of NaCl at a concentration of 8 mg/ml, contains sucrose at a concentration of 95 mg/ml (high density medium), and (b) medium containing 6.4 mg/ml of NaCl and 19 mg/ml of sucrose (low density medium). This provides for isotonicity and for the presence of all medium components (except NaCl) at optimal concentrations throughout the gradient. Gradients are prepared in centrifuge tubes approximately 20 cm long and with an inner diameter of 2 to 4.5 cm, depending on the number of cells to be subjected to the separation procedure.

After the preincubation during 48 hours in spinner culture, the cell suspension (2×10^8 to 8×10^8 cells) is centrifuged and resuspended in 5 to 15 ml of fresh medium. In order to avoid the formation of "streamers" from the concentrated cell suspension into the gradient, the cell suspension is placed on the gradient according to the procedure illustrated in Fig. 2. First the concentrated cell suspension is drawn into a small mixing chamber. Pumping of the cell suspension from the mixing chamber to the bottom of the centrifuge tube is carried out at the same

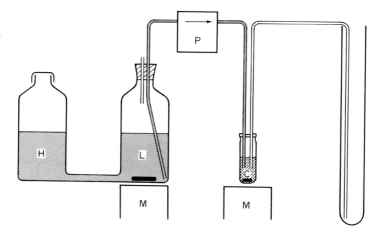

Fig. 2. Experimental design used for placing a concentrated cell suspension on a linear sucrose density gradient. H, high density medium (95 mg/ml sucrose); L, low density medium (19 mg/ml sucrose, 6.4 mg/ml NaCl); P, peristaltic pump; C, cell suspension; M, magnetic stirrer. A linear gradient is formed by mixing of high density and low density medium. While the top portions of the gradient are being pumped into the small mixing chamber, the cell suspension begins to flow from the mixing chamber to the bottom of the centrifuge tube. This results in a gradual dilution of the cell suspension remaining in the mixing chamber by the top portions of the gradient. Continued pumping of the gradient through the mixing chamber into the centrifuge tube lifts the cell layer to the top.

time and at the same rate as pumping of the top portions of the gradient into the mixing chamber. By this means, the cell suspension remaining in the mixing chamber is being gradually diluted by the top portions of the gradient, thus avoiding the formation of a sharp boundary between the cell suspension and the top of the gradient. A linear gradient is formed by gradual mixing of low density medium with high density medium, and the resulting mixture is pumped through the small mixing chamber into the centrifuge tube, thereby lifting the cell layer to the top.

The centrifuge tube containing the cell suspension on top of the gradient is centrifuged during approximately 4 minutes at a centrifuge setting which provides for a speed of 1200 rpm (80g at the top, 400g at the bottom of the centrifuge tube). Acceleration and deceleration of the centrifuge should be gradual in order to reduce to a minimum the formation of turbulences in the centrifuge tube containing the gradient.

After this centrifugation, the cells sedimenting most rapidly have formed a pellet on the bottom of the tube, while the remainder should be distributed over the lower half of the gradient. The fraction of cells sedimenting most slowly and corresponding to 2 to 5% of the original cell number is collected, centrifuged, and resuspended in sucrose-free medium to obtain a cell density of 1×10^5 to 3×10^5 cells/ml. The sucrose-free medium used for resuspension of cells is prepared by mixing fresh medium with an equal volume of "conditioned" medium which has been collected after centrifugation of cells at the end of the second 24-hour preincubation period. The slowly sedimenting cells suspended in this medium are incubated under steady state culture conditions as described above.

V. Methods Used for Characterization of Synchronous Cultures

In order to determine the number of viable cells, the cells were cultured, after appropriate dilution with "cloning medium," in fibrin gels, as described previously (Schindler, 1964). The colonies which developed from the individual cells were counted 5 to 7 days later.

The size distribution of cell populations was determined with the Coulter Counter, by counting the cell number of a sample at increasing threshold settings.

In order to determine mitotic indices, cells were fixed by mixing an aliquot of the suspension culture with an equal volume of a mixture of ethanol, acetic acid, and water (5:2:3, v/v). Subsequently, the cells were centrifuged and resuspended in a small volume of 0.025% crystal violet in 1% acetic acid. In this suspension the percentage of cells in prophase, metaphase, and early anaphase (subsequently termed "mitotic index") was determined by hemocytometer counts. Cells were thus counted as mitoses from the appearance of distinctly visible chromosomes through the entire metaphase and into anaphase, while cells after the onset of furrowing were excluded from the counts.

For determination of cell densities (numbers of cells per milliliter), samples were withdrawn from partially synchronous cultures and mixed with an equal volume of a fixing solution containing 50% ethanol and HCl at a concentration of 0.025 N. The fixed cells were counted with a Coulter Counter.

The proportion of DNA-synthesizing cells was determined as follows: An aliquot of the suspension culture was incubated during 20 minutes with 0.1 μCi/ml of methyl-^{3}H-thymidine (5 Ci/mmole; The Radiochemical Centre, Amersham, UK). Subsequently, the cells were fixed as described above, and smears were prepared, prestained by the Feulgen reaction, and then processed for autoradiography, using NTB-2 Kodak emulsion. After 2 weeks' exposure, the preparations were developed, fixed, and counterstained with Carazzi's glycerin-hemalum through the film. In each sample, the percentage of labeled cells was determined by counting 300 cells.

For measuring the incorporation of thymidine-^{3}H into DNA, aliquots of 1.5 ml were withdrawn from partially synchronous cultures at intervals of 1.5 hours and incubated during 20 minutes with 0.25 μCi/ml of thymidine-^{3}H. Subsequently, the cellular material was fractionated, and the percentage of radioactivity incorporated into DNA was determined as previously reported (Schaer et al., 1971).

VI. Characteristics of Synchronous Cultures

A. Viability of Cells after Centrifugation in Sucrose Gradients

In view of the possibility that a certain proportion of the cell population might be irreversibly damaged by centrifugation in sucrose-containing medium, the capacity of cells to form macroscopically visible colonies in a semisolid medium was determined. As seen in Table II, the number of colonies developing from cells that had undergone centrifugation in a sucrose gradient was rather higher than that developing from untreated control cells. Although the differences may not be significant, the results support the conclusion that the capacity of the cells to multiply is not adversely affected by the centrifugation procedure and/or contact with sucrose-containing medium.

TABLE II

COLONY DEVELOPMENT BY P-815-X2 CELLS SUBJECTED TO SUCROSE GRADIENT CENTRIFUGATION[a]

	Colony yield (%)	
Experiment number	Centrifuged cells	Control cells
1	74	70
2	100	82
3	97	86
Mean	90.3	79.3

[a] From Schindler et al., 1970.

B.   Size Distribution of Slowly Sedimenting Cells as Compared to That of the Original Cell Population

In order to ascertain that centrifugation in a sucrose gradient provided for a separation of small cells from the total cell population, the size distribution of the fraction of cells sedimenting most slowly was determined and compared to that of the original cell population subjected to the sucrose gradient centrifugation procedure. The results of a typical experiment, as illustrated in Fig. 3, indicate that of the fraction of cells sedimenting most slowly, the mean cell size, as well as the variation in size, is considerably smaller than that of the original population. Although the method used does not permit an evaluation of absolute sizes,

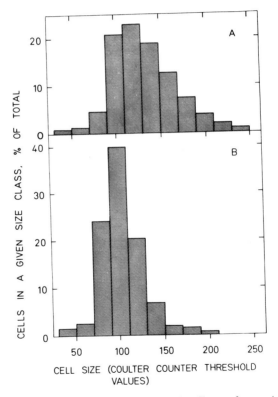

Fig. 3.   Distribution of cell sizes in the total cell population (A) and in the fraction of slowly sedimenting cells re-collected after sucrose density gradient centrifugation (B) (from Schindler *et al.*, 1970). Cell size distribution was determined by measuring cell numbers per milliliter in the cell suspensions with a Coulter Counter at increasing threshold settings. The fraction of slowly sedimenting cells represented 4.2% of the original cell population subjected to the separation procedure.

the results support the conclusion that the fraction of cells sedimenting most slowly is more homogeneous with respect to cell size than the original cell population, and that the average size of the slowly sedimenting cells is inferior to that of all cells.

C. Mitotic Activity of Cultures Derived from Slowly Sedimenting Cells

After collection of the fraction of cells sedimenting most slowly, the cells were incubated under conditions providing for a steady state, and every hour a sample was withdrawn for determination of the mitotic index. The results of a typical experiment, as illustrated in Fig. 4, indicate that the method used provided for a reasonably good synchrony of cell division. During the first few hours after incubation of postmitotic cells, the mitotic index remained very low and then rose to a first peak at 8 hours, which was followed by a second peak at 16 hours, and presumably a third peak at 24 hours.

The observation that the first mitotic peak occurred 8 hours after the onset of incubation supports the conclusion that the majority of cells had

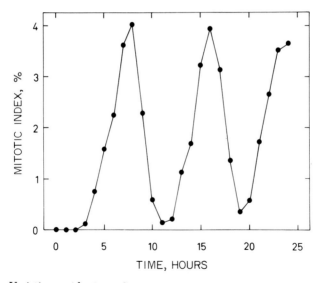

FIG. 4. Variations with time of mitotic activity in cultures derived from slowly sedimenting cells (from Schindler *et al.*, 1970). After preincubation in spinner culture during 48 hours, cells were subjected to sucrose density gradient centrifugation, and the fraction of cells sedimenting most slowly was reincubated under steady state culture conditions. Every hour a sample was withdrawn from the culture, and the cells were fixed and used for determining mitotic indices.

passed through mitosis shortly before they were subjected to the sucrose gradient centrifugation process and collected for reincubation under steady state conditions. Furthermore, the first as well as the second mitotic peak exhibit some degree of asymmetry. This indicates that at the onset of incubation, the population consisting primarily of cells in early interphase also contained some cells in late G_1 and later stages of the division cycle, whereas separation from cells in mitosis and later interphase was nearly complete.

The results presented in Fig. 4 demonstrate that synchrony of cell division was well maintained during the observation period of approximately three cell generations, and in an experiment of longer duration, four successive mitotic peaks were discernible.

D. Pattern of Cell Proliferation in Synchronous Cultures

Since mitotic indices may not represent an entirely reliable measure of rates of increase in cell number, variations in cell density in partially synchronous steady state cultures were determined as follows. Samples were withdrawn at hourly intervals and mixed with an equal volume of a fixing solution containing 50% ethanol and HCl at a concentration of 0.025 N. At the end of the experiment, the fixed cells were counted with a Coulter Counter, and the values were multiplied by the cumulative dilution ratio of the steady state culture as follows:

$$N_t = n_t \times D_t$$

where n_t is the cell number per milliliter at time t when the sample was taken, and D_t is the cumulative dilution of the steady state culture with fresh medium until time t. Thus N_t represents the cell density that would have been attained at time t if the cells had multiplied equally well without any dilution of the culture with fresh medium.

The results of a typical experiment are presented in Fig. 5: The increase in cell number occurred in a stepwise manner with a periodicity of approximately 9 hours. On the other hand, the average cell number doubling time observed in this experiment was 10 hours and thus exceeded the cell cycle time which is reflected by the interval between two successive waves of cell divisions.

Such a difference between cell number doubling time and cell cycle time was consistently observed in the culture system used and was also confirmed in another series of experiments in which cell cycle time determinations were based on the periodicity of the mitotic labeling index after pulse-labeling of steady state cultures with thymidine-^{3}H (Gautschi et al., 1971).

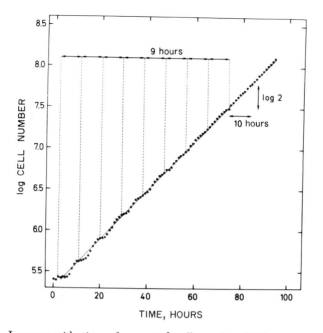

Fig. 5. Increase with time of corrected cell number (N_t) in a partially synchronous cell culture under steady state culture conditions (from Schindler and Hürni, 1971). Samples were withdrawn at 1-hour intervals from a synchronous culture, mixed with an equal volume of fixing solution, and the fixed cells were counted with a Coulter Counter. Measured cell numbers (n_t) were multiplied with the cumulative dilution ratio of the steady state culture, as explained in the text. Corrected cell numbers (N_t) are plotted on the ordinate on a logarithmic scale.

Figure 5 indicates that the increase in cell number attained after the first wave of cell divisions is of the same order of magnitude as that observed during subsequent replication cycles. The procedure used for obtaining cells in early interphase does, therefore, not affect to a significant extent the percentage of cells in the initial population participating in cell proliferation during the first replication cycle. In fact, in a series of 10 experiments, the cell number observed after the first wave of cell divisions was, on the average, 1.87 ± 0.03 (mean ± S.E.) times the original cell number. This closely corresponds to the observed differences between cell cycle time and cell number doubling time in experiments of longer duration.

As seen in Fig. 5, a time lag after incubation of early interphase cells was observed, which may be attributable to manipulation of cells during their separation from the rest of the cell population. A transient retardation of cell cycle progression as a result of the separation procedure was confirmed as follows: after sucrose gradient centrifugation of a cell suspension, all fractions of the gradient including the cell pellet were re-combined, an aliquot of this cell suspension was centrifuged, and the cells were resuspended in sucrose-free medium and reincubated under steady state culture conditions. Every hour a sample was withdrawn for determination of the cell number per milliliter. As seen in Fig. 6, the increase in cell number was somewhat slower during the first 8 to 12 hours after the onset of incubation and then attained a constant rate.

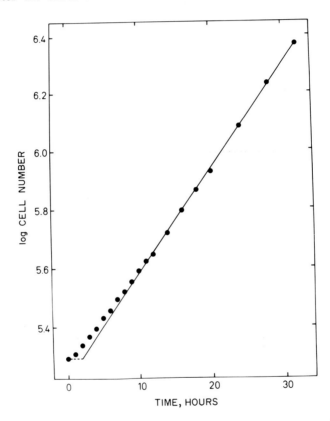

Fig. 6. Retardation of cell multiplication after sucrose density gradient cen-trifugation of cells. A cell suspension was subjected to sucrose density gradient centrifugation. Subsequently all fractions including the cell pellet on the bottom of the centrifuge tube were recombined, an aliquot of the resulting cell suspension was centrifuged and the cells were resuspended in sucrose-free medium and incubated under steady state culture conditions. Samples were withdrawn at 1-hour intervals and mixed with an equal volume of fixing solution, and the fixed cells were counted with a Coulter Counter. Measured cell numbers (n_t) were multiplied with the cumu-lative dilution ratio of the steady state culture, as explained in the text. Corrected cell numbers (N_t) are plotted on the ordinate on a logarithmic scale.

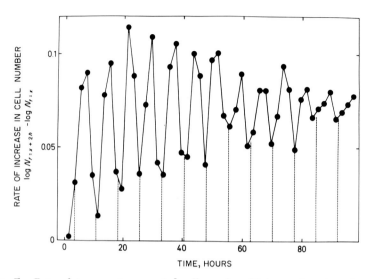

FIG. 7.   Rate of increase in corrected cell number (N_t) as a function of time in a partially synchronous culture under steady state conditions (from Schindler and Hürni, 1971). Samples were withdrawn at 1-hour intervals from a synchronous culture, mixed with an equal volume of fixing solution, and the fixed cells were counted with a Coulter Counter. Measured cell numbers (n_t) were multiplied with the cumulative dilution ratio of the steady state culture to obtain corrected cell numbers (N_t), as explained in the text. Because of the limited precision of Coulter Counter measurements, which is attributable, at least in part, to the relatively small numbers of cells available in individual samples, the data obtained were used to calculate rates of cell multiplication at 2-hour instead of 1-hour intervals.

Progression of cells in the division cycle during the first generation time thus appears to be retarded to a small, but significant extent due to the procedure used for separation of early interphase cells.

The variations with time of the rate of increase in cell number, as observed in a typical experiment, are illustrated in Fig. 7. The synchrony is characterized by relatively broad peaks of cell multiplication rate similar to those described above for mitotic activity. Synchrony is quite well maintained, and rhythmic fluctuations of multiplication rate are evident during ten and possibly twelve cell generations.

E.   Synchrony of DNA Synthesis

Aliquots were withdrawn from partially synchronous cultures at regular time intervals, incubated during 20 minutes with thymidine-³H, and subsequently the cells were fixed and subjected to autoradiography. As

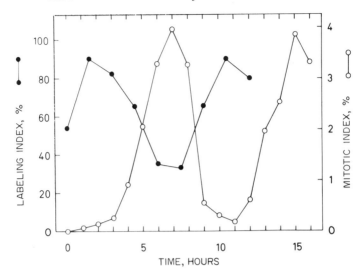

Fɪɢ. 8. Labeling of cells by thymidine-³H (closed circles) and mitotic activity (open circles) in a partially synchronous culture derived from slowly sedimenting cells (from Schindler *et al.*, 1970). Samples were withdrawn from a synchronous culture at regular time intervals and incubated during 20 minutes with thymidine-³H. Subsequently the cells were fixed and subjected to autoradiography. Additional samples were used for determination of mitotic indices.

illustrated in Fig. 8, at the time of incubation of early interphase cells, approximately 50% of cells are already synthesizing DNA. The proportion of cells in DNA synthesis increases to reach a first maximum at 90 minutes and then decreases to a minimum at a time when mitotic activity of the culture is at a maximum. A second maximum of the percentage of DNA-synthesizing cells coincides with a period of minimal mitotic activity. The rhythmic changes in the proportion of DNA-synthesizing cells, which are inverse to those of the mitotic index, correspond well to the cell cycle characteristics of the cell line used: As illustrated in Table I, the period of DNA synthesis is separated from mitosis by a G_1 and a G_2 period which are of similar duration.

In Fig. 9, variations with time of labeling indices are compared with changes in incorporation of thymidine-³H into cellular DNA. Incorporation of the labeled precursor per 10^5 cells during 20-minute periods, as a function of time after incubation of early interphase cells, exhibits marked rhythmic fluctuations, with maximum values which are higher by a factor of 4 than minimum values. On the other hand, fluctuations with time of the labeling index are much smaller. As previously de-

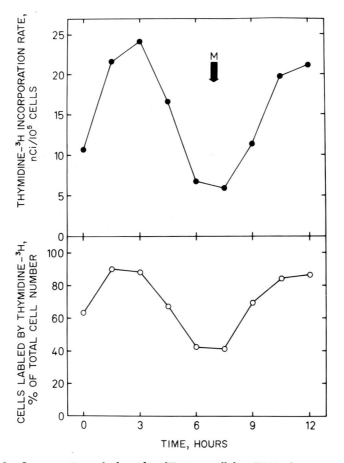

FIG. 9. Incorporation of thymidine-³H into cellular DNA (upper part) and labeling of cells by thymidine-³H (lower part) in a partially synchronous culture derived from slowly sedimenting cells. Samples were withdrawn from the culture at regular time intervals and incubated during 20 minutes with 0.25 μCi/ml of thymidine-³H. Subsequently the cellular material was fractionated, and radioactivity incorporated into DNA of 10^5 cells was determined. Additional samples from the synchronous culture were incubated during 20 minutes with 0.1 μCi/ml of thymidine-³H, and the cells were fixed and subjected to autoradiography. The arrow in the figure indicates the time of maximum mitotic activity.

scribed (Schaer *et al.*, 1971), the fluctuations in thymidine-³H incorporation reflect variations of the relative number of DNA-synthesizing cells as well as changes with time of nucleoside incorporation within the S period.

VII.   Experimental Factors Affecting the Degree of Synchrony

A.   Effect of Preincubation of Cells in Spinner Culture

In Fig. 10 the synchrony of a cell population derived from a spinner culture is compared with that of a cell population prepared by density gradient centrifugation of cells from a nonagitated culture. It is seen that the mitotic activity of the cell population derived from the non-agitated culture is characterized by a longer time interval between in-cubation of early interphase cells and the first mitotic wave. In addition, maxima and minima of the mitotic index are not as pronounced as those of the cell population derived from the spinner culture, and variations with time of mitotic activity appear to be more regular in the latter.

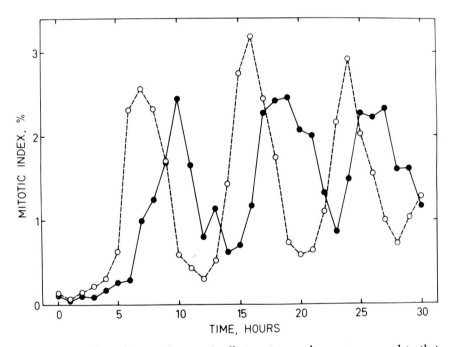

FIG. 10.   Effect of preincubation of cells in spinner culture, as compared to that in nonagitated culture, with respect to the synchrony obtained by sucrose density gradient centrifugation. Cells were incubated during 48 hours in spinner culture (open circles) or in nonagitated culture (closed circles), respectively. Subsequently the cells were subjected to sucrose density gradient centrifugation, and the slowly sedimenting cells were reincubated in sucrose-free medium under steady state cul-ture conditions. Samples were withdrawn from both cultures at 1-hour intervals for determination of mitotic indices.

B. Cell Size Distribution in Different Fractions of the Cell Population Subjected to Sucrose Gradient Centrifugation

The degree of synchrony achieved by density gradient centrifugation evidently is a function of the percentage with respect to the total cell population of the fraction of slowly sedimenting cells used for further incubation. Since position in the division cycle is correlated with cell size, various fractions collected after sucrose gradient centrifugation of a cell population were compared with respect to cell size distribution. In this experiment, differences in cell size distribution thus were used to indicate corresponding differences in distribution within the division cycle between cell populations collected along the gradient. As seen in Fig. 11, the fraction of cells sedimenting most slowly exhibited the smallest average cell size and the highest degree of homogeneity. With

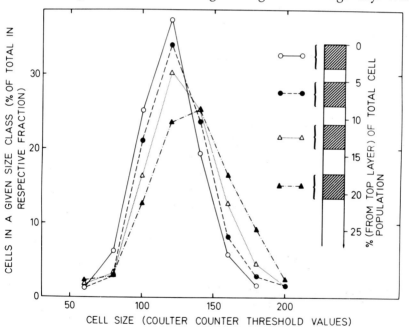

Fig. 11. Cell size distribution in various fractions collected after sucrose density gradient centrifugation of a cell suspension. After preincubation in spinner culture during 48 hours, cells were placed on the sucrose gradient and centrifuged. Subsequently, a series of fractions was collected from the gradient. Fractions containing approximately 3% of the original cell population were recombined, and the size distributions of the resulting cell suspensions were determined, after appropriate dilution in saline, with a Coulter Counter. Size and position of the fractions along the gradient with respect to the total cell population are indicated on the insert at the right side of the figure.

increasing sedimentation rate, an increase in average cell size was observed. Furthermore, the homogeneity with respect to cell size was less pronounced in fractions of cells collected from lower portions of the gradient. It appears, therefore, that the best homogeneity with respect to cell size and, correspondingly, to position in the division cycle, may be achieved by separating the most slowly sedimenting cell fraction from the rest of the cell population. If the relative size of this cell fraction is increased, a lower degree of synchrony may result.

VIII. Discussion

Various criteria may be applied in the evaluation of the degree of synchrony of a proliferating cell system. In general, the passage of cells through a relatively small "window" such as mitosis appears well suited for the characterization of synchrony. Thus, variations with time of mitotic indices permit a reasonably good description of synchronously multiplying cell populations. On the other hand, the duration of mitosis may be affected by the procedure used for preparation of synchronous cultures. Rates of increase in cell number should, therefore, be determined as an independent measure of mitotic activity. In order to thoroughly characterize the synchrony of a proliferating cell system, parameters such as maximum mitotic index, minimum mitotic index, width of the mitotic wave, maintenance of synchrony during successive cell generations, growth fraction of the cell population, and variations with time of cellular DNA synthesis should be combined.

The method presented in this communication is based on the separation by sucrose gradient centrifugation of the smallest cells from the remainder of a cell population grown in suspension culture. Evidently these smallest cells, for the most part at least, represent cells that have passed through mitosis shortly before being collected.

It appears that with regard to the synchrony achieved, collection of the most slowly sedimenting cells is preferable to collection of other cell fractions along the gradient. Separation of late interphase cells on the basis of their rapid sedimentation rate was not attempted because these cells usually are contaminated with cell clumps.

In comparison with methods based on the removal of mitotic cells from monolayers, the initial distribution within the division cycle of early interphase cells obtained by density gradient centrifugation is not as narrow: As seen in Fig. 8, of the population of early interphase cells, about 50% are already in DNA synthesis at the time of incubation. Furthermore, during the period of high mitotic activity, the time required for 80% of the cells to pass through mitosis is approximately 4 hours. Similarly, the data presented in Fig. 5 indicate that the maximum rate of cell multiplication, as observed between 7 and 8 hours, corresponds to a doubling of the cell number within 4.4 hours.

On the other hand, as indicated by Figs. 4, 5, and 7, synchrony is well maintained during successive cell generations. This appears, at least in part, to be a result of (a) the preincubation of the cells in spinner cultures which permits the majority of the cell population to attain a similar, optimal or near-optimal proliferation rate, (b) the use of sucrose gradients in which isotonicity is maintained, and (c) the incubation of the synchronous cell population under the conditions of a steady state suspension culture. Since spinner cultures were used throughout the experiments, the variability in the immediate environment of cells probably was considerably smaller than in monolayer cultures. In addition, the cell line used may exhibit an unusually small variability of generation times which could possibly be attributed to the short duration of the generation time, and, in particular, of the G_1 period.

Cultures of mammalian cells maintaining synchronous growth during prolonged periods of time may be used for further characterization of various kinetic aspects of proliferative behavior, in particular of the variation of generation times within a cell population. In addition, cyclic changes with time of biochemical parameters, such as incorporation rates of labeled nucleic acid precursors (Schaer et al., 1971) and enzyme activities (Schindler et al., 1972), can be detected. The demonstration of cyclic fluctuations in biochemical activities adds strong support to the notion that the observed changes with time reflect the normal progression of cells in the division cycle and are not an artifact attributable to manipulation of cells.

As seen in Table II, cellular viability is not adversely affected by the experimental procedure of collecting early interphase cells. Furthermore, as indicated by the short latency period after reincubation of cells following the separation procedure (Figs. 5 and 6), cellular metabolism apparently is affected to a slight degree only. In addition, the growth fraction of the cell population during the first replication cycle is comparable to that observed during prolonged culture under steady state conditions. On the other hand, it has been reported that in synchronous cultures prepared by Ficoll gradient centrifugation, the cell number increased by a factor of less than 1.4 during the first wave of cell divisions (Warmsley and Pasternak, 1970).

The method of separating early interphase cells from a cell suspension by gradient centrifugation, similarly to methods based on the collection of mitotic cells from monolayers, results in a yield of synchronously dividing cells which represents only a small proportion of the original population. On the other hand, because cell suspensions may be conveniently cultured on a large scale, sucrose gradient centrifugation is suitable for preparing large numbers of synchronously multiplying cells. Synchronous cultures containing up to 8×10^7 cells may be obtained without difficulty if two gradients are used in parallel.

As indicated by the results presented in Fig. 11, the degree of synchrony is inversely related to the fraction of the original cell population collected from the sucrose gradient. The yield of synchronously dividing cells, therefore, may be increased at will if optimal synchrony is not required.

In order to increase the yield of synchronous cells obtained by selective detachment of mitotic cells from monolayers, repeated shaking of monolayer cultures and cold storage of collected mitotic cells until further incubation has been proposed (Lindahl and Sörenby, 1966; Tobey et al., 1967; Lesser and Brent, 1970). Cold storage of cells during several hours may, however, result in changes of cellular metabolism and/or of progression of cells in the division cycle. In fact, partial synchronization of cell populations subsequent to cold storage has been reported (Newton and Wildy, 1959; Miura and Utakoji, 1961). Furthermore, the results of Lesser and Brent (1970) suggest that entry into the G_1 period may be delayed and thymidine nucleotide pool sizes may be decreased after reincubation of mitotic cells subjected to cold storage.

While sucrose density gradient centrifugation also causes some retardation in progression of cells during the first generation time, as seen in Fig. 5, cells exhibit normal proliferation kinetics during subsequent replication cycles. In view of the prolonged maintenance of synchrony, metabolic parameters may, therefore, be studied during more than one cell generation. If cyclic changes of biochemical activities are demonstrated in synchronous cultures, the possibility of artifacts due to manipulation of cells during the separation procedure can be excluded with a high degree of certainty.

In conclusion, because of its simplicity and versatility, preparation of synchronous cultures by density gradient centrifugation of cell populations may be applicable in various studies on the division cycle of cultured mammalian cells. The method appears particularly useful with respect to the large yield of synchronously dividing cells available and to the maintenance of synchronous growth during several cell generations.

REFERENCES

Axelrad, A. A., and McCulloch, E. A. (1958). *Stain Technol.* 33, 67.
Ayad, S. R., Fox, M., and Winstanley, D. (1969). *Biochem. Biophys. Res. Commun.* 37, 551.
Bootsma, D., Budke, L., and Vos, O. (1964). *Exp. Cell Res.* 33, 301.
Bostock, C. J., Prescott, D. M., and Kirkpatrick, J. B. (1971). *Exp. Cell Res.* 68, 163.
Dunn, T. B., and Potter, M. (1957). *J. Nat. Cancer Inst.* 18, 587.
Eidinoff, M. L., and Rich, M. A. (1959). *Cancer Res.* 19, 521.
Gautschi, J. R., Schindler, R., and Hürni, C. (1971). *J. Cell Biol.* 51, 653.
Green, J. P., and Day, M. (1960). *Biochem. Pharmacol.* 3, 190.
Lesser, B., and Brent, T. P. (1970). *Exp. Cell Res.* 62, 470.
Lindahl, P. E., and Sörenby, L. (1966). *Exp. Cell Res.* 43, 424.
MacDonald, H. R., and Miller, R. G. (1970). *Biophys. J.* 10, 834.
Mitchison, J. M., and Vincent, W. S. (1965). *Nature (London)* 205, 987.
Miura, T., and Utakoji, T. (1961). *Exp. Cell Res.* 23, 452.
Morris, N. R., Cramer, J. W., and Reno, D. (1967). *Exp. Cell Res.* 48, 216.
Mueller, G. C. (1969). *Fed. Proc., Fed. Amer. Soc. Exp. Biol.* 28, 1780.
Newton, A. A., and Wildy, P. (1959). *Exp. Cell Res.* 16, 624.
Okada, S. (1967). *J. Cell Biol.* 34, 915.
Petersen, D. F., and Anderson, E. C. (1964). *Nature (London)* 203, 642.
Rueckert, R. R., and Mueller, G. C. (1960). *Cancer Res.* 20, 1584.
Schaer, J. C., and Schindler, R. (1967). *Biochim. Biophys. Acta* 147, 154.
Schaer, J. C., Ramseier, L., and Schindler, R. (1971). *Exp. Cell Res.* 65, 17.
Schindler, R. (1960). *Helv. Physiol. Pharmacol. Acta* 18, C93.
Schindler, R. (1964). *Exp. Cell Res.* 34, 595.
Schindler, R., and Hürni, C. (1971). *Nature (London), New Biol.* 234, 148.
Schindler, R., Day, M., and Fischer, G. A. (1959). *Cancer Res.* 19, 47.
Schindler, R., Odartchenko, N., Grieder, A., and Ramseier, L. (1968). *Exp. Cell Res.* 51, 1.
Schindler, R., Ramseier, L., Schaer, J. C., and Grieder, A. (1970). *Exp. Cell Res.* 59, 90.
Schindler, R., Grieder, A., and Maurer, U. (1972). *Exp. Cell Res.* 71, 218.
Sinclair, R., and Bishop, D. H. L. (1965). *Nature (London)* 205, 1272.
Sinclair, W. K. (1965). *Science* 150, 1729.
Sinclair, W. K. (1967). *Cancer Res.* 27, 297.
Terasima, T., and Tolmach, L. J. (1963). *Exp. Cell Res.* 30, 344.
Tobey, R. A., Anderson, E. C., and Petersen, D. F. (1967). *J. Cell. Physiol.* 70, 63.
Warmsley, A. M. H., and Pasternak, C. A. (1970). *Biochem. J.* 119, 493.
Whitmore, G. F., and Gulyas, S. (1966). *Science* 151, 691.
Xeros, N. (1962). *Nature (London)* 194, 682.

Chapter 3

Production and Characterization of Mammalian Cells Reversibly Arrested in G_1 by Growth in Isoleucine-Deficient Medium

ROBERT A. TOBEY

Biomedical Research Group, Los Alamos Scientific Laboratory,
University of California, Los Alamos, New Mexico

I. Introduction

A. General Comments

An extremely interesting and important class of cells in man exists under normal conditions as a pool of nonproliferating cells, arrested in the postmitotic phase of the cell cycle (i.e., in G_1 arrest or G_0). Under appropriate conditions, these cells can be stimulated either to differentiate—thereby irreversibly losing reproductive capacity—or to resume cycle traverse for a limited number of divisions, ultimately accumulating again in a state of G_1 arrest. Regulation of mammalian cell proliferation *in vivo* is an extraordinarily complex process mediated by a multiplicity of interacting operations of ill-defined nature. Because of the complexity of regulation *in vivo*, it would be desirable to obtain a simpler *in vitro* model system. An ideal model system would be one in which large quantities of proliferating cells could be reversibly induced to enter a nonproliferating, biochemically stable state.

One *in vitro* system currently in use employs nonproliferating lymphocytes cultivated *in vitro* and stimulated to resume cycle traverse and to divide following administration of mitogenic compounds such as phytohemagglutinin (see review by Oppenheim, 1968). Another *in vitro* system consists of low-passage cells sensitive to confluency/density inhibition; following addition of serum, part of the population can be induced to replicate DNA and divide (Nilausen and Green, 1965; Todaro *et al.*, 1965). Both of these *in vitro* systems are useful model systems and have been profitably employed to investigate properties of traversing and nontraversing states. The principal objection to these systems is the difficulty in obtaining sufficiently pure populations containing adequate numbers of cells for biochemical characterization.

We have developed a technique at the Los Alamos Scientific Laboratory in which mammalian cells grown *in vitro* may be accumulated in a stable, reversible state of G_1 by limiting the amount of the essential amino acid, isoleucine, in the culture medium. The effect has been demonstrated in several cell lines, perhaps suggesting a general role for an isoleucine-containing compound in controlling cellular proliferation *in vitro*. By adding back isoleucine, essentially all arrested cells resume cycle traverse, complete G_1, replicate DNA, and divide in synchrony. The number of cells available is only limited by the size of the culture vessel employed. Thus, we have a simple, convenient *in vitro* model system in which we can induce large numbers of cells to enter and recover from a state of nonproliferation merely by manipulating the amount of isoleucine in the medium. This chapter is concerned with production and characterization of cells reversibly arrested in G_1 brought about by growth in culture medium containing varying amounts of isoleucine. It should be stated at the outset that we do not suggest that isoleucine availability plays a regulatory role in cellular proliferation *in vivo*. However, we feel that this isoleucine-dependent phenomenon relating to cellular proliferation observed *in vitro* is of interest because (1) it allows production of large quantities of synchronized cells, not grossly perturbed, suitable for studies of biochemical events in the cell cycle, and (2) it allows one to focus attention on a regulatory product which requires a high level of available isoleucine.

An abundance of data employing numerous different cell types and a variety of analytical techniques suggests that DNA replication and protein synthesis are coordinated in the mammalian cell. That is, gross inhibition of protein synthesis results in inhibition of genome replication (Littlefield and Jacobs, 1965; Young, 1966; Brega *et al.*, 1968; Kim *et al.*, 1968; Weiss, 1969; Hodge *et al.*, 1969). Thus, any manipulation of the culture medium causing a drastic reduction in protein synthetic capacity should be expected to inhibit DNA synthesis in addition to producing an inhibition of other protein-synthesis-requiring operations such as general cell cycle traverse. Reversible inhibition of DNA replication through reversible inhibition of protein synthesis is not intrinsically interesting if accompanied by effects on general cell cycle processes. Rather, it would be extremely useful to devise a system in which genome initiation, but not continued replication, could be preferentially inhibited.

In this chapter we will discuss a system in which limitation of isoleucine results in a specific inhibition of *initiation* of DNA synthesis, while allowing cells already synthesizing DNA at the time of transfer to isoleucine-deficient medium to complete interphase, divide, and accumulate in G_1. It is this specific effect upon initiation but not continuation of DNA synthesis that makes the isoleucine-deficient state interesting. The basic question that arises concerns the nature of the substance(s) required for initiation but not made in cells grown in isoleucine-deficient medium. By employing the isoleucine-deficiency technique, we may be able ultimately to characterize this important regulatory compound; in this sense, we shall refer in this chapter to "isoleucine-mediated regulation of genome replication." Note that the above definition does not preclude other methods for studying this regulatory substance; it merely states that adequate levels of isoleucine must be available for cells to synthesize successfully this product essential for initiation of DNA synthesis. Under appropriate conditions, inhibition of synthesis of this product may be possible by limiting other amino acids, etc.

B. Discovery of the Isoleucine-Mediated Phenomenon

1. Medium Components Deficient in High Cell Concentration Stationary Phase

The role of isoleucine content in genome regulation was discovered by accident. In a study involving a series of monolayer cultures of Chinese hamster cells, line CHO, grown in F-10 medium supplemented with 10% calf and 5% fetal calf sera, several extra cultures remained at the conclusion of the study. The bottle cultures were left in the incubator and, when rediscovered 4 days later, were found to contain viable cell populations. The high degree of viability after this long incubation period at high cell concentration was unexpected, since stationary-phase cultures usually begin dying after short periods at high cell concentrations. Following replacement of the culture medium in the monolayers with fresh, complete medium, the cells divided in synchrony, although only after an initial hiatus of 12 hours during which the cell number remained constant. A similar phenomenon also occurred in suspension cultures of CHO cells grown to high cell concentrations in F-10 medium (Tobey and Ley, 1970).

Further characterization of these stationary-phase cells revealed that the cells were accumulating in the pre-DNA synthetic, G_1, phase of the cell cycle. That the cells were accumulating in G_1 was established by the following criteria: (1) autoradiographic studies which indicated that the nondividing cells were not incorporating labeled thymidine into DNA; (2) microfluorometric measurements which revealed that arrested cells contained the amount of DNA typical for G_1 cells; and (3) data which indicated that synthesis of DNA always preceded cell division (Tobey and Ley, 1970). It was then shown that addition of serum-free medium to stationary-phase cells resulted in synchronous resumption of cell cycle traverse. Serum could also induce partial recovery from arrest, although most of the serum stimulatory activity could be removed by means of dialysis against balanced salt solution (Ley and Tobey, 1970). Since the compounds(s) required for reinitiation of cycle traverse did not appear to be a primary serum factor (with all the inherent difficulties of analysis of serum fractions) but was found in the chemically defined portion of the F-10 medium, attempts were made to identify the limiting compound. Following a series of experiments in which fresh, serum-free medium deficient in various medium components was added back to arrested stationary-phase cells, it was shown that arrested cell populations required both isoleucine and glutamine before resumption of cycle traverse could occur (Ley and Tobey, 1970). Further work (Tobey and

Ley, unpublished observations) indicated that accumulation of cells in G_1 was the result of insufficient quantities of isoleucine only since (1) cells fed glutamine daily during growth to stationary phase (i.e., limited isoleucine only) arrested in G_1, whereas cells fed isoleucine daily during growth to stationary phase did not, and (2) exponentially growing cells transferred to isoleucine-deficient medium accumulated in G_1 arrest, while cells transferred to glutamine-deficient medium stopped throughout the cell cycle. The above results suggested that inability of stationary-phase cells to initiate DNA synthesis and resume cycle traverse was attributable to a deficiency (suboptimal quantity) of available isoleucine in the culture medium.

2. Rapid Induction of G_1 Arrest by Cultivation in Isoleucine-Deficient Medium

The primary effect of isoleucine concentration on G_1 arrest was demonstrated by resuspending cells from an exponentially growing culture in fresh F-10 medium from which isoleucine was omitted, supplemented with dialyzed sera and glutamine to $2 \times 10^{-3} M$ (i.e., twice the normal F-10 content). Within 24 to 30 hours cells accumulated in G_1 and, by merely adding back isoleucine, essentially all cells in the population first initiated DNA synthesis and subsequently divided in synchrony (Tobey and Ley, 1971). Conditions were established for reversibly arresting Syrian hamster and mouse L cells in G_1, indicating that the isoleucine-mediated phenomenon was not peculiar to the CHO cell (Tobey and Ley, 1971). This rapid induction technique is now used routinely in this laboratory in place of stationary-phase cultures.

3. Deficiencies in Other Amino Acids

To ensure that accumulation in G_1 was not the result of general amino acid starvation, exponentially growing CHO cells were placed in F-10 medium supplemented with dialyzed sera and singly deficient in other amino acids [valine, glutamine, leucine, arginine, histidine, or methionine (Ley and Tobey, 1970)]. The culture doubling time immediately slowed to 56 hours or more upon resuspension in the deficient medium, and the total cell number increase was never more than 20%. Upon resuspension in fresh, complete medium, cultures began dividing after a lag period of 7 hours or less with little evidence for synchrony induction. Therefore, it appears that gross depletion of other amino acids resulted in a general inhibition of cycle traverse. More recent studies by Everhart (1972) have confirmed our findings that, under conditions of gross depletion of other amino acids, CHO cells fail to accumulate in large numbers in G_1 arrest. However, Everhart and Prescott (1972) noted that CHO cells could be arrested in G_1 by maintenance in F-12 medium containing 0.1% of the normal concentration of leucine. These results suggest that conditions involving *limitation* (but not starvation) of other amino acid species may yield systems comparable to our isoleucine-deficiency technique.

II.   Culture Conditions Affecting Induction of G_1 Arrest in Isoleucine-Deficient Medium

A.   Cycle Kinetics of CHO Cells in F-10 Medium

A hypodiploid line of Chinese hamster cells [line CHO, originally obtained from Dr. T. T. Puck (Tjio and Puck, 1958)] was grown in F-10 medium as described by Ham in 1963 (purchased as powdered medium from Grand Island Biological Company, Grand Island, New York), supplemented with 10% calf and 5% fetal calf sera (purchased from Flow Laboratories, Rockville, Maryland). In the exponentially growing suspension cultures over the range of 0.75 to 4.5×10^5 cells/ml, the doubling time was approximately 16.5 hours with the following distribution in each phase of the cell cycle determined by standard cell cycle analytical techniques (Defendi and Manson, 1963; Puck and Steffen, 1963; Puck *et al.*, 1964; Tobey *et al.*, 1967a): G_1, 9 hours and 64%; S, 4 hours and 23%; and G_2 + M, 3.5 hours and 13% (Fig. 1). These cycle parameters indicate that, under our growth conditions, the G_1 phase of the cell cycle was extremely long while the duration of S phase was relatively short. The duration of G_1 phase in the CHO cell was dependent upon the batch of serum employed (Tobey *et al.*, 1967a). In all studies described in this report, very large batches of serum (prepared by pooling multiple small batches) have been employed to minimize the variable effects of individual serum lots on cycle kinetics.

The isoleucine-deficient medium was purchased from Grand Island Biological Company and was identical to the regular F-10 medium except that isoleucine has been omitted and glutamine content ($2 \times 10^{-3} M$) was twice that found in normal F-10 medium. The medium was supplemented with dialyzed sera prepared in this Los Alamos Laboratory. Prior to dialysis, the calf serum was heat-inactivated at 56°C for 30 minutes to reduce the degree of cell clumping in isoleucine-deficient medium. Then the calf and fetal calf sera were mixed together in a ratio of 2:1, placed in dialysis tubing, and dialyzed against Earle's balanced salt solution, pH 7.2, at a 1:10 volume for 6 days at 3°C with changes of salt solution on alternate days. Following dialysis, the sera were

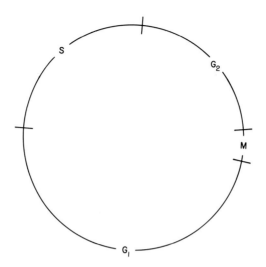

FIG. 1. Schematic representation of line CHO Chinese hamster cell cycle. Cells are grown in suspension culture in F-10 medium (Ham, 1963) supplemented with 10% calf and 5% fetal calf sera. For a culture doubling time of 16.5 hours, the duration of G_1 is 9 hours, S is 4 hours, G_2 is 3 hours, and M is 0.5 hour.

sterilized by filtration and stored at $-20°C$. Dialyzed sera obtained from commercial sources were found to be highly variable from lot to lot and were generally unsuitable due to apparent serum polypeptide breakdown during dialysis; some lots were also found to be contaminated with yeast. Consequently, commercially available dialyzed sera were not employed in any of the experiments described in this chapter.

B. Variable Cell Number Increase Following Transfer to Isoleucine-Deficient Medium Dependent upon Centrifugation Procedure Employed

Following transfer of exponentially growing cells in complete medium to isoleucine-deficient medium (i.e., suboptimal quantities of isoleucine but not total starvation conditions), there was an immediate reduction in rate of cell division. The total number of cells which divided in isoleucine-deficient medium (hereafter referred to as ile$^-$ medium) was dependent upon amount of isoleucine carried over from the exponential culture. [In this and following figures divided fraction represents $N/N_0 - 1$ so that a true population doubling would be indicated by an increase from 0 to 1.0 on the scale provided. N_0 = the initial cell number, and

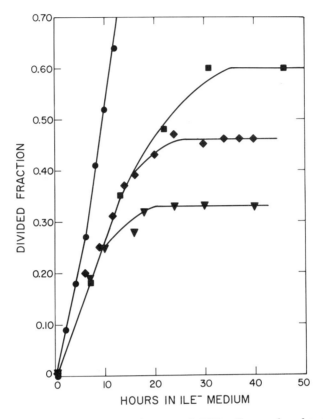

Fig. 2. Effect of method of centrifugation of CHO cells transferred to isoleucine-deficient medium on final cell concentration in the arrested state. Cells from an exponentially growing suspension were either spun down and washed (triangles), spun down in a wide-mouth centrifuge cone (diamonds), or spun down in a narrow-mouth, screw-cap top centrifuge cone (squares) prior to resuspension in isoleucine-deficient F-10 medium containing twice the normal F-10 glutamine content ($2 \times 10^{-3} M$) and supplemented with dialyzed 10% calf and 5% fetal calf sera (ile⁻ medium). An aliquot spun down in a large-mouth centrifuge cone and resuspended in complete medium served as a control (circles). The initial cell concentration in all cultures at $t = 0$ was 150,000 cells/ml. Cell number determinations were made with an electronic particle counter. Divided fraction represents $N/N_0 - 1$ so that a true doubling of the population would appear as an increase from 0 to 1.0 on the ordinate axis.

N = the number of cells at any given point in the experiment. This convention allows direct comparison with autoradiographic analyses of labeled fractions.] For example, in Fig. 2 are results obtained by three separate centrifugation procedures. One culture was spun down and *washed* with ile⁻ medium prior to resuspension. Total cell number increase following resuspension was 33%. In a culture spun down in a wide-mouth centrifuge cone (allowing drainage of the complete medium) and resuspended in ile⁻ medium without an intermediate wash step, some isoleucine apparently was carried along with a resultant cell number increase of 45% following resuspension in ile⁻ medium. Finally, when a culture was spun down in a narrow-mouth centrifuge cone (which did not allow adequate drainage of the medium) and was resuspended in ile⁻ medium without washing, the cell number increase was nearly 60%, implying that even more isoleucine was carried along. Evidence will be provided that appropriate suboptimal quantities of isoleucine added to ile⁻ medium allowed variable numbers of cells to initiate DNA synthesis and divide, dependent upon amount of isoleucine added.

Even though the ultimate level at which cells stopped dividing was different for the various cultures, all three cultures (Fig. 2) attained the arrested state after approximately 20 to 24 hours in isoleucine-deficient medium. Examination of the populations at 24 hours or later indicated that, in all cases, the populations contained 95% or more of cells with a G_1 DNA content. The cell number increase in the washed culture after transfer to ile⁻ medium was 33% or approximately the sum of cells in $S + G_2 + M$ in the initial exponential population. These results suggest that, in cultures appropriately depleted of isoleucine, the cells initially located in S phase or at later positions in the cell cycle could divide ultimately, while cells in G_1 could not commence synthesis of DNA. Stated differently, in washed cells entry of cells into S (initiation of DNA synthesis) is very rapidly inhibited, while cells initially in S, G_2, or M at time of transfer to ile⁻ medium continue traverse of the cell cycle until they accumulate in G_1. Everhart's (1972) data employing double-label autoradiography beautifully demonstrate that washed CHO cells transferred to ile⁻ medium immediately lose the capacity to initiate genome replication, while cells in S phase continue synthesizing DNA. Results in Fig. 2 demonstrate the importance of the method of transfer of cells to ile⁻ medium in predicting the final concentration of arrested cells.

C. Concentrations of Isoleucine Required for Initiation of DNA Synthesis and Cell Division

To obtain an estimate of the amount of isoleucine added back to ile⁻ medium, required for initiation of DNA synthesis, cells were prepared by mitotic selection and were allowed to enter G_1 in ile⁻ medium. These G_1 cells in ile medium were then given thymidine-³H along with different quantities of isoleucine. The fraction of cells labeled with thymi-

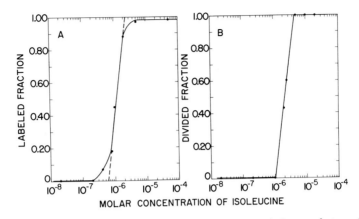

MOLAR CONCENTRATION OF ISOLEUCINE

FIG. 3. Effect of isoleucine concentration on fraction of G_1 population of CHO cells initiating DNA synthesis: (A) Cells prepared by mitotic selection were resuspended in isoleucine-deficient (ile⁻) medium at 150,000 cells/ml containing different amounts of isoleucine and thymidine-³H to a final concentration of 0.05 μCi/ml. The labeled fractions were determined from autoradiographs prepared at 16 hours. Corrections were made, where necessary, for increases in cell number during the 16-hour labeling period; and (B) cells maintained 30 hours in ile⁻ medium were resuspended in fresh ile⁻ medium plus different quantities of isoleucine at $t = 0$ at a concentration of 150,000 cells/ml. Cell number was determined for the various cultures at $t = 28$ hours with an electronic particle counter. Definitions of ile⁻ medium and divided fractions are as in Fig. 2.

dine-³H (fraction synthesizing DNA) was determined autoradiographically after 16 hours. Results in Fig. 3A indicate that it was possible to increase the fraction of cells initiating DNA synthesis from 0 to essentially 100% merely by increasing the concentration of isoleucine approximately fourfold (i.e., from 6×10^{-7} to $2 \times 10^{-6} M$). Similar results were obtained when G_1 cells prepared by growth in ile⁻ medium were given labeled thymidine and different quantities of isoleucine.

To determine the amount of isoleucine required to allow cells to initiate DNA synthesis *and* divide, G_1-arrested cells were given varying quantities of isoleucine and the cell number increase was determined for each culture 28 hours later. The results in Fig. 3B once again indicate that, over a narrow range of isoleucine concentrations, the fraction dividing increased from 0 to 100%. Note that in both Figs. 3A and 3B the initial cell concentrations were 1.5×10^5 cells/ml. As one might expect, changes in initial concentration of cells produce changes in amount of added isoleucine necessary to induce resumption of cycle traverse. Due to uncertainties regarding the concentration of isoleucine in the ile⁻ medium, the values presented do not represent absolute values of isoleucine required for the two cellular processes.

D.   Attempts at Long-Term Maintenance of CHO Cells in G_1 Arrest

If cells are maintained in G_1 arrest in ile⁻ medium in excess of approximately 150 hours, cell viability drops (Tobey and Ley, 1970). It is not clear whether death occurs because of a starvation for isoleucine or for some other medium component. Consequently, attempts were made to maintain cells for prolonged periods in G_1 arrest under conditions where medium components except isoleucine were given to the cells at intermittent periods. This was accomplished by resuspending arrested cells in fresh preparations of ile⁻ medium at intervals. That is, an attempt was made to determine how long cells could survive in medium deficient in isoleucine but sufficient in other nutrients. The results of this experiment are shown in Fig. 4. A control culture incubated in ile⁻ medium increased by the expected increment following transfer to ile⁻ medium (see Fig. 2, diamonds) and thereafter maintained a constant cell number for approximately 6 days (Fig. 4, circles). This long-term maintenance of constant cell number emphasizes the biochemical stability of cells maintained for long periods in ile⁻ medium; no attempts were made to restore isoleucine to this ile⁻ control culture in the experiment in Fig. 4. Instead, an aliquot of cells was prepared in ile⁻ medium and, at varying times thereafter, the cells were resuspended in fresh ile⁻ medium (i.e., refed with fresh ile⁻ medium, see Fig. 4, triangles). After refeeding, there was

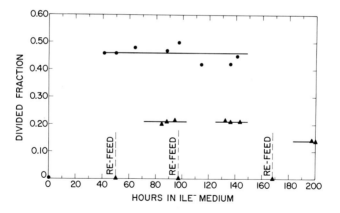

Fig. 4.   Effect of feeding G_1-arrested CHO cells in ile⁻ medium with fresh ile⁻ medium. A culture of CHO cells was resuspended in ile⁻ medium at a concentration of 100,000 cells/ml and maintained in ile⁻ medium thereafter as ile⁻ control (circles). Another culture was placed in ile⁻ medium but, at the times indicated, the cells were resuspended in fresh ile⁻ medium (refed with ile⁻ medium, triangles). All cell number determinations were made with an electronic particle counter. Definitions of ile⁻ medium and divided fraction are as in Fig. 2.

a 21% increase in cell number for both refeedings #1 and #2 and a 14% increase after refeeding #3. The cells apparently were obtaining sufficient isoleucine from the fresh "ile⁻" medium to allow a sizable fraction of cells to divide. Thus, due to technical difficulties, it was impossible to maintain cells in G_1 arrest in ile⁻ for periods of several weeks or more by intermittent refeeding with fresh ile⁻ medium without experiencing increases in cell number.

The explanation for increase in cell concentration following refeeding appears to involve catabolism of serum polypeptides by cells in ile⁻ medium to yield sufficient quantities of isoleucine for the cell division observed (Eagle and Piez, 1960). In this regard, cells maintained for long periods in ile⁻ medium exhibit a very slow increase in protein mass (Enger and Tobey, 1972). Addition of dialyzed serum to G_1-arrested cultures produces approximately the same results as refeeding with ile⁻ medium containing dialyzed serum (R. A. Tobey, unpublished observations). Proteolysis of serum polypeptides during serum dialysis could also liberate additional isoleucine.

E.   Synchronization of Other Cell Lines

The isoleucine-deficiency-induced G_1-arrest technique is not peculiar to the CHO cell. Attempts were made to modify the technique for use in two cell lines which are not readily synchronized by other techniques. For example, it was possible to accumulate mouse L cells in G_1 arrest by culturing them for 90 hours in Eagle's minimum essential medium from which isoleucine was omitted; the medium was supplemented with dialyzed fetal calf serum (Tobey and Ley, 1971). By adding back isoleucine to the arrested population, cell cycle traverse resumed, and cells first initiated DNA synthesis and subsequently divided in synchrony. Another line accumulated in G_1 arrest by growth in isoleucine-limited medium was the BHK21/C-13 line of Syrian hamster cells. When BHK cells were grown for 48 hours in isoleucine-deficient F-10 medium containing dialyzed sera and then were resuspended in fresh, complete medium, the fraction of cells initially synthesizing DNA was essentially zero, then rose rapidly as the cells entered S phase in synchrony (Tobey and Ley, 1971). It should be emphasized that the L and BHK lines employed exhibited no detectable pleuropneumonia-like organisms at the time of experimentation. Since very little effort was expended on determining conditions for induction of G_1 arrest in the L and BHK cell lines, the conditions undoubtedly are far from ideal. However, note the dramatically different conditions required for accumulation in G_1 arrest for the L, BHK, and CHO cell lines.

The conditions for induction of arrest of CHO cells described in this report work well in our hands. However, the protocol given refers only to the CHO cell and should be utilized only as a general guideline when attempting to adapt the technique for use with other cell lines or for use with CHO cells grown under different cultivation techniques than those employed in this laboratory. Among factors that *must* be taken into account in adaptation of the isoleucine-deficiency technique for use with other cell lines are (1) the cell cycle kinetics; (2) the amount of iso- leucine made available to the cells during growth in deficient medium (i.e., the degree of isoleucine insufficiency); and (3) the duration of in- cubation period in the deficient medium. Note that the requirement for isoleucine content for traverse and survival varies greatly among different cell lines. Some cell lines die within hours after transfer to ile⁻ medium supplemented with dialyzed sera. In cases such as these, it is recom- mended that cells be refed with suboptimal quantities of isoleucine dur- ing attempted accumulation in G_1.

F.   Potential Problems in Production of Arrested Cells

1.   CELL CLUMPING

Cell number enumeration is accomplished routinely in this laboratory with an electronic particle counter (Petersen *et al.*, 1969a). Electronic particle counters score cell clumps as a single event and, thus, under- estimate the number of cells in a clumped population. Since in exponen- tially growing suspension cultures of CHO cells the cells exist as an es- sentially monodisperse population, corrections for clumping of cells are not necessary. However, the fraction of cell clumps was highly variable among series of cultures grown in isoleucine-deficient medium supple- mented with different lots of dialyzed sera. In some preparations, the fraction of cells occurring as doublets in ile⁻ medium exceeded 30%. Heat inactivation of calf serum at 56°C for 30 minutes immediately prior to dialysis reduced the clumped fraction considerably and is now a routine step in preparation of dialyzed sera. The effect of heat inactivation prior to dialysis on increase in cell number measured electronically is shown in Fig. 5. Even with this precaution, the fraction of cell doublets is rou- tinely higher (0.08 versus 0.03) and more variable in isoleucine-deficient cultures as compared to exponential cultures in complete medium. We routinely determine the fraction of cells as doublets (higher aggregates are rarely seen) by direct microscopic examination of suspension mate- rial and employ this value to correct the cell number obtained with the electronic particle counter.

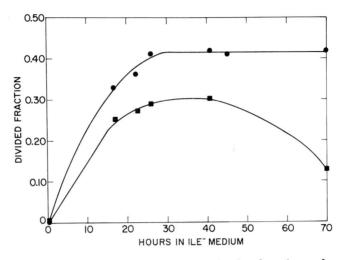

Fɪɢ. 5. Cell clumping in cultures of CHO cells placed in ile⁻ medium supplemented with dialyzed sera ± heat inactivation prior to dialysis. Cells were resuspended in ile⁻ medium supplemented with non-heat-inactivated, dialyzed sera (squares) or in medium supplemented with dialyzed sera heated to 56°C for 30 minutes prior to dialysis (circles). Cell number determinations were made with an electronic particle counter which scores cell clumps as a single event. Definitions of ile⁻ medium and divided fraction are as in Fig. 2.

In order to obtain some information on the nature of clumping, testicular hyaluronidase was added to isoleucine-deficient medium during growth to G_1 arrest (P. M. Kraemer and R. A. Tobey, unpublished observations). The rationale for utilization of hyaluronidase was based on Kraemer's observations (1970) that, in G_1-arrested CHO cells in ile⁻ medium resulting from growth to high cell concentration in stationary phase, the content of surface hyaluronic acid was markedly increased relative to other constituents. Thus, it seemed likely that at least part of the adhesiveness responsible for clump formation might be due to the surface hyaluronic acid. Addition of testicular hyaluronidase to 100 μg/ml to ile⁻ medium containing non-heat-treated dialyzed sera at the time of resuspension of exponentially growing cells in ile⁻ medium yielded results similar to those obtained in medium supplemented with the heated, dialyzed sera (Fig. 6). The apparent increase in cell number, measured electronically in cultures grown in medium containing non-heated sera and no hyaluronidase, was lower; microscopic examination revealed that the fraction of cells occurring in clumps in this culture was sufficient to account for the reduced cell number scored electronically.

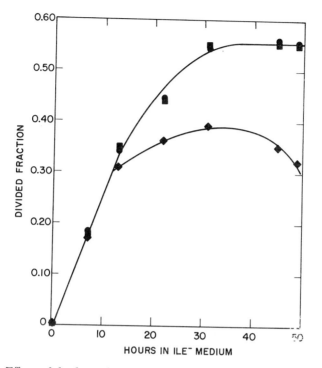

FIG. 6. Effect of hyaluronidase on cell number increase in cultures of CHO cells cultivated in ile⁻ medium containing non-heat-inactivated, dialyzed sera. A control was resuspended in ile⁻ medium containing heat-inactivated sera (circles). Two experimental cultures were resuspended in non-heat-inactivated, dialyzed sera in the presence (squares) and absence (diamonds) of testicular hyaluronidase at a concentration of 100 units/ml throughout the experiment. Cell counts were determined with an electronic particle counter which scores cell clumps as a single event. Definitions of ile⁻ medium and divided fraction are as in Fig. 2.

Treatment of clumped cells with hyaluronidase at the time of cell count determination failed to yield reproducible increases in cell number. We suggest that heat inactivation removes or inactivates a compound which stimulates hyaluronic acid production in cells grown in ile⁻ medium. Alternatively, heat treatment could also activate a repressor of hyaluronic acid synthesis contained in serum. Further experiments obviously will be required before the problem may be resolved. Because of the greater ease and lower cost, heat treatment of serum prior to dialysis is preferred over addition of large quantities of hyaluronidase to the culture medium. However, we strongly suggest that electronic particle data be corrected with data obtained from direct observation of the fraction of clumped cells.

2. *Mycoplasma* CONTAMINATION

In view of the high incidence of contamination of commonly employed cell lines with *Mycoplasma* [pleuropneumonia-like organisms (PPLO)] and documented evidence for competition between cells and PPLO for nutrients (Powelson, 1961; Hakala *et al.*, 1963; Kraemer, 1964; Macpherson, 1966; Levine *et al.*, 1968; Gil and Pan, 1970; Stanbridge *et al.*, 1971), cultures of CHO cells were deliberately infected with the PPLO *Mycoplasma hyorhinis* provided by Dr. Elliot Levine. This organism is a common contaminant of a variety of tissue culture lines. When infected CHO cells were allowed to grow to high cell concentration in stationary phase, the maximum cell concentration attained was only slightly more than half that in an uninfected stationary-phase culture, and infected cells began dying after periods in excess of 50 hours in stationary phase (Tobey and Ley, 1971). Upon resuspension of infected stationary-phase cells in fresh, complete medium, there was little evidence of synchronous cell division. Resuspension of PPLO-infected CHO cells in ile⁻ medium plus dialyzed sera produced a growth pattern similar to that observed in uninfected controls grown in ile⁻ medium. However, following resuspension in complete medium, the degree of synchronization of cell division in infected cultures was extremely low (Tobey and Ley, 1971). Therefore, it is difficult or impossible to demonstrate isoleucine-mediated G_1 arrest in cells infected with PPLO. It is imperative that cells be free of PPLO contamination for studies with isoleucine-deficient medium.

III.   Characterization of Cells in
Isoleucine-Deficient Medium

A.   Entry into G_1 Arrest

1. EVIDENCE THAT CELLS ACCUMULATE IN G_1

Several lines of evidence indicate that cells accumulate in the G_1 phase of the cell cycle as a result of cultivation in ile⁻ medium. In the first approach the mass of DNA in ile⁻ cells was determined directly by means of a modified Schmidt-Thannhauser technique (Enger and Tobey, 1972). Results summarized in Table I indicate that, in cells maintained 30 hours in ile⁻ medium, the DNA mass is identical to that obtained in pure populations of G_1 cells prepared by mitotic selection and allowed to enter interphase in suspension culture. The quantity of RNA per cell in ile⁻

TABLE I

QUANTITIES OF DNA AND RNA IN VARIOUS POPULATIONS OF CELLS[a]

Type of culture	$A_{max}/10^7$ cells	
	DNA	RNA
Exponential	2.01 ± 0.04	5.08 ± 0.24
Ile⁻ medium for 30 hours	1.61 ± 0.05	4.40 to 6.0
Cells in G_1 prepared by mitotic selection	1.62 ± 0.04	4.39 ± 0.10

[a] From Enger and Tobey, 1972.

cultures was much more highly variable than was DNA content but was usually closer to values from exponential cultures than to values from cells in G_1 prepared by mitotic selection. In exponentially growing cultures the modal cell is one in late G_1 due to the predominance of G_1 cells under our cultivation conditions (Anderson et al., 1969, 1970). Thus, the DNA and RNA contents of ile⁻ arrested cells are those expected in cells in mid to late G_1; therefore, we conclude that the ratio of RNA to DNA in the arrested cells is not grossly perturbed.

The second line of evidence indicating arrest in G_1 was provided by examination of the relative DNA contents in ile⁻ populations examined with the Los Alamos flow microfluorometer (Van Dilla et al., 1969; Kraemer et al., 1971, 1972; Tobey et al., 1972). With the flow microfluorometer (FMF) it is possible to determine relative DNA contents in large populations by measuring the amount of fluorescent dyes bound specifically to DNA. Cells for FMF analysis are washed, dispersed, fixed with formalin, and then stained by means of a fluorescent-Feulgen procedure employing fluorescent dyes such as acriflavine in place of the basic fuchsin normally utilized in Feulgen procedures (Tobey et al., 1972). In Fig. 7A are results of FMF analysis of an exponentially growing culture of CHO cells. The major peak at channel 19 represents cells with G_1 DNA content, while at twice the mode of the G_1 peak is a peak at channel 38 representing DNA contents of cells in G_2 or M. Cells in S phase with varying degrees of completion of DNA replication are distributed between the peak representing G_1 and the peak for G_2 and M DNA contents.

As expected from conventional cell cycle kinetic studies, the majority of cells in exponential populations under our cultivation conditions are in G_1. The number of cells in the various phases of the cell cycle can be readily determined from the FMF data (Kraemer et al., 1972), and these values agree extremely well with measurements obtained by standard analytical techniques. In Fig. 7B is the FMF profile of cells maintained

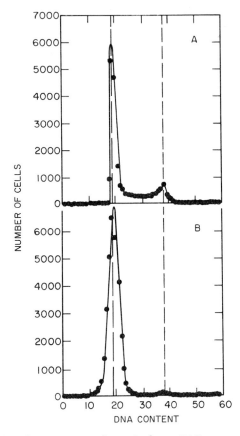

Fig. 7. Flow microfluorometric analysis of relative DNA contents in exponentially growing CHO cultures (A) and in cultures of CHO cells maintained 30 hours in ile⁻ medium (B). Cells were dispersed, fixed with formalin, and Feulgen-stained with the fluorescent dye, acriflavine, as described elsewhere (Tobey *et al.*, 1972). The number of cells examined in the exponential culture was 20,000 and in the arrested culture in ile⁻ medium was 34,000. Ile⁻ medium is defined as in Fig. 2.

for 30 hours in ile⁻ medium. Note that essentially the entire population possessed the G_1 DNA content. There were no cells in S phase; confirmation of S-phase emptying was obtained by the absence of labeled cells in autoradiographs prepared by pretreating cells with thymidine-[3]H. Approximately 1% of the cells appeared to possess the G_2 or M DNA content. These results obtained through FMF analysis agree with the DNA mass data presented above (Table I) in that cells accumulate in G_1 in ile⁻ medium.

FMF analysis yields a reliable means for monitoring a variety of cell cycle kinetics, provided that care is taken to obtain a monodisperse population with a fluorescent stain bound specifically to the DNA and results are obtained with a high-resolution machine. An example of poor quality of data obtained by inappropriate dispersal and staining procedures is illustrated in Fig. 8. Figure 8A represents the DNA distribution pattern of an exponential population employing acriflavine in an appropriate procedure. Figure 8B represents FMF analysis of the same exponential population vitally stained with $10^{-6}\,M$ acridine orange at pH 7.3 and analyzed with the same machine as in Fig. 8A. Note in Fig. 8B the total lack of correlation with DNA content. This is hardly surprising in view of the relatively nonspecific binding of acridine orange to phosphate moieties, both associated and not associated with DNA (Rigler, 1966). In Fig. 8C is the volume distribution spectrum of this exponential population determined with an electronic volume spectrometer. [For a description of the volume spectrometer, see Petersen et al. (1968)]. At best, the acridine orange measurements provide a crude estimate of the volume distribution. These data illustrate the importance of employing a rational preparative procedure.

A third line of evidence suggesting that cells maintained in ile⁻ medium are in G_1 is provided by the observation that, upon adding back isoleucine, all cells first initiate DNA synthesis, then subsequently divide (see Fig. 12).

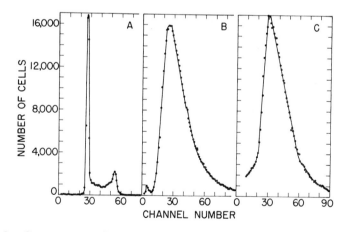

Fig. 8. Comparison of flow microfluorometric analysis of exponentially growing CHO cultures examined by a fluorescent-Feulgen procedure employing acriflavine (A), by a procedure for vitally staining cells with acridine orange (B), and by analysis of volume distribution with a volume spectrometer (C). Cells in (A) were dispersed, fixed in formalin, and Feulgren stained with acriflavine as described previously (Tobey et al., 1972). Cells in (B) were vitally stained with $10^{-6}\,M$ acridine orange at pH 7.3. Cells in (C) were examined without fixation in the volume spectrometer described elsewhere (Petersen et al., 1968).

2. MACROMOLECULAR SYNTHESES

a. Nucleic Acids. The results presented thus far indicate that merely by manipulating the amount of isoleucine in the culture medium, genome replication, in effect, can be turned on or off. The following studies were undertaken to define better the biochemical status of cells accumulating in G_1 arrest in ile⁻ medium (Enger and Tobey, 1972). Upon transfer to ile⁻ medium, the cell number stopped increasing within approximately 24 hours (Fig. 2). Following transfer to ile⁻ medium, the rate of incorporation of labeled thymidine into DNA decreased rapidly, down to 44% of the initial value within the first 2 hours (Fig. 9). In contrast, uridine

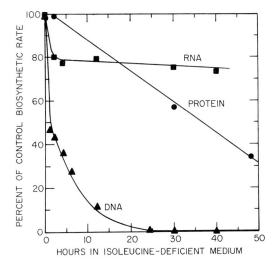

Fig. 9. Rates of DNA, RNA, and protein synthesis in exponentially growing cultures of CHO cells transferred to ile⁻ medium at $t = 0$. Initial cell concentration was 199,000 cells/ml in all cultures. Cultures containing 55 ml of cell suspension were exposed to 6.25 μCi of 5-uridine-³H for 15 minutes, then aliquots of 50 ml were harvested, and the amount of uridine-³H incorporated into RNA was determined by the method of Enger *et al.* (1968), except that 0.3 M KOH treatment at 37°C for 18 hours was substituted for ribonuclease hydrolysis of RNA. Similarly, 60-ml cell cultures were exposed to 2 μCi/ml of methyl-³H-thymidine for 15 minutes, then aliquots of 50 ml were harvested, and the amount of thymidine-³H incorporated into DNA was determined by the method of Enger *et al.* (1968). Polysomes were extracted and analyzed by the techniques described in Enger and Tobey (1972) from aliquots of ~350 ml containing 10^8 cells exposed to 4,5-L-leucine-³H for different time periods ranging from 31.5 to 60 seconds. Translation rate was estimated from an analysis of the increase in specific activity of polysome species as a function of size following incorporation of leucine-³H. Protein synthetic rate was calculated as the product of polysome mass times the translation rate. For details concerning the data in this figure, see Enger and Tobey (1972). Ile⁻ medium is defined as in Fig. 2.

incorporation into RNA was only reduced to 73% of the initial value even after 30 hours in ile⁻ medium (Fig. 9). Experiments in which the mass of stable species of DNA and RNA was determined in ile⁻ cultures yielded a similar striking difference in reductions in DNA and RNA synthetic rates, confirming result obtained with precursor incorporation studies (Enger and Tobey, 1972).

Examination of the various classes of RNA synthesized in arrested cells yielded the results shown in Table II (Enger and Tobey, 1972). Although there was a preferential reduction in rate of synthesis of stable species (ribosomal and 4S RNA) in comparison to unstable species, synthesis of no major RNA species was totally inhibited. Continued synthesis of large quantities of stable species in the absence of an increased rate of turnover would have resulted in a gross imbalance of RNA relative to DNA. From Table I, is obvious that cells in ile⁻ medium did not enter a state of gross nucleic acid imbalance, even though the rates of synthesis of DNA and RNA were decreased in grossly different fashion in ile⁻ medium.

b. Proteins. Since isoleucine is an essential amino acid which must be provided for mammalian cell growth, it might be expected that limiting the amount of isoleucine would result in a rapid, dramatic reduction in protein synthetic capacity. In this regard, restriction of amino acids in the diet of animals *in vivo* or removal from the culture medium *in vitro* results in a depression of biosynthetic capacity due, in large part, to breakdown of polysomes (Staehelin *et al.*, 1967; Sidransky *et al.*, 1967, 1968, 1971; Munro, 1968; Kochhar, 1968; Ward and Plagemann, 1969; Reid *et al.*, 1970; Smulson and Rideau, 1970; Tiollais *et al.*, 1971; Stanners and Becker, 1971; Vaughan *et al.*, 1971; Sidransky and Verney, 1971; Van Venrooij and Poort, 1971). However, cultivation of CHO cells in ile⁻ medium did not result in rapid destruction of polysomes (Enger and

TABLE II

EFFECT OF CULTIVATION OF CHO CELLS IN ISOLEUCINE-DEFICIENT MEDIUM FOR 30 HOURS[a]

RNA species	Percent of control synthetic rate after 30 hours in ile⁻ medium
Ribosomal	25
4S	33
Messenger	50
Heterogeneous nuclear	73

[a] From Enger and Tobey (1972).

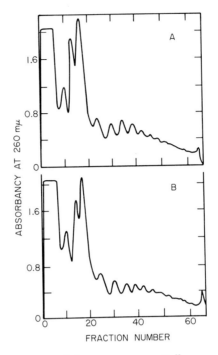

FIG. 10. Polysomes extracted from an exponentially growing culture of CHO cells (A) and from a culture of CHO cells maintained 30 hours in ile⁻ medium (B). Polysome extraction and analysis were prepared as described previously (Enger and Walters, 1970). Ile⁻ medium is defined as in Fig. 2.

Tobey, 1972). The polysome profile in Fig. 10B for cells maintained for 30 hours in ile⁻ medium appears to mimic closely the polysome profile in the exponentially growing control culture (i.e., no gross breakdown of polysomes in isoleucine-deficient medium, see Fig. 10A). The amount of material in the polysome region remained at 90% or more of control value for at least 48 hours with ile⁻ medium (Enger and Tobey, 1972). Zone sedimentation analyses performed on cytoplasm treated with high salt buffer revealed that polysome breakdown during isolation was not occurring (Enger and Tobey, 1972).

The translation rate of cultures maintained for different periods in ile⁻ medium was estimated from an analysis of the increase in specific activity of polysome species as a function of size after incorporation of labeled leucine for time periods less than that required to label fully all species (Enger and Tobey, 1972). Protein synthetic rate was equivalent to the product of synthetic machinery (polysome content) times rate of work (translation rate). As may be seen in Fig. 9, the protein synthetic rate continued at the control rate for at least 2 hours after transfer to ile⁻ medium, *at a time when the DNA synthetic rate was already down to 44% of the initial value.* The protein synthetic rate was 57% and 34% of control values following cultivation in ile⁻ medium for 30 and 48 hours, respectively. These results indicate that synthesis of DNA is much more rapidly and completely inhibited than is synthesis of either RNA or protein species. The observation that protein synthesis continues at significant rates long after initiation of DNA synthesis is completely shut off further suggests that the primary effect of limiting quantities of isoleucine on DNA initiation is not merely the result of a reduction in general protein synthetic capacity. Autoradiographic experiments with monolayer cultures of CHO cells in medium deficient in isoleucine led Everhart (1972) to conclude that the rate of protein synthesis was sufficiently high to allow cells in S at time of transfer to ile⁻ medium to complete S phase but was insufficient to allow entry into S phase from G_1. Thus, protein synthesis essential for completion of late interphase occurs in cells in ile⁻ medium, whereas synthesis of an essential component required for initiation of genome replication does not.

The preceding analyses provide estimates of protein synthetic rate in a relative rather than a net sense. For estimation of the latter value, biuret determination of protein revealed that the protein content per cell increased by approximately 4% during the interval between 30 and 48 hours following transfer to ile⁻ medium (Enger and Tobey, 1972). This indicates that there is a small net increase in protein content in cells arrested in G_1. These results suggest that arrested cells are either obtaining additional isoleucine, perhaps via catabolism of serum polypeptides (Eagle and Piez, 1960), or are synthesizing isoleucine-poor protein species during arrest. As will be shown in a later section, apparently normal isoleucine-containing proteins are made during G_1 arrest.

The slight decrease in polysome content in ile⁻ cultures could arise from insufficient quantities of available messenger RNA species to support full formation of polysomes. Availability of unassociated mRNA was demonstrated by use of low levels of cycloheximide to reduce the rate of translation but not affect the rate of initiation or ribosome attachment. Addition of cycloheximide to cells maintained for 48 hours in ile⁻ medium resulted in a migration of most of the 80S ribosomes and ribosomal subunits into polysomes (Enger and Tobey, 1972). The same phenomenon was observed in ile⁻ cells pretreated with 10 μg/ml of actinomycin prior to addition of cycloheximide. This superformation of polysomes indicates that, even in the absence of *de novo* synthesis of mRNA species, an abundance of mRNA exists in ile⁻ cells. Ile⁻ cells may provide a valuable system for investigation of messenger RNA storage and stabilization.

c. Histones. Histone metabolism was examined in both arrested cells and in traversing cells in different phases of the cell cycle following release from arrest (Gurley *et al.*, 1972a). These studies were of interest because sufficient quantities of cells could be provided in G_1 and S at very high levels of purity to allow a comprehensive comparison of histone metabolism in these phases of the cell cycle. Furthermore, since it was shown that histone synthesis occurred during ile⁻-mediated G_1 arrest, these studies also provided an opportunity for partial characterization of this class of proteins synthesized during arrest. Regarding this latter point, the techniques of analysis of protein synthesis described in the preceding section provided little information regarding the nature of the proteins synthesized in the arrested state. For example, one could envision that the labeled amino acid was incorporated into small polypeptide fragments bearing little or no resemblance to native proteins synthesized in untreated cells.

Among the findings of Gurley *et al.* (1972a) were the following. In agreement with previous studies involving a variety of techniques (Bloch and Godman, 1955; Irvin *et al.*, 1963; Littlefield and Jacobs, 1965; Spalding *et al.*, 1966; Prescott, 1966; Borun *et al.*, 1967; Gurley and Hardin, 1968), it was demonstrated that histone synthesis occurred mainly during S phase. It was possible to obtain sufficient quantities of pure G_1 cells (prepared by either mitotic selection or in ile⁻-induced arrest) so that histone synthetic capacity could also be examined definitively for the first time in G_1 cells. Results obtained indicated that traversing G_1 cells synthesized histones at 3.5% to 5% of the rate in S-phase cells, while arrested G_1 cells synthesized histones at approximately 2% of the rate for S-phase cells (Fig. 11). Thus, histone synthesis was not confined exclusively to S phase and occurred even in nontraversing cells.

The electrophoretic patterns in Fig. 11 suggest that the histones made in arrested cells are not grossly different from histones synthesized in untreated cells. Since the histone fractions of the CHO cell contain isoleucine in concentrations ranging from 1.3 to 5.9 mole % (Gurley and Hardin, 1968), it appears that large isoleucine-containing proteins (as opposed to small polypeptide fragments) are synthesized by arrested cells in ile⁻ medium. The degree of faithfulness of protein synthesis in ile⁻ medium remains to be determined; however, since cells survive long periods in ile⁻ medium and quickly resume traverse following isoleucine addition, the cells have successfully carried out maintenance operations during arrest which would presumably necessitate synthesis of functional protein species. In any event, the above work provides an example of the usefulness of the ile⁻ system in studies of biochemical events in the cell cycle.

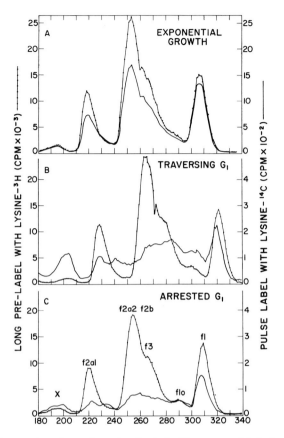

FIG. 11. Preparative electrophoresis of whole histones prelabeled with lysine-³H and pulse labeled with lysine-¹⁴C for 2 hours. Histones were isolated from exponentially growing cells (A), from cells traversing G_1 following preparation by mitotic selection and entry into G_1 in suspension culture (B), and from cells maintained 36 hours in ile⁻ medium. The medium is defined as in Fig. 2. [Reprinted with permission from Gurley *et al.* (1972a, p. 639).]

B. Recovery from Isoleucine-Mediated G_1 Arrest

Arrested cells may be induced to resume cycle traverse either by addition of isoleucine to the deficient medium or by resuspension of cells in fresh, complete medium. For studies of transition between nontraversing and traversing states uninterrupted by chemical/physical manipulation (cf. Tobey, 1972a,b), it is necessary to stimulate traverse by merely adding isoleucine to the arrested culture. Since the ile⁻ medium is partially depleted of a number of nutrients during accumulation of cells in G_1, following addition of isoleucine the growth-supporting capacity of the medium is less than that of an equivalent amount of fresh, complete medium. Consequently, in cultures released from arrest through isoleucine addition, the period prior to resumption of DNA synthesis is increased and the rate of cycle traverse is decreased in comparison to values obtained in cultures placed in fresh medium (R. A. Tobey, unpublished observations). Generally speaking, the degree of synchronization is greater in cells restored to fresh medium. Resuspension in fresh medium is especially preferable in studies of biochemical events during the recovery period (cf. Tobey and Crissman, 1972a).

Resuspension of cells maintained for 30 hours in ile⁻ medium in fresh, complete medium containing isoleucine produced the results shown in Fig. 12 (solid data points). Labeled thymidine was added at the time of resuspension in fresh medium, and samples were removed at intervals thereafter for determination of the labeled fraction via autoradiography.

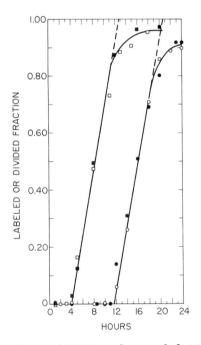

FIG. 12. Rate of initiation of DNA synthesis and division in a culture resuspended in complete medium at $t = 0$ after 30 hours in ile⁻ medium (solid figures) and in a culture prepared by mitotic selection and allowed to traverse the cell cycle in suspension culture (open figures). Thymidine-³H was added at $t = 0$ to a concentration of 0.05 μCi/ml in both cultures. At the times indicated aliquots were removed, and the labeled fractions were determined from autoradiographs (squares). Aliquots were also removed for determination of cell number with an electronic particle counter (circles). Definitions of ile⁻ medium and divided fraction are as in Fig. 2.

Since the label was in the culture continuously from the start of the experiment, the curve representing increase in labeled fraction provided a measure of the entry of cells into S phase. The divided fraction in Fig. 12 again represents $N/N_0 - 1$ so that a true doubling of the population would represent an increase from 0 to 1.0 on the scale provided. Several observations are immediately obvious from consideration of the data in Fig. 12. Cells released from an isoleucine-deficient state first synthesized DNA then divided, providing further proof that they had accumulated in the G_1 phase of the cell cycle during maintenance in ile⁻ medium (see also Table I and Fig. 7B regarding this point). Furthermore, *nearly the entire population initiated DNA synthesis and divided,* indicating that cultivation in ile⁻ medium did not appreciably affect the capacity for cycle traverse upon subsequent restoration of isoleucine.

In regard to this latter point one can obtain a rapid, convenient, and quantitative estimate of the effects of maintenance in ile⁻ medium upon subsequent traverse capacity through determination of the "traverse perturbation index" with the Los Alamos flow microfluorometer (Tobey *et al.*, 1972). The term "traverse perturbation index" is defined as the fraction of cells converted to an abnormally slow or to a nontraversing state as the result of experimental manipulation. For determination of this value, one monitors the capacity of individual cells to increase their DNA content, on the basis of measurements of relative DNA contents at varying times after experimental manipulation (i.e., in this case, at times following restoration of isoleucine to ile⁻ cells). Relative DNA contents in individual cells in large populations of cells were obtained with the flow microfluorometer, as described previously (Figs. 7 and 8). A culture was released from G_1 arrest, and Colcemid was added before any cell had reached mitosis to prevent normally traversing cells from reentering G_1; in the presence of Colcemid, cells arrest in metaphase and maintain the DNA content of mitotic cells. After 18 hours, flow microfluorometric analysis yielded the DNA distribution shown in Fig. 13. The initial ile⁻ culture was a nearly pure G_1 population as in Fig. 7B. The DNA distribution pattern in Fig. 13 reveals that a fraction of the cells representing 12.4% of the total population existed in S phase separate from the bulk of cells containing DNA contents ≥ mitotic cells (see Fig. 13 legend). Separation of the two populations of different traverse capacity could be enhanced by incubation for longer periods, but the fraction of slowly traversing cells remained at approximately 12.4%. Thus, FMF analysis of cultures released from G_1 arrest revealed that, while essentially the entire population (both normal and perturbed cells) could traverse to S or farther, the fraction completing genome replication and advancing to mitosis was approximately 88%.

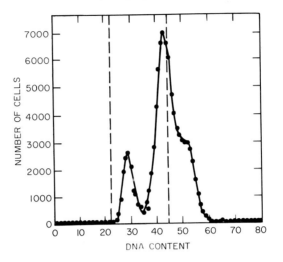

FIG. 13. Flow microfluorometric determination of the traverse perturbation index in a culture recovering from ile⁻-G_1 arrest. A culture of CHO cells maintained 30 hours in ile⁻ medium was resuspended in fresh, complete medium at $t = 0$. Colcemid was added to a concentration of 0.2 μg/ml at $t = 9$ hours, and the flow microfluorometric DNA pattern was obtained at $t = 18$ hours. Cells were dispersed, fixed with formalin, and Feulgen stained with acriflavine as described elsewhere (Tobey *et al.*, 1972). The shoulder in the curve of the major fraction, indicating cells with a greater DNA content than mitotic cells, represents cells which have spontaneously escaped Colcemid-induced metaphase arrest and have reinitiated cycle traverse *in the absence of an intervening division*. These cells have commenced a second round of DNA synthesis without dividing and, therefore, possess more DNA than metaphase-arrested cells. Spontaneous dissolution of the mitotic apparatus in Colcemid-treated DON Chinese hamster cells was observed previously by Stubblefield (1964).

In Fig. 12 the fraction of cells which entered S phase (determined autoradiographically) following reversal of isoleucine-mediated arrest was essentially 1.0, while approximately 90% of the cells divided. Thus, the data obtained by FMF and standard cell cycle analyses are in excellent agreement. Not only do the data in Fig. 13 confirm the low degree of effect on cycle traverse of the isoleucine-deficiency protocol, but they also provide further proof of the usefulness of FMF in population kinetic analysis.

A question arises concerning where, in G_1, cells accumulate as the result of cultivation in ile⁻ medium. Consideration of RNA content relative to DNA content (Table I) might suggest arrest in mid- to late G_1. However, as will be shown in a later section, the pattern of RNA accumulation in cells recovering from ile⁻ medium is unique, suggesting that they do not behave like true late G_1 cells. The cell volume distribution of ile$^-$-arrested cells determined electronically with a volume spectrometer (Tobey and Ley, 1970) is similar to that for exponentially growing cells containing large as well as small cells; in contrast, a highly homogeneous population of small cells is observed in cultures prepared by mitotic selection and allowed to enter G_1 (Anderson *et al.*, 1967a). On the basis of volume distribution alone, one might suspect that ile$^-$ cells are scattered throughout the cell cycle. This obviously is not the case, since several independent analytical procedures described in an earlier section clearly indicate that ile$^-$ cells accumulate in G_1 (Table I, Figs. 7 and 12).

Comparison of the rate of entry into S phase and subsequent rate of cell division in reversed ile$^-$ cultures (Fig. 12, solid points) and in suspension cultures, started from cells prepared by mitotic selection (Fig. 12, open points), revealed that the two populations initiated genome replication and divided in *nearly identical fashion*. Although we know little about the distribution of ile$^-$ cells in the G_1 phase, the distribution of cells in populations prepared by mitotic selection has been extensively studied. The initial selected mitotic population should be minimally perturbed from a biochemical standpoint (Petersen *et al.*, 1969b) and occupy only approximately 1% of the cell cycle. The mitotic fraction decreased from near 100% to 0 in approximately 15 minutes (Tobey *et al.*, 1967b; Petersen *et al.*, 1968, 1969a,b), and these populations initially possessed an Engelberg synchrony index (Engelberg, 1961) of 0.97 where 1.0 is perfect synchrony. These data indicate that mitotically selected cells initially are extremely highly synchronized. On the basis of temporal arguments alone, one would conclude that cells in isoleucine-mediated G_1 arrest were also initially highly synchronized and were situated very close to the M/G_1 boundary. Because of uncertainties as to whether or not traverse commences immediately after addition of isoleucine to ile$^-$ medium, the location of cells at the M/G_1 boundary, deduced from temporal considerations, must be regarded as provisional. Precise localization of cells in G_1 will only become feasible with establishment of a number of clearly defined biochemical event markers scattered throughout the G_1 phase of the cell cycle. No suitable set of G_1 markers is currently available (Mueller, 1969, 1971).

From the data in Table I, it is apparent that cells in isoleucine-mediated G_1 arrest contain an overabundance of RNA relative to cells in G_1 initially prepared by mitotic selection, yet in Fig. 12 the two cultures required the same amount of time to traverse the cell cycle. If the mass of RNA increased at the same rate in both cultures, cells released from arrest would always maintain an excess quantity of RNA and would re-

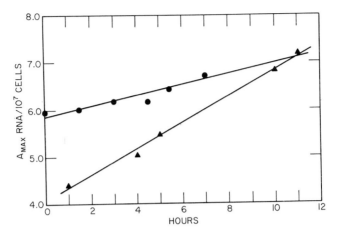

Fig. 14. Increase in RNA mass in a culture resuspended in complete medium at $t = 0$ after 30 hours in ile⁻ medium (circles) and in a culture prepared by mitotic selection and allowed to traverse the cell cycle in suspension culture (triangles). The RNA mass determinations were carried out as described previously (Enger *et al.*, 1968).

main in a perpetual state of biochemical imbalance (cf. Cohen and Barner, 1954; Cohen and Studzinski, 1967; Anderson *et al.*, 1967b). To investigate the rate of RNA accumulation, cultures consisting of refed ile⁻ cells or cells prepared by mitotic selection were allowed to traverse the cell cycle, and aliquots were removed at various intervals for determination of the mass of RNA (M. D. Enger and R. A. Tobey, unpublished observations). Results in Fig. 14 indicate that the rate of increase of RNA was much slower in the reversed ile⁻ culture; however, by 11 hours (the time at which cells began entering mitosis) the RNA contents were equivalent in the two cultures. Resynchronization of glass-grown, reversed ile⁻ cells starting at 11 hours after restoration of isoleucine was accomplished by means of mitotic selection; determination of the accumulation of RNA mass in the subsequent cycle yielded a pattern of increase in RNA identical to that obtained with control cells selected in mitosis in Fig. 14 (M. D. Enger and R. A. Tobey, unpublished observations). Cells released from ile⁻ blockade compensated for their excess RNA content by a reduced rate of accumulation of RNA during the first cycle after release and thereafter possessed a biosynthetic capacity indistinguishable from untreated control cells. These results indicate that, throughout the period of cultivation in ile⁻ medium and also during the subsequent recovery period, RNA metabolism is rigidly controlled and, as a result, cells never enter a state of gross biochemical imbalance.

IV.   Utilization of Cells Reversibly Arrested in G_1

A.   Production at the G_1/S Boundary

In preceding sections information has been presented concerning the method of preparation and biochemical properties of cells reversibly arrested in G_1. In this section a few examples will be presented demonstrating the usefulness of ile$^-$-G_1-arrested cells in studies of mammalian cell cycle events. The long duration of the G_1 phase in CHO cells grown under our cultivation conditions is, at the same time, a blessing and a curse in designing experiments for examination of cell cycle events. Considering for the moment the positive aspects, the preponderance of G_1 cells in an exponentially growing population also implies that the number of cells in S, G_2, and M is relatively low. Had the reverse been true (i.e., a large proportion of cells in S and G_2) during the period of natural depletion of isoleucine as a result of growth to high cell concentration stationary phase, the amount of available isoleucine might have been insufficient to allow a clearing out of late interphase, and the G_1-arrest phenomenon might not have been discovered.

F-10 medium was not designed for continuous support of cell lines in suspension culture but, rather, was intended for maintenance of early-passage cells in clonal growth (Ham, 1963). As such, among other differences, F-10 medium contains only 0.5% to 5% of the isoleucine content of other standard tissue culture media. By merely increasing the isoleucine content in F-10 medium by threefold (still far below other standard media), CHO cells grow to higher cell concentrations but deplete nutrients other than isoleucine, resulting in a failure both to accumulate in G_1 arrest and to remain viable for long periods (Ley and Tobey, 1970). Thus, a subtle alteration in the isoleucine-to-cell ratio drastically alters the capacity for accumulation in G_1 arrest. It is concluded that both the low proportion of non-G_1 cells in exponentially growing populations and the low isoleucine content of the cultivation medium were fortuitous factors contributing to the discovery of isoleucine-mediated G_1 arrest. However, the long duration of G_1 in our cells also contributes greatly to the rapid loss of synchrony prior to entry into S phase.

Loss of synchrony (i.e., synchrony decay) is an unfortunate natural property of mammalian cells in culture, irrespective of the method of synchronization employed, and arises from different rates of cycle traverse by individual cells in the population (Petersen and Anderson, 1964; Engelberg, 1964; Anderson and Petersen, 1964; Tobey et al., 1966a; Whitmore, 1971). From the data in Fig. 12, in cultures prepared either by reversal of ile⁻ cultivation or by mitotic selection and placed in suspension culture (extremely highly synchronized initially), the fastest traversing cells started synthesizing DNA at 4 hours, while the slower moving cells continued to initiate genome replication over the next 8 to 10 hours. The time period required for the entire population to divide was essentially the same as that for entry into S phase. It is not surprising that the bulk of synchrony decay occurred during G_1 phase, in view of the observations that variability in culture doubling times in different nutritional states could be accounted for principally as the result of differences in duration of G_1 phase (Sisken and Kinosita, 1961; Terasima and Tolmach, 1963; Tobey et al., 1967a). The preceding results indicate that populations prepared either by reversal of the ile⁻ state or by mitotic selection are unsuitable for detailed studies of events associated with initiation of genome replication *in the immediate pre-S phase* due to synchrony decay during G_1 traverse. Rather, what is required is a technique that will synchronize cells in late G_1, close to the G_1/S boundary, in sufficient number for biochemical studies.

It should be possible to obtain large quantities of cells at the G_1/S boundary by releasing cells from G_1 arrest and resynchronizing in late G_1 with an agent capable of allowing traverse of G_1 but preventing initiation of genome replication. Hydroxyurea grossly reduces the rate of progression from G_1 to S, preventing initiation of DNA replication and subsequent death due to its S phase cytotoxicity (Sinclair, 1965, 1967; Tobey, 1972a; Tobey and Crissman, 1972a,b). However, hydroxyurea begins to kill non-S-phase cells after exposure times in excess of 10 hours (Sinclair, 1967) so that it is essential to minimize the period of treatment with the drug. Consequently, cells were released from isoleucine-mediated G_1 arrest by resuspension in fresh medium containing isoleucine along with hydroxyurea at a concentration of 10^{-3} M. A sample was removed 10 hours later for determination of the relative DNA content of cells in the population (Fig. 15A). It was readily apparent that, in the presence of hydroxyurea, essentially the entire population still possessed the DNA content of cells in G_1 phase; that is, there was no evidence that cells had initiated genome replication.

To ensure that cells had indeed traversed G_1 in the presence of hydroxyurea, these cells were spun down, washed, and resuspended in fresh, complete medium. At intervals thereafter, aliquots were removed for the determination of cell concentration with an electronic particle counter and of the fraction labeled via autoradiography following exposure to thymidine-³H in a series of 15-minute pulse labeling periods. The results in Fig. 15B indicate that, although the fraction of cells synthesizing DNA initially at the time of removal of hydroxyurea was essentially

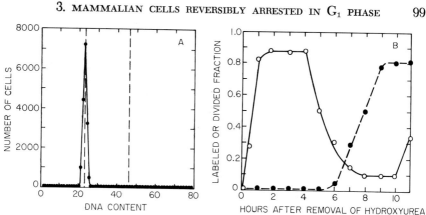

FIG. 15. Evidence that cells recovering from ile⁻-mediated G_1 arrest in the presence of hydroxyurea accumulate at the G_1/S boundary. Cells cultivated for 30 hours in ile⁻ medium were resuspended in fresh, complete medium containing isoleucine and hydroxyurea to $10^{-3} M$ for 10 hours, at which time an aliquot was removed for flow microfluorometric analysis (A). Immediately thereafter the cells were spun down, washed, and resuspended in fresh medium without hydroxyurea. At times thereafter aliquots were pulse-labeled for 15-minute periods with 2 μCi/ml of thymidine-³H, and autoradiographs were prepared for determination of the labeled fractions (open figures). Aliquots were also removed for determination of cell number with an electronic particle counter (solid figures). Ile⁻ medium and divided fraction are defined as in Fig. 2.

zero, the entire population synchronously moved into S phase immediately following resuspension in drug-free medium. Similar results were obtained when reversed ile⁻ cultures were treated with 5 μg/ml of cytosine arabinoside for 10 hours and then washed out, although the time required for resumption of cycle traverse after drug removal was much more variable than in cultures released from hydroxyurea blockade (Tobey and Crissman, 1972a,b). Use of hydroxyurea instead of cytosine arabinoside is recommended, especially for studies of entry into S phase following removal of drug.

The preceding results indicate that extremely large quantities of cells may be synchronized at the G_1/S boundary by combined use of isoleucine-mediated, G_1-arrested cells with hydroxyurea. Such cells are suitable for study of biochemical events associated with initiation of genome replication, although appropriate controls should be employed to correct for possible biochemical alterations such as "unbalanced growth" arising from use of agents such as hydroxyurea. However, note that care should be taken when selecting exposure times to hydroxyurea, since cells slowly enter S phase after prolonged treatment periods with hydroxyurea (i.e., <1% after 10 hours but 4% to 6% after 18 hours) measured via thymidine-³H autoradiography (Sinclair, 1967; Tobey, 1972b).

B. Characterization of Chemotherapeutic Agent Effects on Cell Cycle Traverse

To establish meaningful chemotherapy regimens, it is essential to pre-test new agents to determine not only relative host-tumor cell toxicity but also to discover possible cell cycle, phase-specific toxic effects and effects on cell cycle progression. Concerning this latter point, the effects on cycle traverse are important because agents can cause accumulation of cells in specific segments of the cell cycle with increased susceptibility to phase-specific agents. Conversely, an agent may be self-limiting in that the compound inhibits entry of cells into phases in which cytotoxic effects of the drug are maximal. Effective chemotherapy scheduling must account for factors such as these, whether the drug is administered singly or in combination with other chemotherapeutic agents.

As shown in previous sections of this chapter, CHO cells grown *in vitro* can be reversibly induced to convert from a nontraversing to a traversing state by manipulating the amount of isoleucine in the culture medium. G_1-Arrested CHO cells in ile$^-$ medium also exhibit an enhanced survival following X-irradiation relative to traversing cells, and cells released from isoleucine-mediated G_1 arrest can recover from effects of X-ray-induced division delay significantly faster than similar continuously traversing cells (R. A. Walters and R. A. Tobey, unpublished observations). These properties of CHO cells grown *in vitro* mimic at least superficially properties observed *in vivo* for tumors. That is, following therapeutic treatment, persistent surviving tumor cells in G_0 can be induced to resume cycle traverse (Barendsen, 1970).

The rationale for design of the chemotherapeutic test protocol employing arrested and reversed CHO cells has been described in detail elsewhere (Tobey, 1972a). Cells in G_1 arrest after growth in ile$^-$ medium were given the test agent and 1 hour later (defined as $t = 0$) were given isoleucine to reinitiate cycle traverse along with thymidine-^{3}H to label cells synthesizing DNA (0 sample). As a control, arrested cells were given isoleucine and thymidine-^{3}H at $t = 0$ but no test drug (control-0). Both cultures received Colcemid at $t = 9$ hours to prevent cells from reentering G_1, and autoradiographs were prepared at $t = 18$ hours for determination of the labeled and mitotic fractions. If the agent prevented entry into S, there would be neither labeled nor mitotic cells in the 0 sample. If the agent allowed cells to initiate DNA synthesis but prevented them from completing G_2, the population in the 0 sample would consist of labeled, nonmitotic cells with the labeled fraction approximately equivalent to the labeled fraction in the control-0 sample. In the absence of effects, labeled and mitotic values would be equivalent in both treated and control cultures.

An additional culture was released from G_1 arrest, and part of the population had commenced synthesizing DNA at the time of drug addition. Arrested cells received isoleucine at $t = 0$ to reinitiate cycle traverse, then test agent was added at $t = 6$ hours (at a time when approximately 25% of the cells were in S with the remainder in G_1). Thymidine-^{3}H was then added at $t = 7$ hours to label any cells still synthesizing DNA or continuing to enter S phase. After Colcemid was added at 9 hours, an autoradiograph was prepared at 18 hours for determination of the labeled and mitotic fractions (6-hour sample). A control culture of arrested cells received isoleucine at $t = 0$ and high levels of thymidine-^{3}H between the interval 6¾ to 7 hours, at which time an autoradiograph was prepared to indicate the fraction of cells in S phase at the time of addition of thymidine-^{3}H to the 6-hour sample (designated control-7). If the test agent failed to affect traverse, labeled and mitotic fractions in the 6-hour sample would be equivalent to values from the control-0 culture. If labeled and mitotic fractions in the 6-hour sample were equivalent to values in the control-7 culture, the agent specifically inhibited entry into S but was without effect on cells in S at the time of drug addition, and these completed interphase and accumulated in mitosis. If the labeled and mitotic fractions were 0, the agent prevented initiation of DNA synthesis.

Combining the autoradiographic data with cell number increases observed after addition of test agent to asynchronous cultures yields a variety of distinguishable classes of agents with different effects upon the CHO cell cycle (Table III). On the basis of this very simple test, large classes of compounds may be separated in terms of effects on traverse. [For actual data obtained with this system and a number of different chemotherapeutic agents, see Tobey (1972a,b).] Combined use of data from autoradiographic and cell enumeration techniques with data concerning relative DNA contents measured with the flow microfluorometer adds a new dimension in detailed analysis of the effects of chemotherapeutic agents on cell cycle traverse (Tobey and Crissman, 1972b). Information obtained to date suggests that this simple *in vitro* system is of predictive value in extrapolation to tumor response *in vivo* (Tobey 1972a,b; Tobey and Crissman, 1972a,b).

A variety of additional studies are currently underway in this laboratory employing ile⁻ CHO cells in examination of cell cycle events in traversing and nontraversing populations. For example, Gurley *et al.* (1972b) have recently shown that, while histone F2a2 phosphorylation occurs throughout the cell cycle and also in G_1-arrested cells, phosphoryl-

TABLE III

SUMMARY OF RESULTS EXPECTED FOR AGENTS AFFECTING CHINESE HAMSTER
CELL CYCLE TRAVERSE WITH THE USE OF TEST PROTOCOL

---

0 Sample

Drug at $t = -1$ hour, isoleucine to $4 \times 10^{-5} M$ and thymidine-^{3}H to 0.06 μCi/ml at $t = 0$, Colcemid to 0.2 μg/ml at 9 hours, and an autoradiograph prepared at $t = 18$ hours

---

Control-0

Isoleucine to $4 \times 10^{-5} M$ and thymidine-^{3}H to 0.06 μCi/ml at $t = 0$, Colcemid at $t = 9$ hours, and an autoradiograph at $t = 18$ hours

---

6-Hour sample

Isoleucine to $4 \times 10^{-5} M$ at $t = 0$, drug at $t = 6$ hours, thymidine-^{3}H to 0.06 μCi/ml at $t = 7$ hours, Colcemid to 0.2 μg/ml at 9 hours, and an autoradiograph prepared at $t = 18$ hours

---

Control-7

Isoleucine to $4 \times 10^{-5} M$ at $t = 0$, thymidine-^{3}H at $t = 6\frac{3}{4}$ hours, and an autoradiograph prepared at $t = 7$ hours

Labeled and mitotic fractions were determined in all autoradiographs. Cell number values apply only to line CHO cells cultured in F-10 medium supplemented with 10% calf and 5% fetal calf sera in spinner flasks.

Agent does not affect cycle traverse

    a. Labeled and mitotic fractions in both 0- and 6-hour samples = Control-0
    b. No effect upon cell division

Agent inhibits DNA synthesis and prevents G_1 to S transition

    a. Labeled and mitotic fractions in both 0- and 6-hour samples = 0
    b. Exponential cultures given drug, cell number increase = sum of cells in $G_2 +$ M ($\sim$13%)

Agent inhibits operations throughout the entire cycle

    a. Labeled and mitotic fractions in both 0- and 6-hour samples = 0
    b. Exponential cultures given drug, cell number increase < sum of cells in $G_2 +$ M ($<$13%)

Agent primarily inhibits completion of G_2 but also secondarily inhibits G_1 to S transition

    a. Mitotic fractions in both 0- and 6-hour samples = 0
    b. Labeled fraction in 0- and/or 6-hour sample > 0, but less than Control-0 and Control-7, respectively
    c. Exponential cultures given drug, cell number increases < sum of cells in $G_2 +$ M ($<$13%)

TABLE III (*Continued*)

Agent affects specifically the G_1 to S transition

   a. Labeled and mitotic fractions in 0 sample = 0

   b. Labeled and mitotic fractions in 6-hour sample equal to or slightly less than Control-7 values

   c. Exponential cultures given drug, cell number increase = sum of cells in S + G_2 + M ($\sim$35%)

Agent prevents completion of G_2

   a. Labeled fractions in both 0- and 6-hour samples = labeled fraction in Control-0

   b. Mitotic fraction in both 0- and 6-hour sample = 0

   c. Exponential culture given drug, cell number increase < sum of cells in G_2 + M (<13%)

Agent prevents completion of M

   a. Labeled and mitotic fractions in both 0- and 6-hour samples = Control-0

   b. Exponential cultures given drug, cell number increase $\simeq$ the number of cells in M, and mitotic fraction increases without concomitant increase in cell number ($\simeq$3%)

ation of histone F1 fraction appears to be turned on with DNA replication. Kraemer and Tobey (1972) have employed ile⁻ cells in combination with thymidine blockade to demonstrate that, although the rate of *synthesis* of surface heparan sulfate is maximal during early interphase, *loss* of this material from the cell surface occurs mainly during late interphase and mitosis.

C.   Attempts at Isolation of an Isoleucine-Dependent Genome Regulatory Compound

The most useful aspect of the isoleucine-deficient state ideally is that it should provide a means for isolating a compound (or compounds) playing an essential role in regulation of mammalian genome replication. Stated differently, some compound containing isoleucine or one of its derivatives (or required for activation of a regulatory product) is not synthesized in cultures containing suboptimal quantities of isoleucine. Alternatively, a species inhibiting initiation may be synthesized in arrested cultures. Unfortunately, we have not yet been able to isolate and characterize compounds of this nature. Part of the difficulty in isolating this type of material lies in the ubiquitous distribution of isoleucine and its derivatives throughout the mammalian cell. Not only is isoleucine a normal constituent of protein, but one of the products of isoleucine metabolism in the mammalian cell is α-methylbutyryl-CoA (Meister, 1965) which is incorporated into branched-chain fatty acids and ultimately may appear as a lipid component. Under appropriate catabolic conditions, isoleucine can be degraded to propionyl-CoA or even to acetyl-CoA (Meister, 1965). Therefore, one cannot hope to follow labeled isoleucine into a regulatory product merely by adding labeled isoleucine to ile⁻ cells; the label should be distributed in a multitude of cell subfractions. As expected, in unpublished preliminary experiments conducted in this laboratory shortly after discovery of the isoleucine-deficiency phenomenon involving cellular fractionation of ile⁻ cells fed with labeled isoleucine, the label was indeed distributed among the various fractions.

As discussed earlier, coordination of protein synthesis with DNA replication assures that attempts at reversal of isoleucine effects under conditions of protein synthesis inhibition will be unsuccessful, even if the regulatory compound which ile⁻ cells fail to synthesize is not a protein. Further complicating the biochemical picture is the fact that the first product in the isoleucine degradative pathway, α-keto-β-methyl valeric acid, is produced via transamination (Meister, 1965); resumption of traverse by arrested ile⁻ cells fed with the α-keto derivative of isoleucine would not prove that the missing compounds required for genome replication were nonprotein. In view of the readily reversible nature of transamination reactions in general, the α-keto derivative fed to ile⁻ cells would almost certainly be converted rapidly to isoleucine, with the net effect of feeding ile⁻ cells with isoleucine. The next step in isoleucine degradation is addition of a CoA moiety onto the α-keto derivative of isoleucine (Meister, 1965). Although this step is irreversible, even if one could obtain the α-keto CoA compound, it is extremely doubtful that this CoA compound could enter intact cells. Thus, stimulation of arrested ile⁻ cells to resume cycle traverse by feeding with a derivative of isoleucine does not appear to be feasible technically at the present time.

One is left with the unpleasant alternative of examination of species suspected *a priori* of involvement in regulation of genome replication, either stimulators made in traversing cells or inhibitors synthesized in arrested cells. Even more distressing is the distinct possibility that the isoleucine-containing moiety may represent only a minor species not detectable by standard biochemical analytical techniques. The problem may be likened to a search for a needle buried in one of a series of haystacks; not only is it impossible to decide in which haystack the search should begin, but the searcher is further handicapped because of uncertainties regarding the appearance of the hidden needle (i.e., he might find the needle but not recognize it as such).

Everhart and Prescott's (1972) demonstration that CHO cells accumulate in G_1 arrest through cultivation in medium containing suboptimal concentrations of leucine (as well as in ile⁻ medium) may represent a bright spot in an overall grossly pessimistic problem area. Everhart and Prescott (1972) contend that the differential effect of low quantities of essential amino acids on initiation of genome replication and completion of interphase is the result of differential rates of protein synthesis required for the two processes. Presumably, in the absence of sufficient quantities of an essential amino acid, synthesis of protein required for initiation of DNA synthesis does not occur, whereas synthesis of traverse-essential proteins required for completion of S and G_2 does take place. Beyond suggesting that the factor required for initiation is protein in nature, Everhart and Prescott have not yet characterized a regulatory product.

Perhaps some published data tend to confirm Everhart and Prescott's hypothesis concerning differential rates of protein synthesis required for initiation of DNA replication and general cycle traverse (Tobey, 1972a). Treatment of cultures comprised of cells in G_1 and S with the competitive inhibitor of isoleucine, L-O-ethylthreonine, revealed a strikingly different response, dependent upon the concentration of drug employed. L-O-Ethylthreonine is presumably activated and bound to isoleucyl tRNA species in mammalian cells, although transfer of the inhibitor to the developing polypeptide chain in place of an isoleucine molecule apparently does not occur (Shigeura et al., 1969). With increasing concentrations of L-O-ethylthreonine, progressively larger numbers of isoleucyl tRNA molecules bind the inhibitor and are rendered inactive. At the same time that protein synthetic capacity is decreased through inactivation of isoleucyl tRNA, increasing amounts of isoleucine are available for conversion via nonprotein metabolic pathways. Therefore, it is possible to reduce protein synthetic rate by addition of L-O-ethylthreonine to the culture medium.

At a concentration of $8 \times 10^{-4} M$, L-O-ethylthreonine prevented cells from entering S phase but allowed a large proportion of cells already in S at the time of drug addition to complete genome replication and to enter mitosis (Tobey, 1972a). At concentrations of $2 \times 10^{-3} M$ or greater the drug stopped cells at multiple stages of the cell cycle, and cells began dying within 10 hours [i.e., a classic gross inhibition of protein synthesis (see Tobey, 1972a)]. Thus, partial inhibition of protein synthesis through inactivation of a fraction of the isoleucyl tRNA species under conditions providing excess isoleucine for nonprotein-related reactions produced a differential effect upon entry into S and late interphase traverse highly similar to results in isoleucine-deficient medium.

Similar findings were reported in an earlier study of effects of the protein synthesis inhibitor, cycloheximide, on DNA synthesis in synchronized HeLa S-3 cells. Kim *et al.* (1968) concluded that, although protein synthesis was necessary throughout S phase, the process of initiation of DNA synthesis was more sensitive to the drug than was continuation of genome replication.

If Everhart and Prescott are correct in that the species required for initiation of DNA synthesis involves protein synthetic capacity, the problems concerning isolation and characterization of regulatory species are simplified by at least an order of magnitude. Once again note that the initiator species may represent a minority species of protein that may be extremely difficult to detect with available biochemical techniques. For instance, one could speculatively envision that the initiation species activates genome replication by bringing about a subtle conformational change in the nuclear membrane/DNA polymerase/chromatin complex, causing cells to progress from a non-DNA-replicating state to one in which genome replication could occur. Conversely, if inhibitors are synthesized in arrested cells under conditions of nutrient limitations, it is possible that the species of interest interacts with the chromatin in a subtle way, rendering it nonreactive for replication. Dozens of additional *ad hoc* models could be conjured up in which specialized synthesis of small numbers of low molecular weight polypeptides could grossly alter the capacity of cells to replicate DNA.

Salas and Green (1971) have described a potentially useful technique for isolating proteins which bind to DNA and have determined synthetic rates for several classes of DNA-affinity proteins in both arrested and traversing cultures of mouse embryo line 3T6 cells. Extracts of proline-^{3}H-labeled cells were chromatographed on columns of denatured calf thymus DNA-cellulose by the chromatographic procedure developed by Alberts and co-workers (1968). Proteins bound to DNA were eluted with 0.15 M NaCl, and these proteins were then fractionated into 8 major species by means of polyacrylamide gel electrophoresis. The synthesis of one protein species (designated P6) appeared to be coordinated with DNA replication in traversing cells while an additional species, P1, appeared to be synthesized at an appreciable rate only in cells arrested in G_1 by growth to saturation density. With increasingly stringent arrest conditions brought about by manipulation of the serum content in the medium, the rate of synthesis of P1 species increased and decreased only after cells were released from arrest and initiated genome replication. The presence of tryptophan in all species led Salas and Green to conclude that none of their DNA-binding proteins was histones.

While these results are extremely interesting and ultimately may yield information on genome-regulatory compounds, there are several reservations that must be borne in mind: (1) due to a relatively poor degree of synchrony following transition from a nontraversing to a traversing state in their 3T6 cells, precise kinetic information regarding synthetic patterns of the species of interest could not be obtained; (2) cause and effect relationships have not yet been established (that is, do the various compounds synthesized in arrested and traversing states determine the capacity of cells for proliferation, or are they secondary products which possess no regulatory function?); and (3) the differences in synthetic rate of compounds in traversing and arrested states were quantitative rather than qualitative, although this may be due to the presence of small numbers of contaminating arrested cells in traversing populations, etc. Salas and Green provide the first suggestion that isolation and characterization of genome-regulatory compounds may soon be feasible technically.

Even though the exact biochemical mechanism of regulation of genome replication is unknown, reversible arrest of cells in G_1 by manipulation of the isoleucine (or other amino acid) content in the culture medium should continue to be extremely useful for studies of biochemical events in the mammalian cell cycle.

V.   Summary and Conclusions

A technique has been developed in which extremely large numbers of cells suitable for biochemical studies may be induced reversibly to undergo transition from a proliferating to a nonproliferating state, dependent upon the amount of isoleucine in the culture medium. Cells in isoleucine-deficient medium arrest in G_1 and bear at least superficial resemblance to cells in the G_0 state *in vivo*. Arrested cells remain viable and biochemically competent for several days in a state of biochemical balance without significant growth and remain ready to resume cycle traverse upon addition of isoleucine. Among factors to be considered in establishing isoleucine-dependent G_1 arrest are these: (1) the period of exposure of cells to medium containing suboptimal quantities of isoleucine; (2) the degree of isoleucine deficiency (different types of cells require different levels of isoleucine during accumulation in G_1); and (3) the physiological state of cells shifted to ile⁻ medium (i.e., cells *must* be free of *Mycoplasma* infection). Manipulation of the first two parameters is essential if one wishes to adapt successfully the ile⁻ system to other cell lines than CHO.

Upon resuspension of washed CHO cells in ile⁻ medium there is an immediate cessation of *initiation* of DNA synthesis, with a reduced rate of DNA synthesis for cells in S at the time of medium transfer (Fig. 9). Although the DNA synthetic rate has dropped to approximately 40% of initial value within 2 hours after transfer of ile⁻ medium, RNA synthesis is only slightly affected (Fig. 9). Synthesis of stable RNA moieties is reduced to a greater extent than unstable species, although no major class of RNA is shut off completely (Table II). The amount of available messenger RNA (mRNA) does not appear to be a limiting factor, although it is possible that synthesis of specific mRNA species required for cycle traverse is selectively inhibited. In any event, cells in ile⁻ medium do not enter a state of gross biochemical imbalance.

Protein synthetic rate is unaffected during the first 2 hours in ile⁻ medium, at a time when DNA synthesis has already decreased by 60% (Fig. 9). Furthermore, protein synthesis is still proceeding at 50% to 60% of initial rate after 30 hours in ile⁻ medium, long after DNA synthesis has terminated. Therefore, it appears that synthesis of DNA is much more rapidly and completely inhibited than is protein synthesis. This observation suggests that the effects of limiting amounts of isoleucine on DNA initiation and completion of interphase are *two separate phenomena*. Everhart and Prescott (1972) believe that these two phenomena merely require different levels of protein synthesis and that, under appropriate culture conditions (for example, those described in this chapter or limitations in other amino acids in culture medium), the phenomena are separable.

It is apparent from the data in Fig. 9 that the rates of synthesis of both RNA and protein remain at high levels long after DNA synthesis is grossly inhibited. Continuous RNA and protein synthesis must occur throughout interphase in the CHO cell up to within 114 and 8 minutes of mitosis, respectively, for successful completion of interphase and cell division (Tobey *et al.*, 1966a,b). Therefore, it is not surprising that protein and RNA synthesis remain at high levels during cultivation of cells in ile⁻ medium, since cells in late interphase must synthesize traverse-essential protein and RNA species before dividing and accumulating in G_1.

The gradual reduction in protein synthetic rate with increasing incubation periods in ile⁻ medium is the result of a *reduction in polypeptide translation rate* (Enger and Tobey, 1972) in cells containing a near-normal number of polysomes (Fig. 10). The reduction in translation rate suggests that the rate-limiting step is availability of isoleucine such that partially completed peptide chains must step at isoleucine-requiring sites and await the catabolic liberation of isoleucine from preformed moieties. In turn, this further implies that the rate of polypeptide turnover must increase in G_1-arrested cells and that at least some of the proteins made during arrest must contain isoleucine. In regard to this latter point, recall that apparently normal isoleucine-containing histones are made during ile⁻ arrest (Fig. 11).

The ile⁻ system involving normal polysome size and numbers but a reduced translation rate is very different from results obtained in normal stationary-phase cultures. During stationary phase, macromolecular synthetic rate and quantity of polysomes are grossly reduced (Ward and Plagemann, 1969; Tobey and Ley, 1970; Becker *et al.*, 1971; Stanners and Becker, 1971). The ile⁻ system resembles more closely recent results obtained in contact- or confluency-inhibited cells in monolayer culture in which macromolecular synthetic rates remain high during the period of arrest (Weber and Edlin, 1971).

In summary, the observations presented in this chapter indicate that cells cultivated in ile⁻ medium enter a unique biochemical state in which synthesis of macromolecular species is rigidly controlled throughout. Due to the low degree of biochemical perturbation introduced during cultivation in ile⁻ medium, the recovering cells are extremely useful for studies of biochemical events within the mammalian cell cycle. It also may be possible ultimately to derive from the isoleucine technique described in this chapter insight into fundamental biochemical mechanisms which play an integral role in regulation of proliferation in the mammalian cell.

ACKNOWLEDGMENTS

This work was performed under the auspices of the U. S. Atomic Energy Commission. The author wishes to thank Dr. Harry Crissman, Dr. Duane Enger, Dr. Lawrence Gurley, Dr. Paul Kraemer, Dr. Kenneth Ley, and Dr. Ronald Walters for their contributions in collaborative studies involving the isoleucine technique. The author also wishes to acknowledge the excellent technical assistance provided by Mrs. Phyllis Sanders, Mrs. Evelyn Campbell, Mrs. Susan Carpenter, Mr. Joseph Valdez, and Mr. John Hanners.

REFERENCES

Alberts, B. M., Amodio, F. J., Jenkins, M., Guttman, E. D., and Ferris, F. L. (1968). *Cold Spring Harbor Symp. Quant. Biol.* **33,** 289.
Anderson, E. C., and Petersen, D. F. (1964). *Exp. Cell Res.* **36,** 423.
Anderson, E. C., Petersen, D. F., and Tobey, R. A. (1967a). *Biophys. J.* **7,** 975.
Anderson, E. C., Petersen, D. F., and Tobey, R. A. (1967b). *Nature (London)* **215,** 1083.
Anderson, E. C., Bell, G. I., Petersen, D. F., and Tobey, R. A. (1969). *Biophys. J.* **9,** 246.
Anderson, E. C., Petersen, D. F., and Tobey, R. A. (1970). *Biophys. J.* **10,** 630.
Barendsen, G. W. (1970). "Time and Dose Relationships in Radiation Biology as Applied to Radiotherapy, NCI-AEC Conference, Carmel, California, 1969," pp.

339–351. Clearinghouse for Federal Scientific and Technical Information, National Bureau of Standards, U. S. Dept. of Commerce, Springfield, Virginia.

Becker, H. J., Stanners, C. P., and Kudlow, J. E. (1971). *J. Cell. Physiol.* **77**, 43.

Bloch, D. P., and Godman, G. C. (1955). *J. Biophys. Biochem. Cytol.* **1**, 17.

Borun, T. W., Scharff, M. D., and Robbins, E. (1967). *Proc. Nat. Acad. Sci. U. S.* **58**, 1977.

Brega, A., Falaschi, A., deCarli, L., and Pavan, M. (1968). *J. Cell Biol.* **36**, 485.

Cohen, L. S., and Studzinski, G. P. (1967). *J. Cell. Physiol.* **69**, 331.

Cohen, S. S., and Barner, H. D. (1954). *Proc. Nat. Acad. Sci. U. S.* **40**, 885.

Defendi, V., and Manson, L. A. (1963). *Nature (London)* **198**, 359.

Eagle, H., and Piez, K. A. (1960). *J. Biol. Chem.* **235**, 1095.

Engelberg, J. (1961). *Exp. Cell Res.* **23**, 218.

Engelberg, J. (1964). *Exp. Cell Res.* **36**, 647.

Enger, M. D., and Tobey, R. A. (1972). *Biochemistry* **11**, 269.

Enger, M. D., and Walters, R. A. (1970). *Biochemistry* **9**, 3551.

Enger, M. D., Tobey, R. A., and Saponara, A. G. (1968). *J. Cell Biol.* **36**, 583.

Everhart, L. P. (1972). *Exp. Cell Res.* **74**, 311.

Everhart, L. P., and Prescott, D. M. (1972). *Exp. Cell Res.* **75**, 170.

Gil, P., and Pan, J. (1970). *Can. J. Microbiol.* **16**, 415.

Gurley, L. R., and Hardin, J. M. (1968). *Arch. Biochem. Biophys.* **128**, 285.

Gurley, L. R., Walters, R. A., and Tobey, R. A. (1972a). *Arch. Biochem. Biophys.* **148**, 633.

Gurley, L. R., Walters, R. A., and Tobey, R. A. (1972b). *Arch. Biochem. Biophys.* (in press).

Hakala, M. T., Holland, J. F., and Horoszewicz, J. S. (1963). *Biochem. Biophys. Res. Commun.* **11**, 466.

Ham, R. G. (1963). *Exp. Cell Res.* **29**, 515.

Hodge, L. D., Borun, T. W., Robbins, E., and Scharff, M. D. (1969). In "Biochemistry of Cell Division" (R. Baserga, ed.) pp. 15–37. Thomas, Springfield, Illinois.

Irvin, J. L., Holbrook, D. J., Jr., Evans, J. H., McAllister, H. C., and Stiles, E. P. (1963). *Exp. Cell Res., Suppl.* **9**, 359.

Kim, J. H., Gelbard, A. S., and Perez, A. G. (1968). *Exp. Cell Res.* **53**, 478.

Kochhar, O. S. (1968). *Exp. Cell Res.* **49**, 598.

Kraemer, P. M. (1964). *Proc. Soc. Exp. Biol. Med.* **117**, 910.

Kraemer, P. M. (1970). *J. Cell Biol.* **47**, 110a.

Kraemer, P. M., and Tobey, R. A. (1973). *J. Cell Biol.* **55**, 713.

Kraemer, P. M., Petersen, D. F., and Van Dilla, M. A. (1971). *Science* **174**, 714.

Kraemer, P. M., Deaven, L. L., Crissman, H. A., and Van Dilla, M. A. (1972). *Advan. Cell Mol. Biol.* **2**, 47–108.

Levine, E. M., Thomas, L., McGregor, D., Hayflick, L., and Eagle, H. (1968). *Proc. Nat. Acad. Sci. U. S.* **60**, 583.

Ley, K. D., and Tobey, R. A. (1970). *J. Cell Biol.* **47**, 453.

Littlefield, J. W., and Jacobs, P. S. (1965). *Biochim. Biophys. Acta* **108**, 652.

Macpherson, I. (1966). *J. Cell Sci.* **1**, 145.

Meister, A. (1965). "Biochemistry of the Amino Acids," 2nd ed., Vol. 2, pp. 729–757. Academic Press, New York.

Mueller, G. C. (1969). *Cancer Res.* **29**, 2394.

Mueller, G. C. (1971). In "The Cell Cycle and Cancer" (R. Baserga, ed.), pp. 269–307. Dekker, New York.

Munro, H. N. (1968). *Fed. Proc., Fed. Amer. Soc. Exp. Biol.* **27**, 1231.

Nilausen, K., and Green, H. (1965). *Exp. Cell Res.* **40**, 166.

Oppenheim, J. J. (1968). *Fed. Proc., Fed. Amer. Soc. Exp. Biol.* **27**, 21.

Petersen, D. F., and Anderson, E. C. (1964). *Nature (London)* **203**, 642.

Petersen, D. F., Anderson, E. C., and Tobey, R. A. (1968). *In* "Methods in Cell Physiology" (D. M. Prescott, ed.), Vol. 3, pp. 347–370. Academic Press, New York.

Petersen, D. F., Tobey, R. A., and Anderson, E. C. (1969a). *In* "The Cell Cycle: Gene Enzyme Interactions" (G. M. Padilla, I. L. Cameron, and G. L. Whitson, eds.), pp. 341–359. Academic Press, New York.

Petersen, D. F., Tobey, R. A., and Anderson, E. C. (1969b). *Fed. Proc., Fed. Amer. Soc. Exp. Biol.* **28**, 1771.

Powelson, D. M. (1961). *J. Bacteriol.* **82**, 288.

Prescott, D. M. (1966). *J. Cell Biol.* **31**, 1.

Puck, T. T., and Steffen, J. (1963). *Biophys. J.* **3**, 379.

Puck, T. T., Sanders, P. C., and Petersen, D. F. (1964). *Biophys. J.* **4**, 441.

Reid, I. M., Verney, E., and Sidransky, H. (1970). *J. Nutr.* **100**, 1149.

Rigler, R., Jr. (1966). *Acta Physiol. Scand., Suppl.* **267**, 1.

Salas, J., and Green, H. (1971). *Nature (London), New Biol.* **229**, 165.

Shigeura, H. T., Hen, A. C., Hiremath, C. B., and Maag, T. A. (1969). *Arch. Biochem. Biophys.* **135**, 90.

Sidransky, H., and Verney, E. (1971). *J. Nutr.* **101**, 1153.

Sidransky, H., Bongiorno, M., Sarma, D. S. R., and Verney, E. (1967). *Biochem. Biophys. Res. Commun.* **27**, 242.

Sidransky, H., Sarma, D. S. R., Bongiorno, M., and Verney, E. (1968). *J. Biol. Chem.* **243**, 1123.

Sidransky, H., Verney, E., and Sarma, D. S. R. (1971). *Amer. J. Clin. Nutr.* **24**, 779.

Sinclair, W. K. (1965). *Science* **150**, 1729.

Sinclair, W. K. (1967). *Cancer Res.* **27**, 297.

Sisken, J. E., and Kinosita, R. (1961). *J. Biophys. Biochem. Cytol.* **9**, 509.

Smulson, M. E., and Rideau, C. (1970). *J. Biol. Chem.* **245**, 5350.

Spalding, J. K., Kajiwara, K., and Mueller, G. C. (1966). *Proc. Nat. Acad. Sci. U. S.* **56**, 1535.

Staehelin, T., Verney, E., and Sidransky, H. (1967). *Biochim. Biophys. Acta* **145**, 105.

Stanbridge, E. J., Hayflick, L., and Perkins, F. T. (1971). *Nature (London), New Biol.* **232**, 242.

Stanners, C. P., and Becker, H. J. (1971). *J. Cell. Physiol.* **77**, 31.

Stubblefield, E. (1964). *In* "Cytogenetics of Cells in Culture" (R. J. C. Harris, ed.), pp. 223–248. Academic Press, New York.

Terasima, T., and Tolmach, L. J. (1963). *Exp. Cell Res.* **30**, 344.

Tiollais, P., Galibert, F., and Boiron, M. (1971). *Eur. J. Biochem.* **18**, 35.

Tjio, J. H., and Puck, T. T. (1958). *J. Exp. Med.* **108**, 259.

Tobey, R. A. (1972a). *Cancer Res.* **32**, 309.

Tobey, R. A. (1972b). *Cancer Res.* **32**, 2720.

Tobey, R. A., and Crissman, H. A. (1972a). *Exp. Cell Res.* (in press).

Tobey, R. A., and Crissman, H. A. (1972b). *Cancer Res.* **32**, 2726.

Tobey, R. A., and Ley, K. D. (1970). *J. Cell Biol.* **46**, 151.

Tobey, R. A., and Ley, K. D. (1971). *Cancer Res.* **31**, 46.

Tobey, R. A., Petersen, D. F., Anderson, E. C., and Puck, T. T. (1966a). *Biophys. J.* **6**, 567.

Tobey, R. A., Anderson, E. C., and Petersen, D. F. (1966b). *Proc. Nat. Acad. Sci. U. S.* **56**, 1520.

Tobey, R. A., Anderson, E. C., and Petersen, D. F. (1967a). *J. Cell Biol.* **35**, 53.

Tobey, R. A., Anderson, E. C., and Petersen, D. F. (1967b). *J. Cell. Physiol.* **70**, 63.

Tobey, R. A., Crissman, H. A., and Kraemer, P. M. (1972). *J. Cell Biol.* **54**, 638.

Todaro, G. J., Lazar, G. K., and Green, H. (1965). *J. Cell. Comp. Physiol.* **66**, 325.

Van Dilla, M. A., Trujillo, T. T., Mullaney, P. F., and Coulter, J. R. (1969). *Science* **163**, 1213.

Van Venrooij, W. J., and Poort, C. (1971). *Biochim. Biophys. Acta* **247**, 468.

Vaughan, M. H., Jr., Pawlowski, P. J., and Forchhammer, J. (1971). *Proc. Nat. Acad. Sci. U. S.* **68**, 2057.

Ward, G. A., and Plagemann, P. G. W. (1969). *J. Cell. Physiol.* **73**, 213.

Weber, M. J., and Edlin, G. (1971). *J. Biol. Chem.* **246**, 1828.

Weiss, B. G. (1969). *J. Cell. Physiol.* **73**, 85.

Whitmore, G. F. (1971). *In Vitro* **6**, 276.

Young, C. W. (1966). *Mol. Pharmacol.* **2**, 50.

Chapter 4

A Method for Measuring Cell Cycle Phases in Suspension Cultures

P. VOLPE AND T. EREMENKO

Cell Biology Laboratory of the International Institute of Genetics and Biophysics, Naples, Italy

I. Introduction

This chapter is concerned with the determination of the cell cycle parameters for suspension cultures.

An individual cell exists for a finite time and then divides into two daughter cells. Each newly born cell grows in mass and, after a certain period, divides again. The interval from one cell division to the next is defined as the *generation time*. This parameter varies from one cell to another in a population over a fairly large range, although most of the cells tend to have a generation time near an average value for the popula-

113

tion. In a given culture the length of the cell life cycle can ordinarily be calculated only as an average value.

The interval between two successive cell separations can be subdivided into four main phases, the definitions of which have their foundation in the model proposed by Howard and Pelc (1953) and in the theoretical discussion by Quastler and Sherman (1959). The major cell cycle stages are as follows: (1) The G_1 *phase* is the lifetime of a young cell lasting until the start of the replication of its genetic apparatus; (2) the S *phase* is the period of DNA synthesis during which the genome is duplicated; (3) the G_2 *phase* is the mature stage of the cell lasting until the prophase; (4) the M *phase* is the period of mitotic division during which chromosomes can be seen clearly in the cell (prophase, metaphase, anaphase, and telophase). For mammalian cells cultivated at 37°C, in monolayers or in suspension, the lengths of the S, G_2, and M phases are relatively constant (Watanabe and Okada, 1967). The G_1 phase, however, is extremely variable (Sisken, 1963) and is entirely missing in at least one line of Chinese hamster cells (Robbins and Scharff, 1967).

The causes for these differences are several. Defendi and Manson (1963) have attributed the variations of the generation time of different cell lines to genetically determined differences in the durations of the G_1 phase. Puck *et al.* (1964), on the other hand, comparing HeLa and Chinese hamster cells, concluded that the variation of their generation time is not attributable to the change of one specific stage, but to the variations in the length of all phases in the same proportion. The study of the effects of variations of environmental conditions has, however, confirmed that the G_1 phase is the more variable one. In human amnion cells, it was observed that the G_1 phase is severely affected by temperature variations (Sisken, 1963, 1965). This finding was confirmed for L5178 Y cells, in which Watanabe and Okada (1967) found that the alteration of the generation time by temperature could be attributed mainly to changes in the length of the G_1 and S stages. In pig kidney cells it was also observed that pH changes alter the length of the G_1 phase (Sisken, 1963). In mice Todaro *et al.* (1965) established that contact-inhibited fibroblasts cease growth at the G_1 stage. Finally, phytohemagglutinin-stimulated leukocytes were found to traverse the S phase before division, indicating that circulating leukocytes are arrested in the G_1 stage (Bender and Prescott, 1962; Rubin and Cooper, 1965). Regarding the temperature effect, HeLa cells appear to represent an exception, since it was found by Rao and Engelberg (1965) that the M phase is the most thermosensitive stage of the cell life cycle.

In order to determine the length of each phase of the cell cycle, various methods have been used, some of which have given variable values.

Other methods, although more precise, have proved to require complicated calculations and have been unsuitable biochemical experiments on a large scale. Many of these methods are reviewed in Vols. I, II, and III of *Methods in Cell Physiology* and elsewhere in the literature (Watanabe and Okada, 1967; Petersen *et al.*, 1969).

The *cell life cycle* has been equated to the time for cell number in a population to double (Prescott, 1959; Merchant *et al.*, 1964), as the time for an average of one division per cell (Christensen and Giese, 1956; Nachtwey and Cameron, 1968), as the time between two 50% points of two ascending limbs of labeled mitosis curves after thymidine-^{3}H pulse labeling (Prescott, 1959; Thrasher, 1966; Quastler and Sherman, 1959), as the time required to halve radioautographic grain counts after thymidine-^{3}H pulse labeling (Leblond, 1959), and as the time between two successive divisions determined by time-lapse cinematography (Sisken, 1963). The generation time has also been obtained on the basis of the metaphase accumulation rate in a cell population blocked with colchicine or Colcemid (Leblond, 1959; Hooper, 1961; Puck and Steffen, 1963).

The length of G_1 phase has been estimated from the increase in the percentage of the labeled cells in a Colcemid-treated population continuously labeled with thymidine-^{3}H (Puck and Steffen, 1963), from the increment in the percentage of labeled cells after X-irradiation of a continuously labeled cell population (Kim and Evans, 1963), and from the fraction of unlabeled cells with low DNA content (Mak, 1965). The length of G_1 has also been calculated as the time required for the percentage of labeled cells to reach the first plateau after X-irradiation (Kim and Evans, 1963) or after treatment with colchicine or Colcemid (Puck and Steffen, 1963) in a continuously labeled cell culture. Finally, the G_1 stage has been measured by combining time-lapse cinematography and autoradiography (Sisken, 1963, 1964) and by selection of dividing cells with a micropipette and pulsing of the cells with thymidine-^{3}H at intervals after division (Stone and Cameron, 1964).

The S phase has been estimated from the percentage of labeled cells shortly after pulse labeling (Quastler and Sherman, 1959), from the difference in percentage-labeled cells between a single and a double labeling (Piligrim and Maurer, 1962, 1965), and from the difference in percentage of cells labeled with thymidine-^{3}H and percentage of cells labeled only with thymidine-^{3}H after double pulsing with thymidine-^{3}H and thymidine-^{14}C (Baserga and Lisco, 1963). The phase of DNA synthesis has also been measured by the rate of accumulation of the labeled cells after colchicine or Colcemid action on a continuously labeled cell culture (Puck and Steffen, 1963), from the time of appearance and disappearance of labeled mitotic cells after pulse labeling (Quastler and

Sherman, 1959; Dewey and Humphrey, 1962), and from the time required for labeled metaphase cells to reach a plateau of grain counts after a pulse (Stanners and Till, 1960).

The G_2 phase has been measured by determining the fraction of unlabeled nonmitotic cells with high DNA content (Mak, 1965), by the difference between the total mitotic cells and the labeled mitotic cells after Colcemid action (Puck and Steffen, 1963), and by the difference between the percentage of the labeled cells at the plateau of the X-irradiated cells and the percentage at the plateau of the X-irradiated cells in a continuously labeled culture (Kim and Evans, 1963). It has also been calculated as the time between the accumulation of mitotic cells and the accumulation of labeled mitotic cells after Colcemid treatment in a continuously labeled cell population (Puck and Steffen, 1963) or as the time between the administration of a pulse of thymidine-^{3}H and the appearance of labeled mitotic cells (Defendi and Manson, 1963).

The mitotic phase M has been determined on single cells by time-lapse cinematography or direct observations (Sisken, 1963) and as the time for mitotic index to decrease to zero after X-irradiation (Knowlton et al., 1948). The M phase has also been calculated from percentage of cells in mitosis (Leblond, 1959; Hooper, 1961).

All these methods have been successfully employed for cell cycle analysis both in monolayer and in suspension cultures. In most cases, however, the generation time and cycle phases could not be estimated simultaneously in a single culture. Moreover, as mentioned, in many cases the experimental procedures appear overly sophisticated and the calculations complicated. A simple procedure is described here which allows measurement of the generation time and of the four phases all at the same time in a single, thymidine-synchronized suspension culture. The method is sufficiently precise to be used in biochemical experiments requiring large masses of cells.

II. Methods

A. Cell Cultures and Media

Suspension cultures of HeLa cells (Gey et al., 1952) are subcultured every 4 days using minimum essential medium as modified by Joklik, without Ca^{2+} ions, and containing 10% calf serum (Volpe and Eremenko, 1970a). Cell density is kept between 0.2 and 1.6 × 10^6 cells/ml (Volpe, 1969; Volpe and Eremenko, 1970a). The suspensions are maintained in spinners (Bellco Glass) at $37°C \pm 0.2$ and aerated by a constant flow of 5% CO_2 in air.

B.   Cell Synchronization

Cells are synchronized with a slight modification of the Puck method (1964), as shown in Fig. 4. For each replacement, from one medium to another, the cells are sedimented in the Sorvall GSA rotor at 1000 rpm for 10 minutes in the cold. The experiment starts with a suspension containing 0.5×10^6 cells/ml. The degree of synchronization achieved is indicated by the density of the culture, which, at the end of the second "thymidine shock," remains at 0.5×10^6 cells/ml (Fig. 5). As soon as the cells are returned to medium without thymidine, the length of each phase of the cellular cycle is measured as reported earlier (Volpe and Eremenko, 1970b).

C.   Cell Counting

Aliquots of the cell suspension are withdrawn during stirring and quickly put in ice in siliconized tubes. After mixing, cells are observed with phase contrast and counted using a hemocytometer chamber. Cell number, expressed in 1×10^6/ml, includes single and double cells, each of the latter counted as two. The cell division index is the percentage of these double cells in the population.

D.   DNA Synthesis

DNA labeling is followed by giving a thymidine pulse of 20 minutes at $37°C$ to a constant aliquot of suspension (5 μCi of thymidine-^{3}H/ml). Samples are analyzed for radioactive DNA as follows. They are centrifuged for 5 minutes at 1000 rpm in the Sorvall SS-34 rotor. The harvested cells are resuspended in 1 ml of distilled water and dissolved in 1 ml of a solution of SDS-EDTA (0.01 M NaCl; 0.01 M tris chloride pH 7; 1% sodium dodecylsulfate and 0.05 M EDTA). The samples are then brought to 7% trichloroacetic acid (TCA), left in ice for 20 minutes, filtered onto millipore HA 0.45 μm filters and washed with 5% TCA. The determination of the radioactivity is made in *Bray's* solvent using a scintillation spectrometer.

E.   Chemicals

The Joklik-modified minimum essential medium for spinners (Gibco catalog no. 12-616) and calf serum were furnished by Gibco. Before use, the serum is heated for 2 hours at 50°C and filtered on Seitz EKS II. Unlabeled thymidine was obtained from Sigma. Thymidine-³H (6.7 Ci/mmole) was supplied by New England Nuclear.

III. Results

A. Cell Growth Cycle of HeLa Cells in Suspension

It was found that during 3–4 days of growth in suspension HeLa cells divide stepwise in an apparent natural synchronization (Volpe, 1969; Volpe and Eremenko, 1970a), whereas under the same conditions of growth Chang's liver cells are asynchronous (Strecker and Eliasson, 1966; Volpe, 1969; Volpe and Eremenko, 1970a). Figure 1 shows that the HeLa cell growth curve follows four regular cycles, each with a periodicity of less than a day. During the first cycle the percentage of double cells decreases from 80 to 35; during the second cycle this value increases to 60%; during the third cycle the number of double cells returns to the zero time, high level. DNA in the culture increases concurrently also in a stepwise fashion (Volpe, 1969). The reason for this "spontaneous" synchronization of the HeLa cell growth is still unknown. In our laboratory Chang's liver and HeLa cells can be artificially synchronized with thymidine for different purposes obtaining, however, the

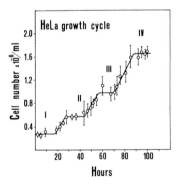

FIG. 1. Stepwise growth curve of "spontaneously" synchronized HeLa cells cultivated in suspension culture. The cells were grown as described in the section on "Methods." A culture is initiated with a density of 0.2 × 10⁶ cells/ml. The cultures achieve maximum population density in 80 hours (1.6 × 10⁶ cells/ml), having completed four cell life cycles, each lasting 20 hours. The points are the mean values for 10 experiments carried out with different volumes of cell suspension (5 with 1 liter, 3 with 2 liters, and 2 with 4 liters).

same results. To start thymidine synchronization, in both cases the cells are withdrawn during the multiplication phase of growth. The experimental data reported in this chapter are those obtained with HeLa cells.

B. Optimum Concentration of Thymidine for Inhibiting the Growth of HeLa Cells

Xeros (1962) found that excess thymidine blocks nuclear DNA synthesis. Puck (1964) used this effect to synchronize eukaryotic cells at the threshold of the S phase. Data are reported here which establish the optimum concentration of thymidine to inhibit the HeLa cell growth in suspension. Figure 2 shows that 2 mmole thymidine is enough to give a minimum cell growth, when the culture has been subjected to the thymidine for 40 hours. A clearer picture is seen in Fig. 3, which indicates that this thymidine concentration is able to inhibit increase in cell number and cell divisions, while a 2.5 mmole-concentration of the nucleoside probably approaches a toxic level.

C. Double Thymidine Synchronization of HeLa Cell Suspensions

A somewhat idealized representation of the distribution of cells throughout the life cycle at the different times of the double "thymidine shock" is shown in Fig. 4. The development of cell number and cell divisions during the course of this synchronization is illustrated in Fig. 5,

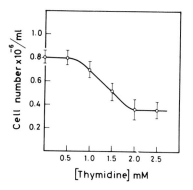

FIG. 2. Effect of the thymidine concentration on the cell number in HeLa suspension cultures. The cells were cultivated as for Fig. 1. They were grown for 40 hours in the presence of various concentrations of thymidine. All cultures, each of 200 ml, were inoculated from the same cell line, and growth was started in parallel with a density of 0.2×10^6 cells/ml. The points represent the average values for 6 experiments.

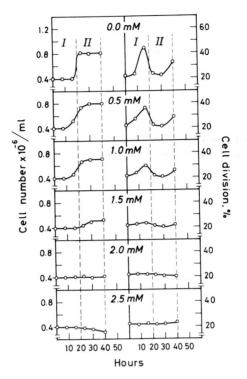

Fɪɢ. 3. Growth and cell division curves for HeLa cells cultivated for 40 hours in the presence of various concentrations of thymidine. The cells, as in Fig. 1, were resuspended in fresh medium every 4 days. Starting with a density of 0.2×10^6 cells/ml, both the newly inoculated control without thymidine, and experimental suspensions with thymidine, were grown for 40 hours (up to the end of the second cell life cycle). Cell numbers and divisions are calculated as described under "Methods."

which shows that under the action of 2 mmole thymidine, during the first and the second shock, the population density stays constant at 0.5×10^6 cells/ml, and the number of double cells is below a level of 10%. In both cases, since cell number does not increase, while a low level of cell division continuously takes place, a certain percentage of old cells must be dying. During the period of growth in the absence of thymidine (8 hours), both cell number and divisions remain constant until the time when the cells, having synthesized DNA, arrive in the M phase (see Fig. 4). Reaching the M phase, those cells that were not synchronized by the first thymidine shock, rapidly divide, and consequently the cell number increases. The increase of cell number, however, is not proportional to the increase in double cells, since some old cells continue to disappear.

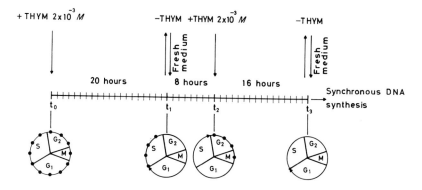

FIG. 4. Thymidine synchronization of HeLa cells grown in suspension. The cells were withdrawn during the multiplication phase (at the third cell life cycle as shown in Fig. 1). Adjusted to density of 0.5×10^6/ml, they were transferred into medium containing 2 mmole thymidine. At this time (t_0) the majority of cells are dividing, while many others are distributed at random throughout the cell cycle. After 20 hours in the presence of 2 mmole thymidine (t_1), some cells are still distributed randomly through the S phase and all the rest of the population is phased at the end of G_1. At this time thymidine was removed and the cells transferred into fresh medium for 8 hours at the end of which (t_2) all pass through the S phase. Then, thymidine was added to the culture for another 16 hours. At this point (t_3) the culture consists of a highly monophasic population concentrated at the end of G_1. Thymidine is finally removed and the cells enter the S phase.

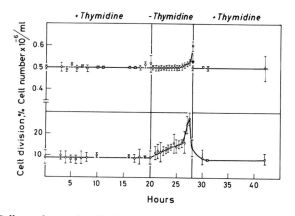

FIG. 5. Cell number and cell division during a double thymidine synchronization of HeLa suspensions. Thymidine (2 mmole) was added to a suspension culture. For 16 hours cell number and cell division remained constant. Thymidine was removed, and the cells grown in the fresh medium for 8 hours. At the end of this time cell division increases appreciably, while cell number increased less. After a new addition of 2 mmole thymidine, the culture remained at a density of 0.5×10^6 cells/ml (the arrow shows the point at which about 20% of cells were discarded). For another 16 hours cell number and cell division were constant. The points represent the average of 30 experiments carried out with different volumes of cell suspension (20 with 1 liter, 6 with 2 liters, and 4 with 4 liters).

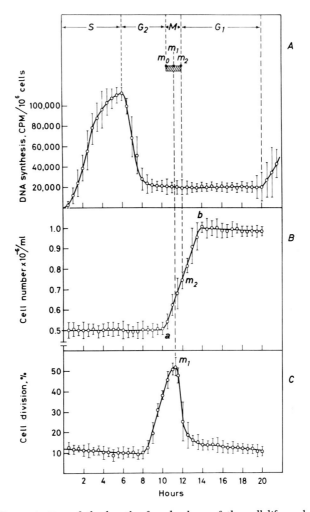

Fig. 6. Determination of the length of each phase of the cell life cycle in HeLa suspensions synchronized twice with thymidine. At the end of the second "thymidine shock," after washing, the culture (entering S phase) was transferred into fresh minimum essential medium and left to grow for 24 hours. Every 30 minutes, DNA synthesis (curve A), cell number (curve B), and cell division (curve C) were estimated. The cycle phases, S, G_2, M, and G_1 are marked at the top; m_0, m_1, and m_2, as shown in Fig. 7, represent respectively the beginning, the middle, and the end of the M phase. The segment ab in the cell number curve B is the experimentally observed linear multiplication phase. The points of the B and C curves represent the mean of 30 experiments. The DNA synthetic curve A is the mean of 10 experiments. The experiments are carried out with different volumes of cell suspension (20 with 1 liter, 6 with 2 liters, and 4 with 4 liters of suspension cultures).

D. Generation Time and Measurement of Its Phases

Starting from the time of return to medium without thymidine (Fig. 4), samples are taken at short intervals to measure (1) DNA synthesis, (2) population density, and (3) rate of mitotic divisions (Fig. 6). Curve A shows that maximum DNA labeling takes place after 6 hours, while a new round of its synthesis starts after 20 hours. Curve B shows that cell number remains constant for 10 hours (0.5×10^6 cells/ml) and then doubles in a short time reaching a new plateau (1.0×10^6 cells/ml). Curve C, finally, shows that the maximum formation of double cells is reached after 11.5 hours. The lengths of the cell cycle and its phases are measured by combining the information from all these curves.

The length of the cell cycle is the period between the starting points of the two peaks of DNA synthesis (Fig. 6A), in these experiments 20 hours. The length of the S phase is given by the time between removal of thymidine and the time of maximum DNA labeling (Fig. 6A), namely 6 hours. After such maximum, DNA synthesis decreases rapidly to minimal values. During this period there is no change in cell number (Fig. 6B) or in cell division (Fig. 6C). The G_2 phase precedes cell division (see Fig. 7); thus, for an appreciable time, the cell number does not change (Fig. 6B). To determine the endpoint of this phase one has to measure first the M phase, which is the shortest in the cell cycle (see

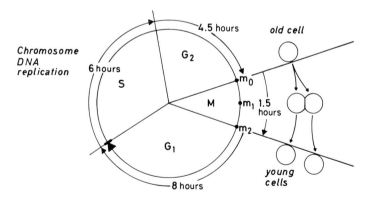

FIG. 7. Representation of the cell life cycle and its phases. The numbers give the lengths of the cell cycle phases determined for HeLa cells grown in suspension. The arrow shows the time when the medium containing 2 mmole thymidine was removed and replaced with one without thymidine. G_1, S, G_2, and M are the four main phases of the cell cycle; m_0, m_1, and m_2 are, respectively, the beginning, the middle, and the end points of the cell division phase M. At the end of the G_2 phase the mature cell, with a double DNA charge, enters the M phase. After mitosis, the resultant double cells divide. The newly born cells, with a normal DNA charge, initiate new life cycle.

Fig. 7). Cell division reaches its maximum 11.5 hours after the beginning of the S phase (Fig. 6C). The top point of the peak corresponds to the time when the suspension contains a maximum number of cells in division. In Fig. 7 this point is m_1. On the other hand, since the linear increase *ab* in cell number (Fig. 6B) is the integral of the sharp peak of cell division (Fig. 6C), its middle point signals the end of the mitotic phase M, which in Fig. 7 is shown as m_2. The M phase, thus is twice the time between m_1 and m_2 going to m_0 counterclockwise, since m_1 is the middle point of the whole phase (Fig. 7). In these experiments the M phase is calculated to be 1.5 hours long. This is in accord with the turning point *a* in the cell number curve B, which takes place slightly later than 10 hours after the beginning of DNA synthesis (Fig. 6). The G_2 phase, thus, is 4.5 hours long, and the G_1 phase, lasting from the m_2 point (Fig. 6B) to the beginning of a new DNA synthesis phase (Fig. 6A), is 8 hours.

IV.   Concluding Remarks

An easy graphic method is described in this chapter for measuring the length of the cell cycle and its phases in suspension cultures. For HeLa cells, the length of the cell cycle is about 20 hours, while the phases S, G_2, M, and G_1 last 6, 4.5, 1.5, and 8 hours respectively. For most mammalian cells grown at 37°C the lengths of S, G_2, and M phases range in intervals of 6 to 9 for S, 2 to 5 for G_2, and 1 to 2 hours for M. G_1 is the most variable phase of the cell cycle. As mentioned, G_1 can last 30 hours or more in some cells and be entirely missing in others (Robbins and Scharff, 1967). The duration of the cell cycle and its phases determined by the present procedure for HeLa cells falls within these limits and corresponds fairly well with the relative values found for the same cells (Puck and Yamada, 1962) under completely different experimental conditions.

The graphic method provides some advantages with respect to other methods. It is readily reproducible; it is also rapid and accurate, since only three easily determined parameters are required, which are all measured in parallel in a single culture. The sensitivity of the method, moreover, does not strictly depend on the background number of doubled cells in the M phase. Looking at Fig. 6, in fact, one can see that the vertical projections, in particular that of the m_1 and m_2 points, are the determining ones. The flattening of the curves, especially that of curve C (Fig. 6), does not affect the measurements in any way, and this is the main advantage of the method. For precise measurement of the length of the M phase, which is the shortest in the cycle, an accurate synchronization of the suspension culture is, however, required. Normally, at the end of the double thymidine synchronization, as mentioned in the section on "Methods," the population density remains at the level of 0.5×10^6 cells/ml. Moreover, when the synchronization is satisfactory, the peak of cell divisions (Fig. 6C) is sharp, reaching the level of 50%, and the cell number exactly doubles in a short time (Fig. 6B).

The method discussed here appears to be very useful in biochemical experiments requiring relatively large masses of cells, since synchronized suspension cultures may be grown at any volume. In our laboratory this method was successfully employed in quantitative studies on cell proteins (Volpe and Eremenko, 1970a), for the analysis of nuclear and mitochondrial DNA synthesis during the HeLa cell life cycle (Volpe and Eremenko, 1972), for studying the control and the structure of different ornithine-δ-transaminase units produced in specific interkinetic phases (Volpe and Eremenko, 1971), and, finally, for investigating the cell cycle dependency of poliovirus RNA replication (Eremenko et al., 1971, 1972a,b; Koch and Volpe, 1972). The method was used as well in biochemical studies on Chang's liver cells (Strecker et al., 1970), for which almost the same lengths of the HeLa cell cycle phases were established.

ACKNOWLEDGMENTS

Thanks are due to Professor H. J. Strecker, who financed the preliminary experiments at the Albert Einstein College of Medicine, New York (SNF, GB 5061 X). Sincere appreciation is also due to Professor E. Pancini, of the Institute of Physics of Naples University, and to Dr. E. Whitehead, of Euratom Organization, for critical discussion of the method. The skillful technical assistance of Mr. C. Buono is appreciated.

REFERENCES

Baserga, R., and Lisco, E. (1963). J. Nat. Cancer Inst. 31, 1559.
Bender, M. A., and Prescott, D. M. (1962). Exp. Cell Res. 27, 221.
Christensen, E., and Giese, A. C. (1956). J. Gen. Physiol. 39, 513.
Defendi, V., and Manson, L. A. (1963). Nature (London) 198, 359.
Dewey, W. C., and Humphrey, R. M. (1962). Radiat. Res. 16, 503.
Eremenko, T., Benedetto, A., and Volpe, P. (1972). In "International Virology" (J. L. Melnich, ed.), Vol. 2, p. 281. Karger, Basel.
Eremenko, T., Benedetto, A., and Volpe, P. (1972a). Nature (London), New Biol. 237, 114.
Eremenko, T., Benedetto, A., and Volpe, P. (1972b). J. Gen. Virol. 16, 61.
Gey, G. O., Coffman, W. D., and Kubicek, M. T. (1952). Cancer Res. 12, 264.
Hooper, C. E. S. (1961). Amer. J. Anat. 108, 231.
Howard, A., and Pelc, S. R. (1953). Heredity 6, Suppl., 261.

Kim, J. H., and Evans, T. C. (1963). *Radiat. Res.* **21**, 129.

Knowlton, N. P., Hempelmann, L. H., and Hoffman, J. G. (1948). *Science* **107**, 625.

Koch, A. S., and Volpe, P. (1972). *In* "International Virology" (J. L. Melnick, ed.), Vol. 2, pp. 277–282. Karger, Basel.

Leblond, C. P. (1959). *In* "The Kinetics of Cellular Proliferation" (F. Stohlman, ed.), pp. 31–47. Grune & Stratton, New York.

Mak, S. (1965). *Exp. Cell Res.* **39**, 286.

Merchant, D. J., Kahan, K. H., and Murphy, W. H. (1964). "Handbook of Cell and Organ Culture." Burgess, Minneapolis, Minnesota.

Nachtwey, D. S., and Cameron, I. L. (1968). *In* "Methods in Cell Physiology" (D. M. Prescott, ed.), Vol. 3, pp. 213–259. Academic Press, New York.

Petersen, D. F., Tobey, R. A., and Anderson, E. C. (1969). *In* "The Cell Cycle" (G. M. Padilla, I. L. Cameron, and G. L. Whitson, eds.), pp. 341–359. Academic Press, New York.

Piligrim, C., and Maurer, W. (1962). *Naturwissenschaften* **49**, 544.

Piligrim, C., and Maurer, W. (1965). *Exp. Cell Res.* **37**, 183.

Prescott, D. M. (1959). *Exp. Cell Res.* **16**, 279.

Puck, T. T. (1964). *Cold Spring Harbor Symp. Quant. Biol.* **29**, 167.

Puck, T. T., and Steffen, J. (1963). *Biophys. J.* **3**, 379.

Puck, T. T., and Yamada, M. (1962). *Radiat. Res.* **16**, 589.

Puck, T. T., Sanders, P. C., and Petersen, D. (1964). *Biophys. J.* **4**, 442.

Quastler, H., and Sherman, F. G. (1959). *Exp. Cell Res.* **17**, 420.

Rao, P. N., and Engelberg, J. (1965). *Science* **148**, 1092.

Robbins, E., and Scharff, M. D. (1967). *J. Cell Biol.* **32**, 303.

Rubin, A. D., and Cooper, H. L. (1965). *Proc. Nat. Acad. Sci. U. S.* **54**, 469.

Sisken, J. E. (1963). *In* "Cinemicrography in Cell Biology" (G. G. Rose, ed.), pp. 143–168. Academic Press, New York.

Sisken, J. E. (1964). *In* "Methods in Cell Physiology" (D. M. Prescott, ed.), Vol. 1, pp. 387–401. Academic Press, New York.

Sisken, J. E. (1965). *Exp. Cell Res.* **40**, 436.

Stanners, C. P., and Till, J. E. (1960). *Biochim. Biophys. Acta* **37**, 406.

Stone, G. E., and Cameron, I. L. (1964). *In* "Methods in Cell Physiology" (D. M. Prescott, ed.), Vol. 1, pp. 127–140. Academic Press, New York.

Strecker, H. J., and Eliasson, E. E. (1966). *J. Biol. Chem.* **241**, 5750.

Strecker, H. J., Hammar, U., and Volpe, P. (1970). *J. Biol. Chem.* **245**, 3328.

Thrasher, J. D. (1966). *In* "Methods in Cell Physiology" (D. M. Prescott, ed.), Vol. 2, pp. 323–357. Academic Press, New York.

Todaro, G. J., Lazar, G. K., and Green, H. (1965). *J. Cell. Comp. Physiol.* **66**, 325.

Volpe, P. (1969). *Biochem. Biophys. Res. Commun.* **34**, 190.

Volpe, P., and Eremenko, T. (1970a). *Eur. J. Biochem.* **12**, 195.

Volpe, P., and Eremenko, T. (1970b). *Exp. Cell Res.* **60**, 456.

Volpe, P., and Eremenko, T. (1971). *In* "Advances in Cytopharmacology" (D. Clementi, ed.), Vol. 1, pp. 257–261. Raven, New York.

Volpe, P., and Eremenko, T. (1972). *Eur. J. Biochem.* (in press).

Watanabe, I., and Okada, S. (1967). *J. Cell Biol.* **32**, 309.

Xeros, N. (1962). *Nature (London)* **194**, 682.

Chapter 5

A Replica Plating Method of Cultured Mammalian Cells

FUMIO SUZUKI AND MASAKATSU HORIKAWA

*Department of Radiation Biology, Faculty of Pharmaceutical Sciences,
Kanazawa University, Kanazawa, Japan*

I. Introduction

Genetic and biochemical studies of single somatic mammalian cells *in vitro* were developed by Puck and his co-workers (Puck and Marcus, 1955; Puck *et al.*, 1956; Puck and Fisher, 1956), who introduced the colony-forming technique as used in the field of microbial genetics. During the 17 years since then, a number of tools and techniques necessary for the genetic analysis of cultured somatic mammalian cells have been developed. For example, the isolation of drug-resistant cell lines (Gartler and

127

Pious, 1966; Chu and Malling, 1968), nutritional auxotrophs (Kao and Puck, 1968, 1969), and temperature-sensitive mutants (Naha, 1969; Thompson et al., 1970) has been reported, and techniques have been developed for genetic analysis with somatic cell hybrids (Barski et al., 1960, 1961; Littlefield, 1964; Ephrussi and Weiss, 1965; Weiss and Ephrussi, 1966).

However, the lack of a replica plating technique, as used in the field of microbial genetics, for cultured mammalian cells has delayed more detailed genetic analysis in somatic mammalian cells. Recently, Goldsby and Zipser (1969) described a technique for the replica plating of a clonal population of mammalian cells. This procedure should be useful for the isolation of useful mutants, for further purification of a mutant cell line, and for the investigation of mutagenesis at the somatic cell level similar to that at bacterial (Lederberg and Lederberg, 1952) and fungal levels (Roberts, 1959). We have established a technique for a simpler and more suitable replica plating method for cultured mammalian cells (Suzuki et al., 1971), by improving the technique described by Goldsby and Zipser (1969). In this chapter we shall describe the technical procedure for replica plating and an example of the application of the technique to the characterization of mutants of cultured mammalian cells.

II.   Procedure for Replica Plating of Cultured Mouse 3T6 Cells and HeLa S3 Cells

The replica plating method described in this section is useful for the replica plating of various cultured mammalian cell lines in vitro as mentioned later. Here we shall describe a new procedure for replica plating of cultured mammalian cells that is simpler and more suitable for genetic analysis than that reported previously (Goldsby and Zipser, 1969), mainly by using cultured mouse 3T6 cells and HeLa S3 cells as materials.

A.   Cells and Medium

A clonal cell line (C1-R2A) derived from mouse 3T6 cells grown in 80% Eagle MEM G medium (Eagle MEM medium supplemented with 0.292 gm/liter L-glutamine) plus 20% bovine serum, and HeLa S3 cells grown in 90% Eagle MEM medium plus 10% bovine serum were used for the present study.

B. Master Plates, Replica Plates, and Hand Replicator

MicroTest II tissue culture plates (Gateway International Inc., 401-03 South Vermont Av., Los Angeles, Calif. 90005; catalog no. 3040), which contain a matrix of 96 flat-bottom wells as shown in Fig. 1 were employed as master plates for single cell culture and as replica plates for replica plating culture.

The hand replicator used for the transfer of cells from 96 flat bottom wells on a master plate to replica plates was prepared as follows: A hole was drilled in the center of each flat bottom in 96 wells of a MicroTest II tissue culture plate. Two vinyl tubes of different sizes (outside diameter 7 mm, wall thickness 1 mm and outside diameter 5 mm, thickness 1 mm, respectively) were fixed in each well with an adhesive, then a small glass tube (outside diameter 3 mm, length 25 mm) was fixed into each small vinyl tube, and a piece of sponge was fixed on the top of each protruding small glass tube as shown in Fig. 2. The capillary attraction of cell suspensions into small protruding glass tubes was thus adapted for the preparation of this hand replicator. MicroTest II tissue culture plates and a hand replicator were sterilized by immersion in absolute ethanol and 10–20 minutes exposure to ultraviolet light of a 10-W germicidal lamp.

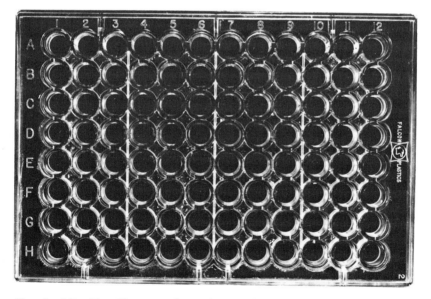

Fig. 1. MicroTest II tissue culture plate employed as master plate and replica plate.

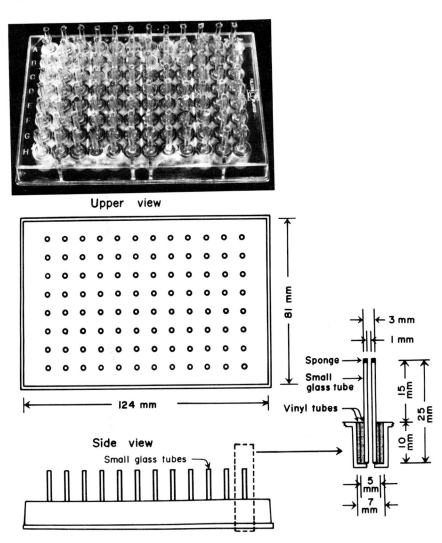

FIG. 2. Hand replicator.

C. Seeding of Single Mouse 3T6 Cells in a Master Plate and Transfer of Cells from Master Plate to Replica Plates (Replica Plating)

Aliquots of 0.2 ml mouse 3T6 cell suspension which contains 8 cells/ml were added to each of 96 wells of MicroTest II tissue culture plates with a 1-ml sterile disposable hypodermic syringe (Jintan Terumo, Tokyo).

TABLE I

FREQUENCY DISTRIBUTION OF WELLS CONTAINING FROM 0 TO 8 MOUSE
3T6 CELLS (CL-R2A)

	No. of cells per well									
	0	1	2	3	4	5	6	7	8	Total
No. of wells	20	34	17	14	7	2	1	0	1	96
Percentage	20.8	35.4	17.7	14.6	7.3	2.1	1.0	0	1.0	100

The number of cells in each well was then checked under an inverted microscope. The frequency distribution of wells with cells from 0 to 8 is shown in Table I. The culture media in wells which received no cells were sucked up by a syringe, and 0.2 ml of cell suspension was added again. The frequency distribution of wells with cells from 0 to 8 after the addition was changed to that shown in Table II.

Thus, a single mouse 3T6 cell can be seeded into 40 to 44 of the 96 wells of a plate. After seeding, the plates were covered and incubated at 37°C for 10 days in a humidified CO_2 incubator which was flushed with a mixture of 5% CO_2 and 95% air. After 10 days of incubation, the culture medium in each well was discarded, and 0.015 ml of 0.25% warmed trypsin and EDTA solution was added to each well. Then 0.2 ml of the culture medium was quickly added to each well and reincubated at 37°C for 3 days in a humidified CO_2 incubator. By this trypsin and EDTA treatment, the clones located in corners of wells were loosely dispersed. These dispersed cells easily re-formed a monolayer on each flat well bottom during reincubation. After 3 days of reincubation, the culture medium in wells was discarded and 0.015 ml of 0.25% warmed trypsin and EDTA solution was added to each well to detach the monolayer cells from the flat bottom surface. The replication process entails detachment, dispersion, and transfer of cells.

TABLE II

FREQUENCY DISTRIBUTION OF WELLS CONTAINING FROM 0 TO 8 MOUSE 3T6
CELLS (CL-R2A) AFTER THE SECOND ADDITION OF 0.2 ML CELL
SUSPENSION IN WELLS WHICH RECEIVED NO CELLS

	No. of cells per well									
	0	1	2	3	4	5	6	7	8	Total
No. of wells	10	42	19	14	7	2	1	0	1	96
Percentage	10.4	43.8	19.8	14.6	7.3	2.1	1.0	0	1.0	100

After 0.2 ml of culture medium was quickly added to each well, 96 protruding small glass tubes of a hand replicator (see Fig. 2) were inserted into 96 wells of the master plate which had undergone cell detachment. Dispersion was accomplished by gentle up-and-down motion of the hand replicator on the master plate. Replication was achieved by pulling out the hand replicator from the master plate, inserting this into 96 wells of the replica plates devoid of culture medium. Thus, this hand replicator, based on the application of capillary attraction of cell suspensions into the small glass tubes, is able to accomplish simultaneously the dispersion and transfer of cells to 10–15 replica plates from a master plate by a single insertion of the hand replicator into the cell suspensions of the master plate.

After 0.2 ml of culture medium was added to wells of the replica plates containing transferred cells, the plates were incubated at 37°C for 5–10 days in a humidified CO_2 incubator. After incubation, the number of clones transferred to identical positions (wells) on the replica plates from a master plate were determined with an inverted microscope.

D.   Replica Plating Efficiency

The results of three experiments which were done independently on different days by using the procedure described above are shown in Table III. In each experiment, as many as 10 replica plates were obtained at one time from a master plate with high replica plating efficiency of 95.3–99.7%.

As mentioned above, single mouse 3T6 cells could be seeded into 40–44 of the 96 wells of a master plate, whereas 37–38 clones originating from a single cell are obtained in a master plate, as can be seen in Table III. Thus, the average plating efficiency in wells of the master plates in these three experiments is 89.3%. Goldsby and Zipser (1969) obtained 6 replica plates at one time from a master plate of Chinese hamster Don lung fibroblasts with 97% replica plating efficiency, using their replicator and their procedure. Their replica plating efficiency is in good agreement with those (95.3–99.7%) obtained with mouse 3T6 cells (C1-R2A) in the present study. However, our replica plating method using a hand replicator which works by the application of the principle of capillary attraction of cell suspensions is simpler and more easily available, and 10–15 replica plates can be obtained at one time from a master plate.

Furthermore, we have found that the present replica plating method can be used for the replica plating of cultured HeLa S3 cells, Chinese hamster Don cells, Chinese hamster hai cells, and many other mammalian

TABLE III

Number of Clones Transferred to Identical Positions (Wells) on the
Replica Plates from a Master Plate Seeded with a Single Cell
of Mouse 3T6 Cells (CL-R2A) and Their Percent Replica
Plating Efficiency. Three Experiments Were Done on
Different Days, Independently

	Master plates		
	Experiment no. 1	Experiment no. 2	Experiment no. 3
Experiment began:	2/2/1971	2/12/1971	2/16/1971
No. of clones on master plate:	38	37	37
Replica plates	Number of clones transferred to identical position on the replica plates		
R1	35	36	36
R2	37	37	37
R3	34	35	37
R4	38	37	37
R5	37	37	37
R6	35	36	37
R7	36	37	37
R8	37	37	37
R9	36	36	37
R10	37	36	37
Mean:	36.2	36.4	36.9
Percent replica plating efficiency:	95.3	98.4	99.7

cell lines *in vitro*. For example, by seeding single HeLa S3 cells into 96 wells of a master plate by the procedure described above and by incubation for 13 days, 36 clones originating from a single cell were obtained. These clones were transferred to three clean glass plates by a hand replicator and then air dried and Giemsa stained. Figure 3 shows Giemsa-stained 36 HeLa S3 clones located in identical positions on a master plate and the three glass plates obtained by a single cell seeding and the replica plating, respectively.

When glass plates with 36 clones, obtained by replica plating, are treated with various reaction mixtures containing a substrate and cofactors at the optimum temperature, specific enzyme-sufficient or enzyme-deficient mutant clones can be detected and isolated. Moreover, if cells transferred to glass plates from a master plate by a hand replicator can easily become attached to the glass surface, and if these glass plates can

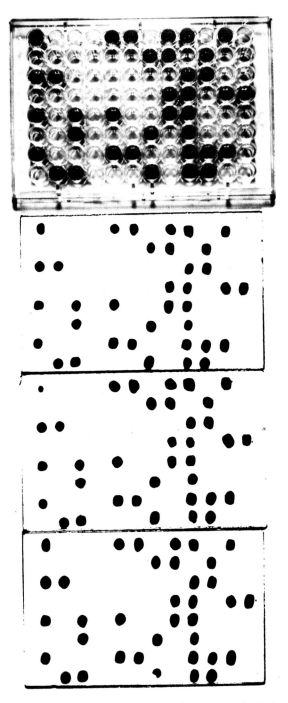

FIG. 3. Giemsa-stained 36 HeLa S3 clones located in identical positions on the master plate and three glass plates obtained by a single cell seeding and replica plating.

be incubated in some vessel containing a culture medium, it will be easier to obtain replica plating cultures than with the use of MicroTest II tissue culture plates as replica plates.

III.   Detection and Isolation of Nutritionally Deficient Mutants from the Original Chinese Hamster *Hai* Cells by a Replica Plating Method

Many kinds of mutant cell lines such as drug-resistant cells (Gartler and Pious, 1966; Chu and Malling, 1968), nutritional auxotrophs (Kao and Puck, 1968, 1969), and temperature-sensitive mutants (Naha, 1969; Thompson *et al.*, 1970) have already been isolated from various cultured mammalian cell lines. These mutant cell lines should be useful for the genetic analysis of cultured mammalian cells and for the investigation of mutagenesis at the somatic mammalian cell level. In this section we shall describe the results of an experiment designed to detect and isolate the nutritionally deficient mutant cells from the original Chinese hamster *hai* cells as an example of the application of the replica plating method.

A.   Cells and Medium

A Chinese hamster *hai* cell line derived originally from the lung of a newborn male Chinese hamster was used in the present study. The cells were grown in a complete medium composed of 90% Eagle MEM + N18 medium (see Table IV) and 10% dialyzed calf serum. Dialyzed calf serum was prepared by dialyses for 24 hours against cold running tap water and against every 800 ml of distilled water in cold five times for a total of 24 hours. In this medium and in a humid atmosphere of CO_2 and air at 37°C, the average generation time of these cells was about 14 hours as determined by cell growth.

For detecting and for isolating alanine-, asparagine-, proline-, aspartic acid-, serine-, glycine-, hypoxanthine-, or thymidine-deficient mutants (auxotrophic mutants) from the original Chinese hamster *hai* cells, various media were prepared, each of which lacked one of the ingredients—L-alanine, L-asparagine, L-proline, L-aspartic acid, L-serine, glycine, hypoxanthine, or thymidine—of the Eagle MEM + N18 medium as listed in Table IV. These media supplemented with 10% dialyzed calf serum were used as selection media.

TABLE IV

COMPOSITION OF EAGLE MEM + N18 MEDIUM

Eagle MEM		N18	
Ingredient	Content (gm/liter)	Ingredient	Content (gm/liter)
L-Arginine HCl	0.126	L-Alanine	0.009
L-Histidine HCl·H$_2$O	0.042	L-Asparagine	0.013
L-Isoleucine	0.052	L-Proline	0.035
L-Leucine	0.052	L-Aspartic acid	0.013
L-Cystine	0.024	L-Glutamic acid	0.015
L-Tyrosine	0.036	L-Serine	0.021
L-Lysine HCl	0.073	Glycine	0.008
L-Methionine	0.015	Sodium pyruvate	0.110
L-Phenylalanine	0.032	Putrescine dihydrochloride	0.0003
L-Threonine	0.048	Choline chloride	0.014
L-Tryptophan	0.010	Vitamin B$_{12}$	0.001
L-Valine	0.046	Inositol	0.016
L-Glutamine	0.292	FeSO$_4$·7H$_2$O	0.0008
Succinic acid	0.075	ZnSO$_4$·7H$_2$O	0.0009
Sodium succinate	0.100	Hypoxanthine	0.004
Choline bitartrate	0.0018	Lipoic acid	0.0002
Folic acid	0.001	Linoleic acid	0.00008
i-Inositol	0.002	Thymidine	0.0007
Nicotinamide	0.001		
Calcium pantothenate	0.001		
Pyridoxine HCl	0.001		
Riboflavin	0.0001		
Thiamine HCl	0.001		
Biotin	0.00002		
NaCl	6.800		
KCl	0.400		
NaH$_2$PO$_4$	0.115		
MgSO$_4$	0.0935		
CaCl$_2$	0.200		
Glucose	1.000		
Phenol red	0.006		
Kanamycin	0.060		

B. Seeding of Single Chinese Hamster *Hai* Cells in a Master Plate and Preparation of Replica Plating Cultures

MicroTest II tissue culture plates were employed as master plates for single Chinese hamster *hai* cell culture and as replica plates for replica plating culture as well, as described above. The hand replicator shown in Fig. 2 was used for replica plating.

By the procedure described above, 0.2 ml of a Chinese hamster *hai* cell suspension containing 8 cells/ml in a complete medium composed of 90% Eagle MEM + N18 medium and 10% dialyzed calf serum was added to each of 96 wells of a MicroTest II tissue culture plate. The culture medium in wells which received no cells was sucked up by a syringe and 0.2 ml of cell suspension was added again. Thus, a single Chinese hamster *hai* cell can be seeded into 42 of the 96 wells of a plate. The master plate was incubated at 37°C for 13 days in a humidified CO_2 incubator, and 38 clones originating from a single cell were obtained.

The culture medium in the wells was discarded and 0.015 ml of 0.25% warmed trypsin and EDTA solution was added to each well to detach the monolayer cells from the flat bottom. After 0.2 ml of Eagle MEM medium alone was quickly added to each well, the cells in the wells were dispersed and transferred, by a hand replicator, to 17 replica plates devoid of culture medium. Then 0.2 ml of complete medium, composed of 90% Eagle MEM + N18 medium and 10% dialyzed calf serum, was added to 38 wells of a master plate containing cell remainders and a replica plate containing transferred cells, and the plates were incubated at 37°C for 10 days in a humidified CO_2 incubator. In addition, 0.2 ml of various selection media omitting L-alanine, L-asparagine, L-proline, L-aspartic acid, L-serine, glycine, hypoxanthine, or thymidine from the Eagle MEM + N18 medium, was added to 38 wells of each of two plates of the remaining 16 replica plates, and the plates were incubated at 37°C for 10 days in a humidified CO_2 incubator. After incubation, the growth of clones transferred to identical positions (wells) on the replica plates from a master plate were determined under an inverted microscope.

C. Detection and Isolation of Nutritionally Deficient Mutants

The growth of 38 clones in each replica plate cultured with a complete medium or the various selection media for 10 days is shown in Table V. As seen in this table, all 38 clones transferred to identical positions (wells) on the replica plates from a master plate grew actively and formed monolayers in a complete medium or in an alanine-free medium. These results indicate that all 38 clones were alanine-sufficient clones which did not require alanine for their growth. On the other hand, more than half of the 38 clones did not grow in glycine- or thymidine-free medium, about 6–10 of the 38 clones did not grow in asparagine-, proline-, aspartic acid-, or serine-free media, and only two clones did not grow in hypoxanthine-free medium.

In Table VI are summarized the results shown in Table V, by a transformed presentation. As can be seen in this table, 38 clones can be divided into seven classes, for the sake of convenience, according to their

TABLE V
THE GROWTH OF 38 CLONES IN EACH REPLICA PLATE CULTURED IN COMPLETE MEDIUM OR VARIOUS SELECTION MEDIA FOR 10 DAYS

Clone no.	Complete medium	Substance omitted from complete medium							
		Ala	Asn	Pro	Asp	Ser	Gly	Hyp	TdR[a]
1	+	+	+	+	−	+	−	+	+
2	+	+	+	+	+·	+	−	+	−
3	+	+	+	+	+	+	−	+	−
4	+	+	+	+	+	+	−	+	−
5	+	+	−	+	+	+	−	+	+
6	+	+	+	+	+	+	−	+	+
7	+	+	+	−	−	+	−	+	−
8	+	+	+	−	+	−	+	+	+
9	+	+	+	−	+	−	−	+	−
10	+	+	+	+	+	+	+	+	−
11	+	+	+	+	+	+	+	+	−
12	+	+	−	+	+	+	−	+	+
13	+	+	−	+	+	−	−	+	−
14	+	+	−	−	−	−	−	−	−
15	+	+	+	−	−	−	−	−	−
16	+	+	+	+	+	+	−	+	−
17	+	+	+	+	+	+	+	+	+
18	+	+	+	+	+	+	+	+	+
19	+	+	−	+	+	+	−	+	−
20	+	+	+	+	+	+	+	+	+
21	+	+	−	+	−	−	+	+	−
22	+	+	+	+	+	+	−	+	−
23	+	+	+	+	+	−	−	+	−
24	+	+	+	−	+	+	−	+	−
25	+	+	+	+	+	+	+	+	+
26	+	+	+	+	+	+	+	+	+
27	+	+	+	+	+	+	+	+	−
28	+	+	+	+	+	+	+	+	+
29	+	+	+	+	+	+	−	+	+
30	+	+	+	+	+	−	−	+	−
31	+	+	+	+	+	+	−	+	−
32	+	+	+	+	+	−	−	+	+
33	+	+	+	+	+	+	−	+	+
34	+	+	+	+	−	−	−	+	−
35	+	+	+	−	−	+	−	+	−
36	+	+	+	−	+	+	+	+	+
37	+	+	+	+	+	+	+	+	+
38	+	+	+	+	+	+	+	+	+

[a] Thymidine.

TABLE VI

PROPERTIES OF 38 CLONES CLASSIFIED INTO 7 CLASSES ACCORDING TO NUTRITIONAL REQUIREMENTS

Clone no.	Required substances						
17							
18							
20							
25							
26							
28							
37							
38							
36		Pro					
6					Gly		
29					Gly		
33					Gly		
10							TdR[a]
11							TdR
27							TdR
2					Gly		TdR
3					Gly		TdR
4					Gly		TdR
16					Gly		TdR
22					Gly		TdR
31					Gly		TdR
5	Asn				Gly		
12	Asn				Gly		
1			Asp		Gly		
8		Pro		Ser			
23				Ser	Gly		TdR
30				Ser	Gly		TdR
32				Ser	Gly		TdR
19	Asn				Gly		TdR
24		Pro			Gly		TdR
21	Asn		Asp	Ser			TdR
13	Asn			Ser	Gly		TdR
7		Pro	Asp		Gly		TdR
35		Pro	Asp		Gly		TdR
9		Pro		Ser	Gly		TdR
34			Asp	Ser	Gly		TdR
15		Pro	Asp	Ser	Gly	Hyp	TdR
14	Asn	Pro	Asp	Ser	Gly	Hyp	TdR

[a] Thymidine.

nutritional requirements. Clones 17, 18, 20, 25, 26, 28, 37, and 38, which belong to class 1, were alanine-, asparagine-, proline-, aspartic acid-, serine-, glycine-, hypoxanthine-, and thymidine-sufficient clones which did not require any of these eight nutritional substances for their growth (prototrophic clones). Simultaneously, it was found that such prototrophic clones represent about one-fifth of the 38 clones grown in a complete medium composed of 90% Eagle MEM + N18 medium and 10% dialyzed calf serum.

On the other hand, clones 36, 6, 29, 33, 10, 11, and 27 (class 2) were proline-, glycine-, or thymidine-deficient clones which required proline, glycine, or thymidine for their growth (one-substance-auxotrophic clones). Clones 2, 3, 4, 16, 22, 31, 5, 12, 1, and 8 (class 3) were two-substance-deficient clones which required any two among asparagine, proline, aspartic acid, serine, glycine, and thymidine (two-substance-auxotrophic clones). Clones 23, 30, 32, 19, and 24 (class 4) were three-substance-deficient clones which required any three among asparagine, proline, serine, glycine, and thymidine (three-substance-auxotrophic clones). Clones 21, 13, 7, 35, 9, and 34 (class 5) were four-substance-deficient clones which required any four among asparagine, proline, aspartic acid, serine, glycine, and thymidine (four-substance-auxotrophic clones). Finally, clone 15 (class 6) was proline-, aspartic acid-, serine-, glycine-, hypoxanthine-, and thymidine-deficient (six-substance-auxotrophic clone), and clone 14 (class 7) was asparagine-, proline-, aspartic acid-, serine-, glycine-, hypoxanthine-, and thymidine-deficient (seven-substance-auxotrophic clone).

From these results, it can be concluded that the majority (about 58–63%) of the 38 clones were glycine- and thymidine-auxotrophic mutants, and that only two clones were hypoxanthine-auxotrophic mutants which required hypoxanthine for their growth. At the same time, these results suggest that the original Chinese hamster *hai* cells grown in a complete medium consisted of nutritionally sufficient cells and various kinds of nutritionally deficient cells. In any case, we have described here how easily we could detect and isolate the nutritionally deficient mutant cells from the cultured Chinese hamster *hai* cells, with the use of our replica plating method.

IV. Concluding Remarks

Studies of biochemical genetics and mutagenesis in somatic mammalian cells *in vitro* are among the most important future problems in mammalian cell biology. These studies require a large number of mutant cells which have stable and well-defined genetic markers. Biochemical mutants in somatic mammalian cells have been obtained either by establishing cell lines from mutants known in the whole organisms that are also expressed at the cellular level or by means of various methods to isolate cell clones that have developed phenotypes different from those of the cultured parental cell line. There are at least three different useful methods of isolating mutant cells from cultured mammalian cell lines. The first is the mass selection method which has been conveniently and widely used by many workers to select mutant cells resistant to extrinsic agents such as chemicals and radiation. The second is the thymine starvation method which was first proposed and practiced by DeMars and Hooper (1960) to obtain a glutamine-requiring mutant line from cultured HeLa cells. The third is the 5-bromodeoxyuridine and visible light method which was proposed by Puck and Kao (1967) for isolation of nutritionally deficient mutants from Chinese hamster cell cultures. By these methods, a number of genetically useful mutant cells such as temperature-sensitive, drug-resistant, or nutritionally auxotrophic cell lines, etc., have so far been isolated.

However, the difficulty involved in the study of single gene phenomena in diploid cells in which simple recombination techniques, as in bacterial systems, are not available remains to be solved. Furthermore, the lack of replica plating techniques such as those used in the field of microbial genetics for cultured mammalian cells has delayed accurate genetic analysis in somatic mammalian cells. The application of the replica plating method described in this chapter should be useful for the isolation of useful mutants, for further purification of a mutant cell line, and for obtaining a much broader spectrum of somatic cell mutations. Especially, employment of this replica plating method combined with the thymine starvation method (DeMars and Hooper, 1960) or the 5-bromodeoxyuridine and visible light method (Puck and Kao, 1967) mentioned above, will facilitate the further purification of an isolated mutant cell line, and the investigation of mutagenesis in cultured somatic mammalian cells.

ACKNOWLEDGMENTS

The authors wish to thank Dr. I. Yamane, Tohoku University, for his kind gift of Chinese hamster *hai* cells, and Dr. T. Sugahara, Kyoto University, for his many helpful discussions and reading of the manuscript.

REFERENCES

Barski, G. S., Sorieul, S., and Cornefert, F. (1960). *C. R. Acad. Sci.* **251**, 1825.
Barski, G. S., Sorieul, S., and Cornefert, F. (1961). *J. Nat. Cancer Inst.* **26**, 1269.

Chu, E. H. Y., and Malling, H. V. (1968). *Proc. Nat. Acad. Sci. U. S.* **61**, 1306.

DeMars, R., and Hooper, J. L. (1960). *J. Exp. Med.* **111**, 559.

Ephrussi, B., and Weiss, M. C. (1965). *Proc. Nat. Acad. Sci. U. S.* **53**, 1040.

Gartler, S. M., and Pious, D. A. (1966). *Hum. Genet.* **2**, 83.

Goldsby, R. A., and Zipser, E. (1969). *Exp. Cell Res.* **54**, 271.

Kao, F. T., and Puck, T. T. (1968). *Proc. Nat. Acad. Sci. U. S.* **60**, 1275.

Kao, F. T., and Puck, T. T. (1969). *J. Cell. Physiol.* **74**, 245.

Lederberg, J., and Lederberg, E. M. (1952). *J. Bacteriol.* **64**, 399.

Littlefield, J. W. (1964). *Science* **145**, 709.

Naha, P. M. (1969). *Nature (London)* **223**, 1380.

Puck, T. T., and Fisher, H. W. (1956). *J. Exp. Med.* **104**, 427.

Puck, T. T., and Kao, F. T. (1967). *Proc. Nat. Acad. Sci. U. S.* **58**, 1227.

Puck, T. T., and Marcus, P. I. (1955). *Proc. Nat. Acad. Sci. U. S.* **41**, 432.

Puck, T. T., Marcus, P. I., and Cieciura, S. J. (1956). *J. Exp. Med.* **103**, 273.

Roberts, C. F. (1959). *J. Gen. Microbiol.* **20**, 540.

Suzuki, F., Kashimoto, M., and Horikawa, M. (1971). *Exp. Cell Res.* **68**, 476.

Thompson, L. H., Mankovitz, R., Baker, R. M., Till, J. E., Siminovitch, L., and Whitmore, G. F. (1970). *Proc. Nat. Acad. Sci. U. S.* **66**, 377.

Weiss, M. C., and Ephrussi, B. (1966). *Genetics* **54**, 1095.

Chapter 6

Cell Culture Contaminants

PETER P. LUDOVICI AND NELDA B. HOLMGREN

Department of Microbiology and Medical Technology, University of Arizona,
Tucson, Arizona and W. Alton Jones Cell Science Center, Lake Placid, New York

I. Introduction

Most laboratories using cell and tissue cultures are plagued with the problem of contaminations. Initially the problem was thought to be one of only bacteria, yeast, and mold contaminations, and the advent of antibiotics appeared to be the "cure all." Unfortunately this has not proved to be the case and the use of antibiotics has compounded the problem rather than lessened it. In fact the evidence is clear that the use of antibiotics in cell cultures has developed a false sense of security in workers which has led to haphazard cultural techniques. Such relaxation in rigid aseptic procedures is believed to be chiefly responsible for other types of contamination including mycoplasmas, viruses, protozoa, and cells of one type with cells of another. In many cases such contaminations go undetected and therein lies the danger, because data obtained in such studies cannot be properly evaluated and thus lead to erroneous conclusions.

The purpose of this chapter is to make workers aware of these contamination problems and to present some of the methods used to detect, eliminate, control, and prevent such contaminants in cell culture. No attempt is made to cover the subject completely or to include all techniques. The reader is referred to several excellent articles on the subject that have appeared recently or are presently in press (Coriell, 1962; Hayflick, 1965; Macpherson, 1966; Brown and Officer, 1968; Fogh and Fogh, 1969; Fogh et al., 1971; Stanbridge, 1971; Fogh, 1973; Stulberg, 1973; Kruse and Patterson, 1973).

II. Mycoplasma Contamination of Cultured Cells

A. Introduction

There is mounting evidence that mycoplasma contamination of cultured tissue cells causes profound alterations in the nutritional, biochemical, cytogenetic, and immunological characteristics of the cell. Consequently a reevaluation of the possible role of mycoplasma contamination in the interpretation of many reported studies, in which tests for the possible presence of mycoplasma were not done, may be necessary. It is a type of contamination that is easily ignored. In the first place, it is possible for a cell culture to contain as many as 10^7 or 10^8 mycoplasma/

ml without causing noticeable turbidity or overt cytopathology. Second, because of their minute size and the requirement for special stains, these organisms cannot be resolved or identified with the light microscope. Last, knowledge of the biology of mycoplasma and their interaction with the host cell has grown so rapidly that investigators have not kept pace with information concerning a type of contamination considered by many to be annoying but not necessarily serious. The purpose of this section is to emphasize the deep-seated alterations caused by mycoplasma and to summarize the present status in regard to detection, control, and prevention of this form of contamination in cell cultures. Several excellent reviews on the subject have recently appeared (Hayflick, 1965, 1972; Macpherson, 1966; Fogh and Fogh, 1969; Fogh et al., 1971; Stanbridge, 1971; Barile, 1973).

B.   Incidence of Mycoplasma Contamination in Cell Culture

Following the first report in 1956 (Robinson et al., 1956) a number of studies showed that a high percentage of cell lines from laboratories in various parts of the world were contaminated with *Mycoplasma* sp. Table I, taken from an excellent comprehensive review by Barile (1973), shows the incidence found in several different studies. Although both primary and continuous cell cultures were found contaminated with

TABLE I

INCIDENCE OF MYCOPLASMA CONTAMINATION IN CELL CULTURES[a]

Cell cultures	Number positive / Number tested	Percent	Investigators
Primary cells	0/37	0	Rothblat and Morton, 1959
	12/>1500	0.8	Barile, 1962, 1968; Barile and Schimke, 1963; Barile and Del Giudice, 1972; Barile et al., 1962
	0/26	0	Hederscheê et al., 1963
	?	3–4	Levashov and Tsilinskii, 1964
Continuous cells	22/37	59	Rothblat and Morton, 1959
	94/166	57	Pollock et al., 1960
	140/243	57	Barile, 1962, 1968; Barile and Schimke, 1963; Barile and Del Giudice, 1972; Barile et al., 1962
	32/55	58	Hederscheê et al., 1963
	45/49	92	Rakavskaya, 1965
	52/60	87	Ogata and Koshimizu, 1967

[a] From Barile, 1973.

TABLE II

INCIDENCE OF PPLO DETECTION IN RELATION TO THE NATURE OF
THE CELL CULTURES[a]

Nature	Source (number of laboratories)	Number positive / Number of cultures	Percent positive
Primary cells	6	0/10	0
Continuous cells	15	48/92	52
With antibiotics	13	46/64	72
Without antibiotics	3	2/28	7
With serum	13	48/86	56
Without serum	3	0/6	0

[a] From Barile et al., 1962.

mycoplasma, the risk with primary cultures (0 to 4%) was considerably lower than with continuous cell lines (57 to 92%). However, contamination of primary cell cultures does occur (Girardi et al., 1965a; Hayflick and Stanbridge, 1967). Tables II and III, also taken from a publication by Barile et al. (1962), relate the incidence to the nature of the cell culture and to the nature of the laboratory work. From these results, one can infer in retrospect possible sources of contamination. A definite correlation was found between the incidence in cell cultures maintained with antibiotics (72%) versus those cultured without antibiotics (7%) and again with tissue cells propagated in medium supplemented with serum (56%) versus medium containing no serum (0%). Barile (1966) summarized these results as follows: "The lowest incidence of contamination in continuous cells was found in cells grown in small volumes and in antibiotic-free media and used for cell culture metabolic studies. The highest incidence of contamination was found in cells used to propagate

TABLE III

INCIDENCE OF PPLO DETECTION IN CONTINUOUS CELL CULTURES
IN RELATION TO NATURE OF LABORATORY WORK[a]

Nature of laboratory work	Source (number of laboratories)	Number positive / Number of cultures	Percent positive
All laboratories	15	48/92	52
Physiological studies	4	0/31	0
Microbial propagation	9	38/50	76
Tissue-cell propagation	2	10/11	91

[a] From Barile et al., 1962.

microorganisms or in cells supplied by cell culture producers. These cells were grown under conditions that provide a high contamination risk. They were produced on a large scale—grown in large volumes and in large numbers of containers."

In Table IV are listed the identity and number of isolates from uninoculated cell cultures (Barile, 1973). Human, bovine, and swine species of mycoplasmas were responsible for 99% of the contaminations. Early isolates up to about 1964 proved to be human oral and genital strains. This suggested that one source of contamination originated from the hands or oropharynx (through mouth pipetting) of the individuals

TABLE IV

PREVALENCE AND IDENTITY OF MYCOPLASMAS ISOLATED FROM CELL CULTURES[a]

Species	Natural habitat	Isolations	
		Number	Totals[b]
Mycoplasma orale, type 1	Human: oral	370 (35%)	—
M. hominis	genital	67 (6%)	442 (42%)
M. fermentans	genital	5 (0.5%)	—
M. arginini	Bovine: oral, genital	239 (22%)	—
Mycoplasma sp.[c]	oral, genital	80 (7.5%)	—
	conjunctiva	—	—
Acholeplasma laidlawii	genital	75 (7%)	400 (38%)
Acholeplasma sp.[d]	genital	6 (0.6%)	—
M. hyorhinis	Swine: nasal	204 (19%)	204 (19%)
M. arthritidis	Murine: respiratory	5 (0.5%)	9 (0.9%)
M. pulmonis	respiratory	4 (0.4%)	—
M. gallisepticum	Chicken: respiratory	3 (0.3%)	5 (0.5%)
M. gallinarum	respiratory	2 (0.2%)	—
M. canis	Canine: oral, genital	1 (0.1%)	3 (0.3%)
Mycoplasma sp.[e]	genital	2 (0.2%)	—
Total		1063	

[a] From Barile, 1973.

[b] Presents total number and percent of mycoplasmas.

[c] These 80 mycoplasmas were related to 3 distinct bovine serogroups: 67 were related to strain HRC 70-159 and 12 to strain HRC 213 (R. A. Del Giudice and M. F. Barile, unpublished data) and 1 to strain M165/69 of R. H. Leach (unpublished data).

[d] These 6 strains were unrelated to the 3 established species and may represent new species.

[e] These 2 strains were related to a distinct serogroup (canine strain HRC 689) (Barile et al., 1970).

handling the cultures. The ubiquitous use of antibiotics which inhibited obvious contamination with bacteria created conditions selective for *Mycoplasma* sp. since these organisms are absolutely resistant to penicillin and, for the most part, to streptomycin as well. Furthermore, the false security engendered by the use of antibiotics led to relaxation of aseptic techniques. Then, from about 1964 to 1968 the prevalent species isolated was a porcine strain, *M. hyorhinus*. Although trypsin which is prepared from hog pancreas was suspected as the possible source, all attempts to isolate mycoplasma from trypsin have proved futile. Since 1968, the predominant species again changed. Contaminant species isolated from cultured cells were identified as *M. arginini* and various saprophytic and unidentified strains. A source of contamination for these species has been demonstrated by Barile and Kern (1971) and Barile and Del Giudice (1972) to be commercial bovine serum used in the culture medium. Since the mycoplasma are filterable, they were also isolated from so-called "sterile" processed lots of serum when large sample inocula were tested. Barile (1973) concludes that contamination of bovine sera by mycoplasmas is probably due to changes in the processing of the sera which have resulted because of the increased demands for large volumes of the product. Commercial sera which are still collected by sterile intracardiac punctures are rarely contaminated with mycoplasmas (Barile, 1973).

C.   General Properties of the Genus Mycoplasma

Mycoplasma have been isolated from many species of domestic and laboratory animals including man and also from plants (Hampton, 1972). Some of the parasitic species are pathogenic causing diseases predominantly of the mucous membrane, lymph nodes, and joints. Certain other species, originally found in sewage and which do not require sterol for growth, are referred to as saprophytes. There are many similarities between the mycoplasma and the stable L forms of bacteria. At the present time the consensus of opinion appears to be that the mycoplasma constitute a natural biological class. The opposing opinion is that the mycoplasma are the stable L forms of bacteria (Smith, 1971; Neimark, 1964).

Organisms of the genus *Mycoplasma* are the smallest forms capable of growth in a cell-free medium. They differ from bacteria in not possessing a rigid cell wall but instead are limited by a trilaminar surface similar to that of a tissue cell. The parasitic mycoplasma also differ from bacteria in their requirement for cholesterol as a growth factor. The cholesterol is incorporated into the cell surface where it functions possibly in permeation or as a carrier of fatty acids (Smith, 1971). This requirement is satisfied by a serum supplement to the medium. The nature of the cell wall explains the plasticity and pleomorphism of the individual mycoplasmal forms in response to their physical and chemical environment. The characteristic colonial morphology on agar has been likened to that of a "fried-egg," a form which it assumes because of the tendency for the mycoplasma to grow into the agar in the center and then to spread out peripherally over the surface (Figs. 1, 3, 4). Granular colony forms on agar are also seen (Fig. 2). Colonies range from 20 μ (T-strains) to 500 μ in diameter. The mycoplasma cell is pleomorphic and varies in shape from spherules to filaments with branching forms. Mycoplasmas, unlike viruses, have both DNA and RNA. Their genome size varies from 0.45×10^9 to 1×10^9 daltons and they can code for 600 to 1000 proteins (McGee et al., 1967; Neimark, 1967, 1970; Williams et al., 1969). Some species of mycoplasma apparently reproduce by binary fission (Kelton, 1962; Morowitz and Maniloff, 1966; Maniloff and Morowitz, 1967). They resemble viruses in a number of properties (Hayflick, 1965) among which is their size. The smallest reproductive elements range from 125 to 150 mμ and have been reported to pass through a 0.22-μ Millipore filter (Morowitz et al., 1963).

Like viruses, they are absolutely resistant to penicillin and to antibiotics whose mode of action is interference with cell-wall synthesis. Specific antiserum incorporated into the culture medium inhibits the homologous species of *Mycoplasma* but not the organisms of serologically different and distinct species (Edward and Fitzgerald, 1954). This property, which is analogous to the neutralization of viruses by specific antiserum, is the basis of a highly specific method for *Mycoplasma* identification (Clyde, 1964).

D. Classification

Because sufficient information on species of *Mycoplasma* was not available earlier, the first accepted system of nomenclature did not appear in *Bergey's Manual of Determinative Bacteriology* until 1957. In this, the *Mycoplasma* were placed in the class Schizomycetes because of their possible relation to bacteria and their L forms, and all known species were placed in the genus *Mycoplasma*, family Mycoplasmataceae, order Mycoplasmatales. Recently, Edward and Freundt (1970) have proposed that the Mycoplasmatales which do not require sterol be placed in a separate family, Acholeplasmataceae, and genus *Acholeplasma*. This would include the saprophytic species formerly named *Mycoplasma laidlawii*. Since new species and properties are being identified, this order is subject to change. For anyone interested in the history and problems

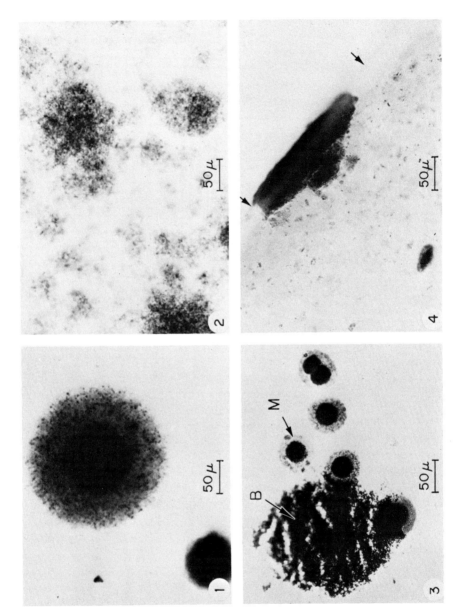

of classifying the mycoplasma, reference can be made to an excellent discussion by Edward and Freundt (1969).

E. Mycoplasma–Cell Interactions

Many investigators are aware and concerned about the risks involved in the interpretation of data obtained experimentally with cell and tissue cultures contaminated with an unsuspected or ignored *Mycoplasma* sp. Studies are increasingly emphasizing the close relationship between the host tissue cell and mycoplasma contaminant whereby some *Mycoplasma* sp. not only compete successfully with the tissue cell for an amino acid, arginine, essential to both tissue cell and mycoplasma, but also for nucleic acid precursors which they require for the synthesis of their own species of procaryotic DNA and RNA. As a consequence of the firm attachment of the mycoplasmas, primarily to the cell surface, and of several enzymatic activities contributed by the mycoplasma, alterations have been reported in tissue cell morphology and growth, cell metabolism, chromosome number and morphology, and in disturbed virus-cell relationships.

One of the first effects of mycoplasma contamination to be reported was the rapid depletion of arginine in the culture medium (Powelson, 1961; Kenny and Pollock, 1963). The mycoplasmas are characterized by their biochemical heterogeneity, but can be divided into two broad classes determined by the pathway utilized to derive energy; i.e., nonfermenting and fermenting species. The nonfermenting species for the most part obtain their energy by degrading arginine via the arginine dihydrolase pathway. For each mole of arginine, one mole of ATP is produced (Schimke, 1967). The fermenting group derive their carbon and energy by the dissimulation of hexoses (Smith, 1971).

Freed and Schatz (1969) showed that when Chinese hamster cells (pseudodiploid line, C14FAF28) were deprived of any single essential amino acid, growth of the cells was inhibited in the S phase of the cell cycle with a decreased uptake of thymidine-^{3}H into nuclear DNA. Chromosomal aberrations, seen as chromosome and chromatid breaks and exchanges, endoreduplication, and severe fragmentation of the chromo-

FIGS. 1–4. Mycoplasma colonies. Fig. 1., *M. orale*, type 1, "fried egg" appearance. Fig. 2., *M. pneumoniae*, "granular" appearance. They grow beneath the surface of the agar, *M. hominis*, Figs. 3 and 4. In a sagittal section preparation, the central button of the colony grows more deeply embedded in agar than the periphery (Fig. 4). Compare the general appearance of the bacterial colony (B) (*Corynebacterium* sp.) with the mycoplasma colonies (M), Fig. 3. All photographs from Barile, 1973.

somes, appeared in the mitoses subsequent to the completion of the cell cycle, which occurred following the addition of the omitted amino acid.

The authors proposed that the chromosome aberrations were a result of the inhibition of protein synthesis following the inhibition of DNA synthesis by deprivation of a single amino acid. Thus, chromosomal aberrations reported for cell lines contaminated with nonfermenting mycoplasma may result from the rapid depletion of arginine from the medium. However, nuclear and chromosomal aberrations have also been shown to result from contamination with fermenting *Mycoplasma* sp. which do not deplete the medium of arginine. Fogh and Fogh (1965, 1967b) reported chromosomal alterations in the FL human amnion cell line which were not completely reversed on elimination of the myco-plasma from the culture. The *Mycoplasma* sp. was identified as *M. fermentans*. Another fermenting sp., *M. laidlawii* has been reported to in-duce chromosomal aberrations in human diploid fibroblasts (Stanbridge *et al.*, 1969).

Other morphological nuclear changes resulting from mycoplasma con-tamination or experimental infection have also been noted. Stanbridge and co-workers (1969) correlated interference of host cell DNA with an alteration ("leopard" nuclei) in the morphology of the nucleus, as seen in May-Grunwald Giemsa-stained cover slips of mycoplasma-contami-nated tissue cells. Jezequel *et al.* (1967) described the appearance of clear perinucleolar halos (also noted by the above authors) and the phenomenon of "nucleolar capping," a segregation of the nucleolar com-ponents, in cultured cell lines (human embryonic lung, Chang liver, human embryonic liver, and rabbit lymph node) from which they iso-lated *M. orale* type I, *M. hominis* type I, and an unidentified *Mycoplasma* sp., which they were unable to grow in subcultures. The addition of excess arginine to the culture medium neither prevented nor reversed the nucleolar changes whereas kanamycin suppressed the development of these changes. They also propose that the morphological changes may reflect the direct effect of mycoplasma contamination on DNA metabolism.

That mycoplasma contamination causes instability of DNA was shown by Randall *et al.* (1965) who recovered thymidine[3]H label in the acid soluble fraction of HeLa and L cell culture medium. Nardone *et al.* (1965) found that the presence of mycoplasma inhibited the incorpora-tion of tritiated thymidine and uridine. Autoradiograms showed exposed silver grains over the margins of the cells rather than over the nucleus. Kenamycin was effective in eliminating the mycoplasma from the culture and in restoring nucleoside incorporation. Holland *et al.* (1967) demon-strated a mycoplasma nucleoside phosphorylase (which cleaved thymi-dine to thymine) in a HeLa cell line and in 38 of 42 cell cultures from which mycoplasma were isolated. The nucleoside phosphorylase contributed by the mycoplasma was similar to that of purified horse liver thymidine phosphorylase.

In addition, a number of investigators have shown changes in pyrimidine deoxynucleoside metabolism in cell cultures caused by contamination with mycoplasma (Hakala et al., 1963; Paton and Allison, 1970). The presence of a pyrimidine deoxyribonuclease in the supernatant of cell-free extracts of broth cultures of M. [Acholeplasma] laidlawii A (Niemark, 1964) and of similar activity for M. hominis (Stock and Gentry, 1969) demonstrated that the activity was associated with the mycoplasma contaminant. The nature of the activity of the pyrimidine deoxyribonuclease was that of an endonuclease, cleaving DNA to oligonucleotide fragments. The latter authors stated that interpretation of viral-induced deoxyribonuclease activity could be due to mycoplasma contamination of either the virus inoculum or of the host cell unless both sources had been tested for mycoplasma.

The presence of such enzymes and their suggestive role in the profound alterations of nucleic acid metabolism is consistent with the results of other investigators. New species of procaryotic RNA and DNA have been described in mycoplasma-contaminated tissue cells in culture. Levine et al. (1967, 1968) demonstrated the synthesis in the cytoplasm of human diploid fibroblasts infected with M. hyorhinis, of a "light" RNA (14S and 20S) with a guanine plus cytosine content of 33% and similarly, of a "light" DNA. In addition, in such cells a 90–95% overall decrease in uridine-^{14}C incorporation into the cell nuclear 35–45S ribosomal precursor RNA and into the ribosomal RNA content per cell was found. Markov et al. (1969) were still able to demonstrate the persistence of two species of mycoplasmal RNA in HeLa cells after culturally confirmed elimination of the mycoplasma contaminant by treatment with Tylan (tylosin tartrate). In a report by Todaro et al. (1971), the authors stated that they were conducting studies to define the specificity and sensitivity of a method which utilizes sucrose gradient sedimentation to detect the altered RNA and DNA in mycoplasma-infected cell cultures labeled with radioactive precursors of either RNA or DNA.

The ability of both nonfermenting and fermenting species to cause chromosomal aberrations which resemble those caused by viruses and by inhibitors of DNA synthesis led Stanbridge (1971) and Stanbridge et al. (1971) to suggest that the chromosomal damage is due to interference with cellular DNA synthesis either by competition for nucleic acid precursors or by degrading the host cell DNA. These authors demonstrated further that depletion of arginine from the culture medium by the nonfermenting mycoplasmas was not the sole cause of chromosomal aberrations in the host cell since these are caused by both fermenting and nonfermenting strains of mycoplasmas (Stanbridge et al., 1969). Furthermore, only the nonfermenting strains depleted the medium of arginine (Table V). They further showed that both types of mycoplasma could grow in Eagle's basal medium (BME) supplemented with 10% calf serum and a variety of nucleic acid precursors but not in fresh BME supplemented with 10% calf serum alone or in this medium conditioned by growth of the human diploid fibroblast cell strain, WI-38. Thus, the essential nucleic acid precursors do not diffuse from the tissue cells into the medium during growth. These results substantiate their hypothesis that mycoplasmas require a close relationship with the cell, either by a firm attachment (Thomas, 1968) to the surface of the cell or in an intracellular position.

Macpherson and Russell (1966) have suggested that the ability of the cell-associated mycoplasma to degrade cell DNA and to cause chromosomal aberrations may explain the "transformation" of euploid to aneuploid cells capable of indefinite growth in vitro. They demonstrated that mycoplasmas were capable of inducing heritable transformation of hamster fibroblasts (BHK21) as manifested in alterations of cell and colonial morphology and in their ability to grow in suspension in dilute agar. Mycoplasma attach very firmly to the surface of tissue cells in culture and to erythrocytes and lymphocytes (Thomas, 1969). The close attachment of the mycoplasma to the surface caused inhibition of lymphocyte response to phytohemagglutinin (PHA). The Mycoplasma species inhibiting the PHA response in lymphocytes were found to be arginine-dependent and the inhibition was reversible on the addition of excess arginine to the cell cultures. The receptor site has been demonstrated to be neuraminic acid in some cases. Other types of receptor sites have also been indicated (Sobeslavsky et al., 1968).

Both Mycoplasma pneumoniae and M. gallisepticum, which cause respiratory tract infections in humans and fowl respectively, have been shown to produce hydrogen peroxide and this is believed to contribute to the virulence of the organisms. Three other species, M. arthriditis, M. pulmonis, and M. neurolyticum when injected in large quantities (10^9 CFU) have toxic properties and induce neurological symptoms and death in the appropriate animals (rats and mice) (Thomas, 1968).

These profound effects of Mycoplasma contamination on the metabolism of the tissue cell are reflected in disturbed virus host-cell relationships. They interfere with virus growth in tissue culture by reducing the number of plaque-forming units (Brownstein and Graham, 1961; Rouse et al., 1963). Rouse and Schlesinger (1967) demonstrated that the factor involved in the maturation step of type 2 adenovirus was arginine-depen-

TABLE V

AMINO ACID CHANGES IN BME DUE TO THE GROWTH OF MYCOPLASMA SPECIES[a]

| | | | Amino acid concentrations (calculated as μmole/0.5 ml) | | | | | | |
| | | | Fermenters | | | | Nonfermenters | | |
Amino acid	Fresh control	Incubated control[b]	Mycoplasma gallisepticum	M. pulmonis	M. hyorhinis	A. laidlawii	M. salivarium	M. arthritidis	M. hominis
Threonine	0.066	0.066	0.058	0.066	0.041	0.066	0.079	0.073	0.0534
Aspartic acid	—	—	—	—	0.0412	—	—	—	trace
Serine	—	—	—	—	—	—	0.143	—	—
Glutamine	0.828	0.367	0.339	0.348	0.330	0.349	0.327	0.374	0.202
Glutamic acid	0.007	0.036	0.063	0.065	0.053	0.070	0.366	0.072	0.109
Proline	—	—	—	—	—	—	—	—	—
Glycine	0.029	0.033	0.041	0.045	0.0371	0.045	0.045	0.0525	0.044
½ Cystine	0.051	0.051	0.0765	0.079	0.0615	0.097	0.042	0.0872	0.109
Alanine	—	—	—	—	—	—	—	—	—
Valine	0.108	0.112	0.110	0.121	0.106	0.125	0.344	0.137	0.126
Methionine	0.031	0.034	0.031	0.088	0.027	0.034	0.082	0.036	0.029
Isoleucine	0.107	0.111	0.108	0.111	0.094	0.117	0.278	0.131	0.119
Leucine	0.109	0.117	0.122	0.127	0.108	0.134	0.395	0.148	0.135
Tyrosine	0.051	0.058	0.055	0.062	0.046	0.059	0.145	0.064	0.057
Phenylalanine	0.054	0.059	0.063	0.070	0.052	0.067	0.185	0.074	0.068
Lysine	0.082	0.078	0.086	0.096	0.077	0.088	0.244	0.098	0.0938
Histidine	0.031	0.029	0.028	0.031	0.025	0.033	0.063	0.032	0.031
Ornithine	trace	0.0159	0.023	0.0216	0.0231	0.019	0.015	0.0266	0.025
Ammonia	0.68	0.638	0.907	1.041	0.719	1.02	0.986	1.327	0.886
Arginine	0.029	0.027	0.026	0.026	0.023	0.022	—	—	—
Tryptophan	—	—	—	—	trace	—	trace	—	—
Citrulline	—	—	—	—	—	—	—	0.086	—

[a] From Stanbridge et al., 1971.
[b] Samples incubated for a period of 13 days at 37°C.

dent and that the inhibition in medium depleted of arginine by mycoplasma could be reversed by the addition of arginine. Inhibition of Rous sarcoma virus (RSV) by *M. orale* type I was demonstrated by Somerson and Cook (1965) and also the suppression of Rous-associated virus (RAV). Ponten and Macpherson (1966) confirmed the finding of Somerson and Cook that the presence of *M. hominis* as a contaminant suppressed the formation of transformed foci by RSV. In contrast, however, the presence of mycoplasma have also been reported to enhance vaccinia virus titers (Hargreaves and Leach, 1970).

Mycoplasma have been isolated from cases of leukemia and other neoplastic tissues, sometimes directly and at other times indirectly by passage through mycoplasma-free tissue cells (Barile, 1967; Hayflick and Stanbridge, 1967). Their role in neoplasia has been assumed to be similar to the other infections to which immunosuppressed patients are subject. However, their virus-like properties and ability to cause chromosomal aberrations and genetically transmissible changes should keep them on the "suspect" list as carcinogenic agents (Macpherson, 1966; Russell, 1966). Fogh (1969), however, points out that long-term mycoplasma-infected FL human amnion cells had a reduced capacity to produce tumors in weanling gold Syrian hamsters.

F.   Detection of Mycoplasma in Cell Culture

Mycoplasma can often be detected in tissue cells grown on cover slips when fixed and stained with either an intensified Giemsa (Marmion and Goodburn, 1961) or the May-Grunwald Giemsa stains (Jacobson and Webb, 1952). They stain with the specificity of DNA and can be seen as minute rounded forms attached to the surface of the cell and in the intercellular spaces. Fogh and Fogh (1964) have described a method where the small round forms can be seen distinctly in the extended cytoplasm of the cell and between the cells as shown in Fig. 5. The method involves a hypotonic treatment, air drying, and orcein staining of cells grown on cover slips. The authors increased the sensitivity of the method by inoculating mycoplasma-free FL amnion cells with the suspected material. In general, staining methods are useful as preliminary screening methods. However their adequacy for detecting low-grade contamination has been questioned (Macpherson, 1966). Serological identification of mycoplasma can be made with fluorescein-labeled specific antiserum (Malizia *et al.,* 1961; Barile *et al.,* 1962; Clark *et al.,* 1963; Del Giudice *et al.,* 1967). Since the mycoplasma are a serologically heterogeneous group and species are being isolated which cannot be identified with antisera to the known human and animal strains, this method has practical limitations.

The morphology of the individual forms can best be seen with the electron microscope. In addition to the sampling problems, it is not always possible by this means to distinguish viruses and mycoplasma (Barile, 1967). There are, moreover, reports of erroneous identification of mycoplasma as viruses (Negroni, 1964; Girardi et al., 1965b).

At the present time the method of choice for detecting mycoplasma in tissue cultures is by culture and the demonstration of the typical colonial morphology on agar. At the same time, there are good reasons for believing that the present media formulations that have been commonly and successfully used still may not contain all growth factors essential for the primary isolation of some species of *Mycoplasma*. Evidence for this lies in the difficulty encountered in isolating some strains directly on agar or without a previous passage in cultured tissue cells or, after primary isolation on agar, the impossibility of maintaining them in subculture on agar (Hayflick and Stanbridge, 1967).

A widely used medium on which many strains can be isolated and subcultured is the formula of Hayflick (1965) on which the agent (*M. pneumoniae*) of atypical pneumonia in man was first demonstrated to be a mycoplasma. Basically, the media formulations consist of 70 parts basal medium (brain-heart infusion, peptone, and NaCl), 20 parts of horse serum, and 10 parts of a 25% autoclaved extract of Fleischmann's type 2040 dry yeast (Standard Brands, Inc.) as a broth or modified to contain 1% final concentration of agar for a solid medium. Antibacterial agents (1:2000 thallium acetate and 1000 units/ml of penicillin) or an antifungal agent (amphotericin B at 1 mg/100 ml) can be incorporated when the inoculum is suspected of being contaminated with microorganisms other than mycoplasma. Although adequate for the growth of many strains, the trend today is toward the development of media which will permit the growth of the more fastidious members of this nutritionally complex group. For example, Barile and Kern (1971) added 0.5% L-arginine hydrochloride and dextrose as energy sources to the broth medium, as described by Hayflick (1965), for isolation of mycoplasma from samples of bovine serum. On the premise that the absolute requirement for nucleic acid precursors may be involved in the mycoplasma-host cell interaction, Stanbridge et al. (1971) tested the addition of certain nucleic acid precursors (0.002% diphosphopyridine nucleotides) to cell-free BME supplemented with 10% calf serum. Because certain species (notably those of the GDL group to which *M. hyorhinus* belongs) frequently fail to grow on agar, Zgorniak-Nowosielsak et al. (1967) obtained growth of these species in an agarose medium containing BHK-21 cells in suspension. Shepard (1969) added 0.5% urea, eliminated the dextrose and arginine, and adjusted the pH to 6.0 for the growth of T-

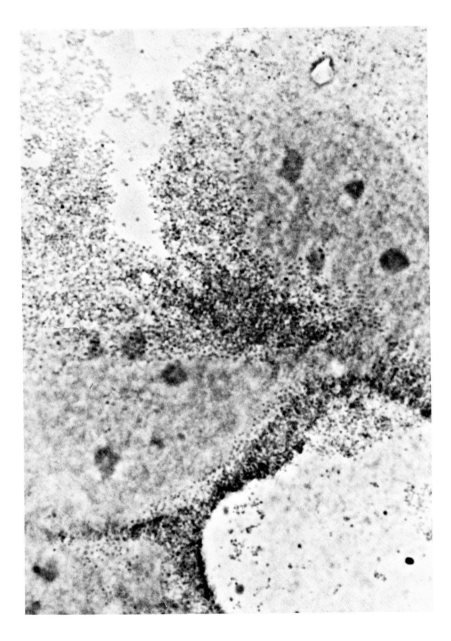

strains. Obviously media more sensitive to the needs of some mycoplasma are needed and probably forthcoming.

The sensitivity and limitations of several chemical methods for detecting mycoplasma have been discussed by Stanbridge (1971) and Macpherson (1966). The most promising, on the basis of recent developments, is the detection of mycoplasma DNA and RNA by radioactive label and banding in a sucrose gradient at a density of 1.20 to 1.24 gm/cm^3 (Todaro et al., 1971). Though these "lighter" RNA's and DNA's are not specific for mycoplasma alone, it should be possible to distinguish those of mycoplasma from those of other procaryotic contaminants.

The inoculum from cultured cells suspected of, or being tested for, the presence of mycoplasma should contain both culture medium and cells from a culture in the logarithmic phase of growth (3 to 5 days after subculture) since 80% of the mycoplasma are cell-associated at this time (Fogh and Fogh, 1967a). Agar plates should be inoculated in duplicate with 0.1 ml of sample, one to be incubated at 37°C in a moist chamber aerobically and the other in an atmosphere of 5% CO_2 in nitrogen. The latter frequently permits a more luxuriant growth and higher incidence of isolations (Barile, 1962; Zgorniak-Nowosielska et al., 1967; N. B. Holmgren, unpublished data). At the same time broth tubes should be inoculated with 1 ml of sample and incubated in a similar manner. Broth cultures serve to enrich the number of mycoplasmas and should subsequently be subcultured to agar media when the culture becomes turbid, or at weekly intervals for 4 weeks (Barile, 1972).

Agar plates are examined after 4 to 6 days incubation at 100× magnification by focusing through the agar onto the surface. Mycoplasma can usually be identified by their typical morphology and growth centrally into the agar. Pseudocolonies, consisting of Ca and Mg soaps (Brown et al., 1940; Hayflick, 1965) which form on the surface of the agar, and small clumps of tissue cells can be confusing but can be distinguished by scraping the suspected colony with a bacteriological loop. Pseudocolonies, bacterial colonies, and tissue cells will be completely removed whereas the central growth of a mycoplasma colony will remain in the agar. When the colonial morphology is atypical, a small square of the agar cut out aseptically and rubbed gently but firmly over the surface of a second plate or a subculture from broth will often give typical growth (Figs. 1–4).

The Dienes Method (Hayflick, 1965) of staining mycoplasma colonies

---

FIG. 5. FL human amnion cells infected with the HT strain of *M. fermentans* as observed under phase optics after hypotonic treatment and orcein staining. ×2000. Photograph furnished through the courtesy of Dr. Jorgen Fogh.

helps in identification. In this method, small squares of cover slips coated with the dried Dienes stain are gently placed over the suspected colony either on the plate or on a small block of agar, transferred colony side up to a glass slide. Mycoplasma stain an intense blue in the center with a lighter blue periphery. Young bacterial colonies also take up the stain but completely decolorize it in 30 minutes. Tissue cells take on a lavender hue. If specific antisera are available, serological identification can be made by the growth inhibition test according to the method of Clyde (1964). Stanbridge and Hayflick (1967) have described a further simplification of the disc technique and Purcell *et al.* (1969) have discussed various factors for the successful performance of the test.

To detect low levels of mycoplasma contamination, 25 ml of fluid samples (e.g., commercial bovine sera) or pieces of unminced tissue specimens are inoculated into 100 ml of broth media (Barile and Kern, 1971). Tissue is not minced to avoid release of mycoplasmacidal factors (Barile, 1973). Larger inoculum may be used to increase sensitivity of detecting mycoplasma (100 ml sample to 400 ml of broth). The broth cultures are incubated at 37°C and subcultured to agar if the broth becomes turbid or at weekly intervals for 4 weeks (Barile, 1973). Barile and Kern (1971) have detected low levels of mycoplasma contamination in commercial sera using such large inocula.

A semisolid broth culture procedure has recently been developed by Barile (1973) for screening cell cultures for bacterial and fungal contamination as well as mycoplasma contamination. To prepare semisolid broth medium, a 0.1% concentration of Noble's agar is added and penicillin as well as thallium acetate are omitted. In practice 1 ml of cell culture specimen is inoculated into 10 ml of medium. Cultures are incubated at 37°C aerobically and periodically checked for microbial growth for at least 4 weeks. According to Barile (1973) the agar provides an oxygen gradient and permits mycoplasmas to produce dense layers of turbid growth with microcolony formation. The arginine-using mycoplasmas produce an alkaline shift while the fermentors produce an acid shift in pH. This semisolid broth medium grows most mycoplasmas, bacteria, and fungi (Barile, 1973).

G.   Sources of Contamination

Several sources or causes of contamination in cultured cells have become evident with the development of tissue culture methods and their applications. Some of the original contaminations arose as a result of contact or droplet infection from the hands or nasopharynx of the tissue culturist. The recent demonstration (Barile and Kern, 1971) that certain species identified as *M. arginini, Acholeplasma laidlawii,* and unidentified species can be isolated from bovine serum established this as another source. The relaxation of aseptic techniques following the false security engendered by the ubiquitous use of antibiotics led to the horizontal spread of the mycoplasma contaminants to other cultured cells. It has been demonstrated that in aerosols created during various procedures (O'Connell *et al.,* 1964; Kundsin, 1968) the mycoplasma, like bacteria, can remain viable at certain temperatures and relative humidities and are capable of infecting cell cultures.

Other suspected sources have been the tissue of origin (van Herick and Eaton, 1945; Hayflick and Stanbridge, 1967) and the induction of bacterial L forms *in vitro* by antibiotics which are active in preventing cell wall synthesis. The latter is still controversial. The origin of the cytopathogenic species *M. hyorhinus* has never been established since no one has been able to isolate mycoplasma from the one suspected source, trypsin. However, an initial infection from a rare happening could have contributed to its widespread prevalence within a laboratory and between laboratories through the exchange of both cultured cells and mycoplasma cultures.

H. Control and Eradication of Mycoplasma Contamination

Several general preventive procedures have evolved through findings suggestive of possible sources of contamination. In general, the trend has been toward the reestablishment of effective but practical aseptic techniques and the elimination of antibiotics wherever possible. Separate rooms or isolated areas where traffic can be kept to a minimum and where dust and the presence of bacteria in the environment can be reduced by frequent washings with antiseptic solutions or, alternatively, the use of laminar flow hoods or rooms outfitted with laminar flow are recommended. An excellent review by McGarrity and Coriell (1971), discusses the sources of contamination and their control. Mouth pipetting is avoided and the cross-contamination between cell lines is prevented by handling only one line of cells at a time. Insofar as the problem of serum contamination is concerned, several procedures, such as inactivation at 56° or 60° or refiltration of the commercially obtained medium and serum followed by storage at refrigeration temperature for 10 days has proved successful in preventing contamination by mycoplasma (Fogh and Fogh, 1969; Fedoroff *et al.,* 1972). Barile (1973) recommends that commercial bovine sera be either heated at 56°C for 30 minutes or filtered twice through 0.20 or 0.22 μ membrane filters. He also recommends that trypsin be filtered either once through a 0.10 μ filter or twice through the usual 0.20 or 0.22 μ membranes. Pretesting of a sample of serum from a given lot and the purchasing and storage of lots found mycoplasma-free at temperatures of $-70°C$ or below are recommended. Finally, constant surveillance of techniques and routine testing procedures for contaminating microorganisms including mycoplasma should be adhered to.

I. Methods for Elimination of Mycoplasma from Cell Cultures

There are a number of procedures which have been tried and found to be quite successful. Among these in the order of successful application are the use of antibiotics (Stanbridge, 1971), treatment with specific antiserum (Pollock and Kenny, 1963), heat treatment (Hayflick, 1960), and exposure of the cell culture to hypotonic treatment (Gori and Lee, 1964). There does not appear to be a "universal" antibiotic that will eliminate any species of mycoplasma from every cell line. Mycoplasma not only vary in susceptibility to the various antibiotics but the toxicity of the antibiotic varies in degree with different cell lines. Perlman *et al.* (1967) have published a table (Table VI) showing antibiotics useful in controlling PPLO contamination in tissue cultures. Some of the essential information necessary to a successful cure is given. The stability of the antibiotic in tissue culture medium is important since this determines the time of medium changes and subculture during the treatment period. The toxic level to the tissue cells is essential to know since, at times, it is better to use a high level of the chosen antibiotic for as short a period of time as possible to avoid the development of resistance. Sometimes, it is best to isolate the mycoplasma and determine its sensitivity to suitable antibiotics incorporated in broth or agar. Frequently the minimal lethal dose to the mycoplasma in bacteriological media is not effective when tested on the tissue cells and should be used in higher concentration. After terminating treatment by removal of antibiotics, the cell culture should be tested at regular intervals to determine whether all mycoplasma were eliminated by the treatment. Though most of the mycoplasma have been shown to be associated with the cell membrane, a few have been found intracellularly by thin-section electron microscopy (Edward and Fogh, 1960). These conceivably may not make contact with the antibiotic, and a delayed reappearance of the infection has been noted. Kanamycin has been found to be very helpful (N. B. Holmgren, unpublished data). On the other hand, the tetracyclines were found to be toxic at low levels. For example, in one such test, mycoplasma isolated from an established amnion cell line, RP Am-1, were not inhibited in agar by 5 μg/ml but were destroyed at 10 and 25 μg/ml of tetracycline.

TABLE VI
ANTIBIOTICS USEFUL IN CONTROLLING PPLO CONTAMINATION IN
TISSUE CULTURES[a]

Antibiotic	Stability in tissue culture media[b]	Concentration showing marked cytotoxicity (μg/ml)	Minimal concentration inhibiting PPLO in agar streak test[c] (μg/ml)	Concentration recommended for controlling PPLO in tissue cultures[d] (μg/ml)
Chloramphenicol	High	30	15	30
7-Chlortetracycline	Very low	80	40	100
6-Demethyl-7-chlorte-tracycline	High	15	5	10
Erythromycin	Moderate	300	15	50
Fusidic acid	High	40	20	20
Gentamicin	High	3,000	1	200
5-Hydroxytetracycline	Moderate	35	5	10
Hygromycin B	Moderate	300	15	50
Kanamycin	Very high	10,000	25	200
Neomycin B	Very high	3,000	15	50
Novobiocin	Low	200	10	50
Paromomycin	High	5,000	20	50
Spiramycin	Moderate	1,000	1	50
Tetracycline	Moderate	35	2	10
Tylosin	Moderate	300	1	10

[a] From Perlman et al., 1967.

[b] Stability scale: half-life of 2 days, very low; half-life of 4 days, low to moderate; half-life of 8 days, very high.

[c] As determined in twofold agar dilution test.

[d] Recommended on basis of 3-day incubation period between medium changes.

However, 25 μg/ml was toxic to the cell and 10 μg/ml proved ineffective. Elimination of the mycoplasma from this cell line was effected after treatment for 10 days with kanamycin at 200 μg/ml. The RP Am-1 line has remained free of contamination for over a period of 5 years.

Treatment with specific antiserum has been successful but requires the isolation and identification of the mycoplasma species in order to procure the specific antiserum, if available, or to prepare a hyperimmune serum to the isolate. Exposure of the contaminated cell line to 41°C for 18 hours (Hayflick, 1960) or increasing the permeability of the cell by 2- to 3-minute exposures to distilled water containing high concentrations of aureomycin, kanamycin, and chloramphenicol, followed by cultivation in complete medium containing one-fifth the concentration previously used in the distilled water, proved successful with some cell lines (Gori and Lee, 1964). Both of these latter methods, however, cause extensive damage to the cells and incur the high probability of selecting cells from the population.

There is some evidence that some tissue cells show greater resistance to mycoplasma infection. The presence of a mycoplasmacidal factor has been demonstrated in HeLa and L cells (Kaklamanis et al., 1969), and it has been suggested that the transformation of the BHK-21 cell line (Macpherson and Russell, 1966) by mycoplasma was attributable to the selection of cells in the population resistant to the mycoplasma contaminant (Russell et al., 1968).

The prophylactic use of low levels of antibiotic or combined antibiotics to which mycoplasma are sensitive is not recommended because of the ease with which these organisms develop resistance. It is recommended that storage in liquid nitrogen of cell lines free of contamination be a part of the routine of cell culture, since, at times, it seems easier to discard the contaminated cell line when a source of noncontaminated tissue cells is available.

III.　Virus Contamination of Cultured Cells

A.　Incidence and Sources of Virus Contamination in Cell Cultures

The general suspicion that viruses are common contaminants of cell and tissue cultures has been held for many years. It was shown by Rivers et al. (1929) that vaccinia virus could persist in tissue cultures of immune cells. Gey and Bang (1951) and Bang and Gey (1951) reported that several viruses persisted for months in established cell lines while Ackerman and Kurtz (1955) showed the persistence of poliovirus in HeLa. Several strains of adenovirus were shown by Rowe (1953) to persist in a latent form in human adenoid tissue. Such tissues also contained herpes simplex and cytomegalic viruses in a latent form. Hull et al. (1956) demonstrated the presence of many strains of adenovirus, foamy agent virus, and vacuolating virus from primary monkey kidney cells.

Much of our present knowledge on virus contamination problems comes from the Conference on Cell Cultures for Virus Vaccine Production, edited by Merchant (1968). The message from that Conference is clear: the problem of virus contamination in cell and tissue cultures is a serious one and improved surveillance is required.

Similar to other types of cell culture contaminants, the sources include: (1) tissues from which cells are cultured, (2) sera, (3) reagents, and (4) laboratory workers. The tissue *per se* appears to be the most important source of virus contamination.

1. PRIMARY CULTURES

Cultures prepared directly from animal organs are commonly designated primary cultures. When such tissues contain indigenous viruses, then problems arise. As pointed out by Kniazeff (1968) such viral contaminants can be agents which: (1) show high incidence of infection in the host, (2) produce a systemic infection and disseminate to tissues commonly used in cell culture, or have a specific tropism for such tissues, (3) remain in the tissues of the host for a sufficiently long time to make isolation likely, (4) produce silent or at most a mild disease, and (5) will replicate or at least survive under *in vitro* cell culture conditions.

a. Human Viruses. Human tissue, because of its limited supply, has not been cultivated as frequently as tissue from other primate and non-primate sources. Usually human fetal tissue is used most frequently whereas adult tissue is rarely used for routine purposes. It is generally believed that the occurrence of contaminating viruses in human tissues is limited. The viruses found most frequently, yet only on rare occasions, include cytomegalovirus and rubella (Kalter and Heberling, 1968).

Also, occasionally herpesviruses and adenovirus have been recovered from nonfetal human tissues. Recent interest in herpes simplex type 2 as a possible etiological agent in carcinoma of the cervix has been heightened because of its latent character as shown recently by its isolation from spontaneously degenerating cultures of carcinoma of the cervix grown *in vitro* for 6 months (Aurelian *et al.*, 1972). Duplicate cultures arising from the same seed remained unaffected. This is similar to an observation made by one of the authors on genital tissue from squamous cell carcinoma of the vulva in which herpes simplex manifested itself in 1 out of 15 culture flasks after 10 days in culture (Christian *et al.*, 1965).

A long list of viruses have been recovered in isolated cases, generally as a result of intrauterine infection, from the fetus, stillbirths, newborns, and malformations. These include rubeola, vaccinia, variola, varicella, poliovirus, Coxsackie virus, ECHO virus, mumps, influenza, Western encephalitis, lymphocytic choriomeningitis, rabies, and infectious hepatitis (Kalter and Heberling, 1968).

Melnick (1968) listed the known latent viral infections of man as herpes simplex, herpes zoster, cytomegalo, adeno, a new herpes type (Burkitt lymphoma), as well as measles virus associated with subacute inclusion body encephalitis (Shaw *et al.*, 1967; Freeman *et al.*, 1967) and congenital rubeola (Rawls and Melnick, 1966). Other viruses causing undetected infections in man are the satellites of adenovirus, whose growth depends on a late state in the replication of the helper adenovirus (Casto et al., 1967; Parks et al., 1967; Hoggan et al., 1966; Blacklow et al., 1967).

b. Simian Viruses. Several species of monkeys have been used as sources of primary tissue cultures for the production of virus vaccines. These include the rhesus (Macaca mulatta), the cynomolgus (M. irus), and the African green (Cercopithecus aethiops) monkeys. Recently other simians have been introduced for large-scale virologic studies. These include the patas (Erythrocebus patos), the marmoset (Saquinus sp.), and the baboon (Papio sp.).

Simian kidney cells have seen large-scale use in the production of virus vaccines for a number of years despite the rather large number of simian viruses isolated from such cells (Hull et al., 1956; Hsiung, 1968a). Adenoviruses make up the largest group of such viruses recovered from tissue culture preparation. The adenoviruses have received the greatest attention because certain members have shown oncogenic potential (for example, SV20, SV33, SV34, SV37, SV38, and SA7) (Hull et al., 1965). Another oncogenic virus, SV40, a papovavirus has been isolated with increasing frequency from primary kidney cultures of rhesus, cynomolgus, and patos monkeys. SV40 is noteworthy because it produces tumors in newborn hamsters and causes transformation of hamster, mouse, monkey, and human cells in tissue culture.

Other viruses which have been isolated from monkey kidney cultures include reoviruses, enteroviruses, herpes B virus, cytomegalo virus, paramyxovirus, SV3, SV5, and foamy viruses. In all, more than 50 viruses have been recovered from several nonhuman primate tissues—chiefly kidney cell cultures (Kalter and Heberling, 1968; Hsiung, 1968a). Of most concern is the herpes B virus which is present primarily in Asian species of monkeys and is fatal for humans. As in cases of infection with many other simian viruses, monkeys infected with herpes B virus show only mild and inapparent symptoms. The same may be said of a fatal hemorrhagic disease agent transmitted from vervet monkeys to man (Hennessen, 1968). It was the apparent cause of 7 fatalities out of 27 cases who came into contact with organs, blood, or tissues and were wearing protective clothing during surgical procedures but not during other handling of material contaminated with monkey blood (Hennessen, 1968). The sources of such latent virus infections in monkeys are somewhat obscure. Some agents (like SV40) are obviously viruses of primates while others may be acquired by the animals after contact with man. This appears to be the case for measles and paramyxoviruses, SV3, and SV5 (Hsiung, 1968b).

c. Bovine Viruses. Bovine adenoviruses 1, 2, and 3, bovine herpesvirus 1, parainfluenza 3, certain enteroviruses, and bovine virus diarrhea are potential contaminants of cell cultures (Kniazeff, 1968). The viruses that are especially significant since they may be present both in fetal and in adult tissues include herpesvirus 1, parainfluenza 3, and bovine diarrhea virus. Newborn and day-old calves may be carriers of bovine diarrhea virus (Burki and Germann, 1964; Romvary, 1965), and thus contamination of cell cultures can occur not only by using fetal bovine tissues but also through the use of fetal bovine serum (Malmquist, 1968; Kniazeff *et al.,* 1967). The latter is a much more common practice in cell culture work. Also the fact that many strains of bovine diarrhea virus do not produce cytopathogenic effect further complicates the problem of detecting this viral cell contaminant.

d. Chicken and Duck Viruses. Avian adenoviruses, leukosis viruses, avian encephalomyelitis viruses, avian infectious bronchitis viruses, and Newcastle disease viruses are potential contaminants of cell cultures made from chicken tissue sources. Lymphoid leukosis viruses and avian adenoviruses are the most likely cell culture contaminants while the other viruses may be contaminants if the flock is in the active stages of infection (Luginbuhl, 1968).

Most chicken flocks are infected with leukosis viruses, which are usually introduced early in life through contaminated respiratory vaccines (Melnick, 1968). The virus is transmitted horizontally through saliva and feces and vertically from viremic hens. Leukosis virus acts as a helper virus for Rous sarcoma virus, producing the protein coats. It also can interfere with Rous sarcoma virus.

There is evidence of a herpes-like virus being associated with Marek's disease. The virus grows in chick cell cultures, and should be considered a potential contaminant of chick embryos and a likely contaminant of adult cell cultures (Witter, 1968). Other chicken viruses less likely to be encountered, yet which can be transmitted via the embryo include avian encephalomyelitis, infectious bronchitis, and Newcastle disease viruses.

Duck plague and eastern encephalitis viruses are possible cell contaminants but insufficient research has been done with duck tissue. The same is true for the viral flora of duck embryo cells (Luginbuhl, 1968; Witter, 1968).

e. Rabbit Viruses. Eleven viruses are associated with various genera of rabbits and hares (Wilner, 1968). Six viruses infect only wild species under natural conditions while five viral agents infect domestic or laboratory rabbits. Two, myxoma and rabbitpox viruses, produce fulminating disease with high mortality while the other three, oral papilloma, virus

III, and reoviruses are at most responsible only for minor or inapparent infection (Wilner, 1968).

Incidence of isolating contaminating viruses in rabbit cell cultures appears to be extremely rare. Isolation of herpes virus apparently has been noted (Duff, 1968).

f. Canine Viruses. The 12 viruses isolated from dogs produce disease (Gillespie, 1968). In this group, canine distemper and infectious canine hepatitis (ICH) occur as relatively common diseases which most dogs incur during their first year of life. This may also be true with parainfluenza and canine herpes virus, although insufficient information is available at this time. Incidence of rabies and pseudorabies in dogs is low but usually fatal (Gillespie, 1968). Mumps, reovirus I, and ECHO virus type 6 isolated from dogs are probably identical with human viruses. Four viruses are of interest because of their oncogenic potential. These include the Lederle strain 255 of ICH virus which produces tumors in hamsters while the other three—canine papilloma, canine venereal tumor, and mast cell leukemia—produce tumor masses in dogs. At present little is known about the oncogenic potential of these agents in man or in human cell cultures. Mumps, ECHO virus type 6, reovirus I, pseudorabies, and rabies are known to produce diseases in man. ICH has been repeatedly isolated from canine kidney cultures (Melnick, 1968).

The frequent occurrence of canine lymphosarcoma suggests the possibility of a latent virus being involved. This is especially true in view of Huebner's findings with murine sarcoma (Huebner, 1967). The same situation is involved for murine sarcoma as for avian sarcoma. In both these sarcomas, virus cannot be detected by ordinary methods, but the addition of the proper helper virus supplies the protein coat to the sarcoma virus genome which in turn allows the virus to be detected. Whether these findings apply to dog lymphosarcoma will depend on finding the correct helper virus (Melnick, 1968).

g. Miscellaneous Viruses. An apparent latent cytomegalo-like virus was isolated from spontaneously degenerating equine kidney cell cultures only after prolonged cultivation of the cells. Cytopathic effect was noted after some 80 days cultivation and 3 subcultures (Hsiung *et al.*, 1969).

A myxovirus-like latent virus designated feline syncytia-forming virus (Fe SFV) has been detected in 90% of the adult feline population of three counties in California (Hackett *et al.*, 1970). Frequently no cytopathic effect in the form of syncytia appear until the cells have been trypsinized and subcultured 3 or 4 times. The Fe SFV does not produce giant cells and inclusion bodies in feline kidney cells as does the feline rhinotracheitis virus (Crandell and Despeaux, 1959). The question of whether leukemia virus stocks are contaminated with Fe SFV remains to be determined (Hackett and Manning, 1971).

2. CELL LINES AND ESTABLISHED CELL LINES

A cell line arises from a primary culture at the time of the first subculture. It remains a cell line as long as it can be subcultured only for a limited period of time (human approximately 70 population doublings in 1 year) and during this time it retains the normal diploid chromosome constitution of the tissue from which it was derived. Obviously cell lines vary in the presence or absence of virus contaminates depending on the original tissue source. If the primary culture comes from human or animal tissues contaminated with virus then it is quite likely that the cell line will likewise contain the contaminant. In some cases a sufficiently prolonged period of culture *in vitro*, with several subcultures, is necessary before the virus will manifest itself by showing overt cytopathic effect. This has been demonstrated both in monkey (Hsiung, 1968a) and in equine cultures (Hsiung *et al.*, 1969).

For some latent virus-infected cells it is quite possible that the cells will not show any cytopathic effect even with prolonged cultivation and frequent subcultures. The presence of viruses in these cells can only be detected by special procedures which will be discussed later.

Diploid cell lines derived from human embryonic organs are relatively free of virus contaminants, both cytopathic and latent viruses. Experimental vaccines have been prepared in diploid cell cultures of WI-38 and administered to more than 250,000 patients as vaccines against poliomyelitis, adenovirus 4, rhinoviruses, and rubella. Extensive laboratory and clinical tests over 5 years show that these cells carry no known human virus or other microorganism (Coriell, 1968; Hayflick, 1968). As stated by Coriell (1968) there are many advantages in using human diploid cells for virus production. These include: (1) they are of embryonic origin and therefore free from many contaminants acquired after birth; (2) they do not harbor any known latent viruses, as do monkey kidney and chick embryo cells; (3) spontaneous alteration has not been known to occur in human embryonic fibroblasts, (4) since they are of homologous origin, they should not sensitize man as readily as cells from another species; (5) they have the widest virus spectrum of any culture system so far described and support growth of some viruses that cannot be grown in any other cell system; (6) they can be tested exhaustively; (7) they can be preserved in large quantity at liquid nitrogen temperature for future use; (8) they can be recovered for mass production of vaccine at will; (9) once the cell line has been proved safe, it can be used for producing a wide variety of viral vaccines; and (10) all of the above data permit the "seed virus" concept to be applied to a "seed cell" system.

Established cell lines are cell lines which have demonstrated the potential to be subcultured indefinitely *in vitro*. Usually less than 75% of the cells retain the normal diploid chromosome constitution of the tissue from which the established cell line was derived. They consist of two types: (1) those derived from normal tissue, and (2) those derived from neoplastic tissue, e.g., HeLa originating from a human carcinoma of the cervix. Evidence has been presented to suggest that all established cell lines derived from normal tissue have been replaced with HeLa by cell contamination (Gartler, 1967) (see Section VI). In any case, established cell lines are generally considered free from viruses since repeated subculture greatly increases the chances that latent viruses harbored by the original tissue will manifest themselves. It is possible however, that the virus, or a part of it has integrated with the genome of the cell and if this occurs then ordinary unmasking procedures are ineffective. It is theoretically possible that an inoculated virus could hybridize with such foreign material in the cell and emerge as a potentially oncogenic human agent in vaccine preparations. It is for such a hypothetical reason that established cell lines are expressly forbidden for use in human vaccine production. Their use in dog vaccine preparations since 1963 without adverse effects (Brown, 1968) as well as in laboratory model experiments in rats (Hull, 1968) would seem to negate this objection.

3. SERUM

Molander *et al.* (1968) examined commercial sera and specially procured batches of fetal calf serum for the presence of virus contaminants. One out of 13 samples of the commercial serum produced cytopathogenicity in bovine tracheal cells. A follow-up report by the same group (Molander *et al.*, 1969) indicated that 14 out of 148 lots of bovine serum contained cytopathic or noncytopathic viruses. The noncytopathic viruses were demonstrated by interference procedures. Two isolates were identified as infectious bovine rhinotracheitis virus. Also there were four documented cases of a bovine cell line being contaminated with bovine diarrhea virus from fetal calf serum used in the medium. Bovine diarrhea virus does not produce any cytopathic effect but its presence was revealed by using fluorescent antibody procedures. The most recent data from this group (Molander *et al.*, 1972) reveals 1 out of 10 lots of specially procured fetal calf serum (collected under sterile conditions and not filtered) and 1 out of 16 lots of commercial fetal calf serum yielded viral contaminants. The virus isolated from the commercial lot caused cell destruction of embryonic bovine trachea cells during initial passage (150 $TCID_{50}/1$ ml) and was identified as bovine herpes virus type 1. The other contaminant from a lot of specially procured fetal calf serum required one blind passage before producing CPE in embryonic bovine trachea cells. This virus tentatively identified as bovine diarrhea virus was inconsistent and incomplete in its cell destruction effect. In other samples of serum sediments tested by interference tests, there were delays in the appearance of CPE caused by the challenge of several known bovine viruses in embryonic bovine trachea cells but it could not be determined whether these delays were due to noncytopathic agents. Further work is needed to refine the interference assay system.

In a recent workshop on serum for tissue culture purposes, organized by Federoff *et al.* (1972), Kniazeff reviewed the types of viruses that have been isolated from fetal calf serum. This includes bovine herpes virus, bovine diarrhea virus, parainfluenza 3, and bovine enteroviruses. Several isolates as yet have not been characterized. Of 150 lots of bovine serum tested, viruses were isolated from 18 lots or 12% of the pools tested. The incidence of viral contamination of serum depends on the age of the animal when bled. It was noted that viruses were infrequently isolated from newborn and agamma sera, while none were detected from sera of older calves and only a small percentage from adult animals (Federoff *et al.,* 1972).

Most commercial suppliers of calf serum are presently screening their sera for viruses by the methods described in Section III,D.

4. LABORATORY TECHNICIANS AND INVESTIGATORS

Although no reports have appeared relative to virus contamination through laboratory workers, it is conceivable that virus contamination could occur in this manner if infected individuals do not take the proper laboratory precautions of adhering to strict aseptic techniques and eliminating all mouth-pipetting procedures.

B. General Properties of Viruses

Viruses are agents that can cause infectious disease. They are submicroscopic particles consisting of a nucleic acid core enclosed in a protein or a protein-lipid coat. The nucleic acid is a single molecule of either RNA or DNA, either single- or double- stranded. Viruses are obligate intracellular parasites and can replicate only in actively metabolizing cells. They are unique in that they direct an infected cell to make numerous virus parts and then assemble them into finished virus particles. Viruses can be reproduced entirely from their nucleic acid.

In shape, viruses differ widely. They may be elongated, rounded, polyhedral, or cuboidal. Their size varies from 20 mμ for foot and mouth disease virus to 300 mμ in diameter for vaccinia virus. The life cycle of a virus in a host or host cell proceeds in a series of orderly steps which include: (1) adsorption of virus to a susceptible host cell, (2) penetration of the virus into the interior of the cell, (3) eclipse phase, in which the virus cannot be detected in the cell, (4) replication phase, in which viral nucleic acid and protein are synthesized within the host cell, (5) assembly and maturation of the newly formed virus particles, and (6) release of the viruses from the cell. In many cases little or no harm to the host occurs by the life cycle of a virus, since many viral infections are silent, inapparent ones both *in vivo* and *in vitro*.

Conversely if the cells are damaged by the viral replication, then disease exists and the pathology of the viral disease relates primarily to the visible effects of intracellular parasitism. Such cellular damage by viruses in tissue culture is known as a cytopathic effect (CPE). The mechanism of viral injury is not well understood, but is believed to be related to the release of enzymes contained in the lysosomes of the cell. In some cases cells infected with virus show no CPE because the virus is deprived of an essential metabolite. If the deficient metabolite is added to the medium, the virus replicates and cell infection changes from an inapparent to an apparent one. Conversely there are other inapparent or latent infections of viruses that never display overt signs of infection or only when induced to do so by unusual procedures. Melnick (1968) has restated the definition of latent infections caused by viruses which was agreed to at the Madison Conference in 1957: "An inapparent infection is a primary infection evolving without symptoms and usually leaving a lasting immunity. Regardless of whether the primary infection evolves into a disease or remains inapparent, the virus in some cases may persist. If the virus persists in the absence of symptoms, the infection is *latent*. In a latent infection, the virus might or might not be in a proviral or in a vegetative state, and virus particles might or might not be produced."

Satellite viruses are morphologically and serologically unrelated to the viruses with which they may be found. They may grow independently of the companion virus, or they may not be able to replicate alone and therefore remain undetected unless the replicating companion virus is present (e.g., adenoviruses). Replication of the associated satellite virus interferes with the replication of the companion virus.

Helper viruses are usually not found with the virus that they help under natural conditions. An exception is the Rous-associated virus–Rous-sarcoma-helper system. In this case the Rous-associated virus is needed to form the viral coat. Newcastle disease virus needs the helper parainfluenza virus type 3 in calf kidney cells, and human adenoviruses undergo an abortive infection in simian cells and need SV40 as a helper virus. A step in the transcription or translation of late events in the virus cycle is apparently not performed efficiently in simian cells in the absence of the helper virus (Rapp, 1968).

In cases of viral infection with some oncogenic agents instead of CPE there is a stimulation of the host cells to increase their number. Such an increase in the number of cells leads to tumor formation *in vivo* and to transformation of the cells *in vitro* giving rise usually to an established cell line with the capacity to grow indefinitely.

C.   Detection of Viruses in Cell Culture

There are a number of rather sophisticated biochemical procedures to monitor increases in viral macromolecules and viruses in cells. However, for the detection of virus contaminants in cell culture there are several simpler methods that are more commonly used in diagnostic laboratories.

1.   CYTOPATHIC EFFECTS (CPE)

Many, but not all, viruses destroy the cells in which they multiply. Such infected cell monolayers gradually develop evidence of cell destruction, as newly formed viruses spread to infect more and more cells in the culture. The destructive changes are known as cytopathic effects (CPE) and they can be readily observed under the low power (5 to 10×) of the light microscope using the condenser racked down and the iris diaphragm partially closed to obtain sufficient contrast. When viral contaminants produce such a distinct cytopathic effect, then the problem of detection is simple. In some cases for latent viruses, a prolonged period of cultivation with intervening subculture of cells is necessary before viral CPE become evident. Also the serum concentration of medium used may inhibit virus replication and therefore prevent detection by CPE. Lowering the serum concentration of the medium to 1 or 2% and using fetal or agamma sera in place of adult sera usually eliminates this problem.

2.   PASSAGE TO MORE SUSCEPTIBLE CELLS

If a distinct viral CPE does not become visible even after prolonged cultivation then the passage of cell material to other more susceptible cells may reveal CPE. Many workers maintain certain susceptible lines in the laboratory, e.g., for potential human virus contaminants they may passage the cells on WI-38, a cell line derived from human embryonic lung, or HEp-2, HeLa, or KB, all established cell lines derived from human carcinoma or primary human amnion. Primary monkey kidney cells have also been used for this purpose but there is a greater chance the monkey kidney cells may carry latent viruses than the cell types mentioned above. For animal tissue with potential animal virus contaminants, corresponding animal cell lines and established cell lines would be used.

3.   HEMADSORPTION

For certain viruses, that do not readily produce CPE but do bud from the cytoplasmic membranes of cells, e.g., myxoviruses, paramyxoviruses, and arboviruses, it is possible to detect their presence because they adsorb red cells when they are added to the medium. This phenomenon, known as hemadsorption, occurs because the newly synthesized viral protein incorporated into the cytoplasmic membrane of cells has an affinity for erythrocytes. This hemadsorption of red cells is readily visible under low power of the light microscope.

4.   INTERFERENCE

The replication of one virus in a cell usually interferes with the multiplication of a subsequently added second virus. Rubella virus was first discovered by demonstrating that cultures of infected monkey kidney cells, showing no CPE, were resistant to challenge with a known CPE-producing ECHO virus. This phenomenon, known as interference, was used to isolate rhinoviruses but it is no longer widely used because susceptible cell lines have been discovered in which these viruses produce CPE. Interference is still a useful technique to use however, when searching for new viruses or for latent viruses that do not produce CPE. In this respect, the avian leukosis complex and Rous sarcoma viruses provide an excellent example of a typical interference technique now used to detect latent viruses in cell cultures. The avian leukosis viruses multiply to high titer in chick-embryo fibroblasts without producing CPE or cell transformations. The presence of the avian leukosis viruses can be detected and measured in these chick embryo fibroblast cultures on the basis that the leukosis viruses prevent the formation of transformation foci in such cells challenged with Rous sarcoma virus. Rous sarcoma viruses consistently "transform" normal chick embryo cells.

Another example of the use of interference to study latent infections of viruses was Henle and Henle's (1965) observation that Burkitt's lymphoma cells were resistant to several myxoviruses because the lymphoma cells harbored the latent herpes-type virus that caused the cells to produce interferon.

5.   IMMUNOFLUORESCENCE

The fluorescent antibody procedure is a qualitative test which is used to identify cells in culture that are making virus protein. The main value of the technique has been as a research tool used to localize the site of intracellular viral antigens. However, it is also valuable in demonstrating the presence of latent viral antigens in apparently normal tissue cells that are, in fact, latently infected with virus. The test is based on the fact that homologous antibody labeled with a fluorescent dye, specifically binds to viral antigen in cells and the complex can be detected by its fluorescence under ultraviolet (UV) light (Coons, 1958).

In practice, test cells are grown as monolayers on cover slips, then fixed with acetone, covered with fluorescein- or rhodamine-tagged antiviral serum. The cover slip is washed off with balanced salt solution to remove nonadsorbed antibodies and the cells are examined using a UV-light microscope. This procedure is the direct immunofluorescence technique. It depends on preparing antibody against different viral antigens from a spectrum of latent viruses usually found in a given host and individually labeling each antibody with fluorescent dye.

Alternatively, the indirect immunofluorescence procedure requires labeling only a single immunoglobulin to the animal species in which the different viral antibodies were produced. Thus, if rabbits were immunized with viral antigens to produce antibodies, then only an anti-rabbit globulin made in goats need be labeled with fluorescent dye. Another advantage of this method is that fluorescein-conjugated immunoglobulins to various animal species are readily available commercially.

In practice for the indirect immunofluorescence the cells grown as monolayers on cover slips are fixed with acetone and covered with non-labeled antiviral serum. After washing the cover slips, fluorescein-labeled antiglobulin to the animal species in which the viral antibodies were produced is added. If the viral antibodies combine with the viral antigens in the cells, then the tagged antiglobulin will also be fixed at those particular sites and they will show fluorescence under UV light.

Smith (1968) uses such an indirect fluorescent antibody system to immunologically monitor dog cell cultures for the presence of viruses. In the system, cover slip cultures of primary dog cells are prepared and then tested with a battery of six sera. One is the serum from the cell donor while the others include a pooled globulin from dogs and four street dog sera which have a broad reactivity to known canine viruses. In this manner one can check each lot of donor cells that come from an animal for adventitious agents before the cells are even used. Rowe (1968) recommends testing primary, secondary, and tertiary passage cells by Smith's procedure since, in addition to prolonged cultivation, an important variable in obtaining growth of latent agents is serial cultivation.

The above procedures can be carried out by most laboratories and are usually used routinely to detect the presence of viral contaminants. If none of these tests demonstrate the presence of virus, it is still not possible to be absolutely certain that a given cell culture is free of virus—for it is possible that a latent, persisting virus infection may still be present which requires complicated procedures for unmasking and which cannot be carried out in routine practice. Several of the more sophisticated techniques for detecting such latent viruses have been listed by Melnick (1968): (1) Direct examination of cultured cells and tissue culture fluids in the electron microscope. Sensitivity of the technique can be increased by density gradient ultracentrifugation. (2) Fusion of test cells with susceptible cells to determine whether this leads to expression of infectious virus. (3) Analysis of the infective dose-response curve to see whether it fits single or double hit kinetics This would determine whether the virus seed has become contaminated with another genome from a known or unknown source in the tissue culture system in which it was grown. (4) Addition of helper viruses to the test cultures to determine whether a new virus appears (satellite) or whether a double hit kinetic curve can be changed into a single hit curve (PARA- and MAC- adenoviruses, murine sarcoma-leukemia viruses). (5) Tests of the donor tissue for the presence of group-specific antigens (such as those of avian leukosis virus) with selected sera that do not react with coat protein antigens. (6) Homology studies to determine whether messenger RNA (mRNA) in the test cell hybridized with nucleic acid of unknown viruses. This method has been applied for detecting mRNA of adenovirus and papovavirus in virus-induced tumors, and should also be applicable to cells carrying other incomplete viral genomes.

Several other sophisticated methods to detect the presence of oncogenic viruses in mammalian cell cultures have been described by Hackett (1973). These include besides electron microscopy procedures, isotope labeling techniques and monitoring for reverse transcriptase. The isotope labeling techniques depend upon RNA-containing tumor viruses characteristically banding at a density of 1.16 gm/cm^3 and forming a peak of radioactivity in that region of the gradient (Todaro et al., 1970). DNA-containing tumor viruses can also be detected as a broad peak of thymidine-^{3}H which encompasses the 1.22 to 1.14 gm/cm^3 density region of the gradient (Hackett, 1973). These techniques have been designated "detection of isotope labeled particles by isopycnic centrifugation" (DIPIC) and they appear to show considerably promise as monitoring methods for detecting oncogenic viruses in cell cultures. An added advantage of the DIPIC technique is that it can also be used as an indicator for mycoplasma contamination of tissue culture (Todaro et al., 1971) and is useful when the mycoplasma strain is difficult to detect by bioassay. Mycoplasmas band below the density of the oncogenic viruses, at 1.22 gm/cm^3 (Hackett, 1973). The detection of the reverse transscriptase (RNA-directed DNA polymerase) enzyme as an indicator of the presence of oncogenic viruses may be more useful for monitoring cells other than human, since the results to date with human tumor cells grown in cell culture have been negative (Hackett, 1973).

D. Prevention of Virus Contamination

Control measures relating to viral contamination from the tissue of origin depend in part on proper handling and testing procedures of the animals donating the tissue. For example with monkeys it has been shown that housing different species in separate quarters reduced the incidence of virus cross-contamination between species (Hsiung, 1968b). Also monkeys should be quarantined for 30 to 90 days before tissues are used for cell culture. This permits time for any overt disease-producing agents to reveal themselves. Contact between monkeys and man during the quarantine should be kept to a minimum (Hennessen, 1968). Where possible similar procedures should be followed for other animals including dogs, chickens, rabbits, cats, etc. The animal can be bled and the serum isolated and tested for antibodies against various known viruses which produce diseases in the species using various serological tests. Subsequently, this serum can be used in the indirect immunofluorescent test with cells grown on cover slips from the same animal (Smith, 1968). When human tissues are obtained for cell cultures, one should be aware of the clinical diagnosis of the patient and if possible any previous history of viral diseases.

Preventive measures to avoid virus contamination of cell cultures through the use of serum in the medium include: (1) using virus-screened serum, (2) testing and eliminating batches of serum contaminated with viruses, or (3) sterilization of sera. Many commercial tissue culture suppliers now supply virus-screened sera and others will do so in the future. The extra cost for screened sera is well worth the price when one considers the time and cost in screening as well as the frustration and economic loss suffered by workers when contaminated cultures are used. Most commercial suppliers will permit investigators to pretest a batch of serum before purchase and then reserve that lot exclusively for their future use. In such cases the worker may wish to test the serum for viral contaminants. The usual procedures employed by commercial suppliers generally follow the techniques of Molander et al. (1972) and are as follows:

1. A sample of the serum (10% of the volume) is concentrated by ultracentrifugation at 120,000g for 1½ hours.

2. The pelleted material is resuspended in standard cell culture

medium (10% fetal calf serum, minimum Eagle's medium [MEM] using a mortar and pestle to effect a ten-fold concentrate).

3. Portions are inoculated in cell cultures of embryonic bovine trachea (EBTr), primary bovine embryonic kidney, and WI-38, human embryonic lung for detection of bovine and human viruses.

4. Inoculated cultures are observed for cytopathic effect over a 7- to 21-day period.

5. If no cytopathic effect develops the cultures are blind passed 2 or 3 times on fresh cultures of the same cell type.

6. Companion negative cultures are challenged with cytopathic viruses, bovine virus diarrhea (BVD), and infectious bovine rhinotracheitis (IRB) virus to demonstrate possible interference.

7. Other negative cultures are treated with red cells to detect the possible presence of noncytopathic hemadsorbing bovine parainfluenza type 3 in serum.

8. Positive cultures are further analyzed for virus type by serological neutralization procedures.

9. A few suppliers also examine the tenfold serum concentrate by electron microscopy for presence of viruses.

Coriell at the workshop on serum questioned the general practicability of monitoring bovine sera for the presence of viruses since these viruses do not appear to grow and are not detected in human cell cultures. Federoff et al. (1972), at the same workshop, concluded that the usefulness of bovine virus testing methods for an investigator depends on the intended use of the serum. For research studies or vaccine production with bovine viruses, such monitoring should be done. For propagation of human cell lines at present, the hazards are not defined and even though experimentally bovine herpes virus may infect human cells, there are no data on the incidence of such infections, if any, during the usual propagations of human or nonbovine cells. Until definitive studies are done, it is left for the investigator to decide how seriously he should consider such viral contaminants (Federoff et al., 1972).

Sterilization of serum to eliminate not only viruses but other contaminating microorganisms as well has been under consideration for years. Such methods include heat inactivation, ionizing radiation, ultraviolet irradiation, and chemical sterilization. Membrane filtration of course has been used effectively for most microorganisms but the 0.20- or 0.22-μ pore size generally used will not remove Mycoplasma or most viruses. Several commercial methods for serum sterilization have been advanced recently. Lo Grippo's method of combining ultraviolet irradiation and β-propiolactone treatment of serum is effective but it is not licensed in the United States because β-propiolactone may have carcinogenic properties (Federoff *et al.*, 1972). Gray Industries' method of serum sterilization using nonionizing radiation apparently permits sterilization of glass-packaged serum without alteration of serum proteins. This method shows considerable promise but it may be too early to assess its effectiveness in terms of assuring sterility without affecting the growth-promoting qualities of serum. Another method using high energy electrons was effective in eliminating even the smallest picornaviruses but the levels of energy needed produced obvious alterations in serum proteins (Federoff *et al.*, 1972).

Prevention of viral contamination of cell cultures by laboratory workers requires only a rigid adherence to aseptic techniques with the additional precaution that no mouth pipetting be permitted.

If a primary culture, cell line, or an established cell line becomes contaminated with a virus, either a cytopathic, a noncytopathic, or a latent virus, there is no certain method to eliminate the contaminant. It is best to discard such cultures rather than try to eliminate the viral infection by hyperimmune serum or by chemical means.

IV. Bacteria and Fungi Contamination of Cultured Cells

A. Incidence and Sources of Bacterial and Fungal Contamination in Cell Cultures

Bacterial, yeast, and mold contaminations occur in most cell culture laboratories in spite of or because of the use of antibiotics. When antibiotics are omitted from the media, contaminations with these microorganisms are usually easily visible by the turbidity or aggregates formed and the subsequent destructive effects on the tissue cells. Conversely when antibiotics are present in the media, they may destroy many contaminants or they may only suppress others leading to a chronic or latent and inapparent infectious state (Coriell, 1962). An inapparent infection is one in which no microscopically detectable change can be seen in the cell cultures and in which routine bacteriological cultures on blood-agar plates or thioglycollate broth remain sterile. Coriell (1962) was able to detect the presence of three different diphtheroids in several established cell lines by: (1) inoculating cells into mycoplasma broth and agar, (2) using a centrifuged pellet of cells as an inoculum, and (3) repeated subculture in antibiotic-free media. One diphtheroid was detected as an inapparent infection in 13 established cell lines. Similar low-level bacterial contaminations (100 to 1000 organisms per milliliter) of cell culture media were detected in 20 of 73 cell cultures examined (27%) from 22 laboratories by inoculation into mycoplasma broth and agar media (Barile and Schimke, 1963). The bacteria were identified as members of *Corynebacterium, Micrococcus,* and *Bacillus* genera as well as several gram-negative enteric rods. There was no turbidity of the cell culture media or cytopathic effect on the cells. None of the investigators previously working with the lines knew they were contaminated. The cultures had been checked for sterility only with thioglycollate broth without taking into account the inhibitory effect of the antibiotics being used or the need for special nutrients to offset the antibiotic-induced damage.

The seriousness of the problem of contamination has not lessened in recent years as illustrated by the fact that 18% of the cell lines submitted to one of the cooperating laboratories of the Cell Culture Committee during 1968 were contaminated with various microorganisms and 27% of the lines submitted in 1969 were similarly contaminated (McGarrity and Coriell, 1971). It is estimated that approximately 50% of cell cultures in current use are probably contaminated with a microorganism of some type (McGarrity and Coriell, 1971).

Pseudomonas aeruginosa is a common contaminant of cell cultures. This is especially true when penicillin and streptomycin are used, since many strains of this organism are resistant to these antibiotics. At low levels *P. aeruginosa* produces plaques in cell monolayers which are indistinguishable from viral plaques (Fig. 6) (Ludovici and Christian, 1969). Other bacteria which produce similar plaques on cell monolayers include *P. fluorescens* and *Alcaligenes fecalis* while many other organisms including *Staphylococcus aureus* and *Escherichia coli* do not. The similarity of the bacterial plaques to viral lesions can also be demonstrated using a solid agar overlay procedure (Fig. 7) (Wexler *et al.,* 1970). McGarrity and Coriell (1971) noted that *Pseudomonas* organisms were the most frequently isolated contaminants over the last 4 years when antibiotics were used.

Yeasts are common contaminants of cell cultures (Coriell, 1962). Many forms grow slowly in cell cultures and are relatively nontoxic to the cells. McGarrity and Coriell (1971) observed that a lot of specially prepared, commercially supplied nonfiltered calf serum contained a slow growing yeast of the *Torula* sp. which was undetectable by their routine quality control checks or those of the supplier. It was first detected when it was used in insect cell cultures incubated at only 28°C. A few yeasts grow best in the presence of living cells and fail to grow on aerobic blood-agar plates (Coriell, 1962). They can usually be detected by direct observation which is sometimes facilitated by using a cell culture medium without

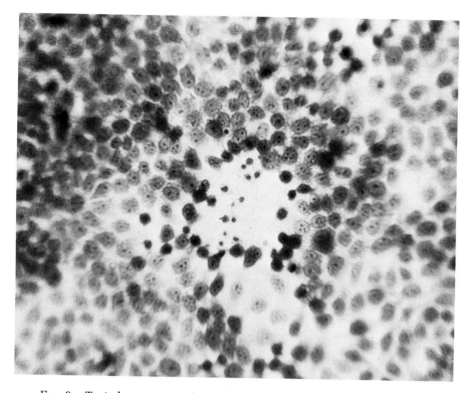

Fig. 6.   Typical appearance of an early *P. aeruginosa* plaque in HeLa monolayer 12 hours after infection. ×70.

Mycostatin or Fungizone. Passage of cells to Sabourand dextrose broth or agar may be necessary in some cases to detect the presence of yeast grown in an antifungal medium.

Molds in cell culture tend to grow slowly and for short experiments are not a serious problem. They can usually be detected by simple microscopic examination of the easily identifiable mycelium. Confirmation can be obtained by inoculating Sabouraud dextrose broth or agar medium. Such infections are usually the result of airborne contamination or faulty technique. High humidity increases the risk of mold contamination. Cell culture reagents placed in cold rooms seem to be especially vulnerable to contamination. Also petri dish cultures requiring a 5% CO_2 and 95% air incubator with humidity are much more susceptible to mold contamination. It is advisable to avoid such cultures for long-term experiments, and to carry established cell lines in closed culture containers.

The source of contamination by bacteria and fungi may be from the

Fig. 7. HeLa monolayer infected with *P. aeruginosa* and overlaid with solid agar medium showing typical virus-like appearance of bacterial plaques 24 hours after infection with the clear central zones of dead cells not staining with neutral red.

original tissue, the cell line, the established cell line, the medium employed, the air, or the technician. Furthermore all three types of organisms are frequently present in clinical specimens of saliva, urine, stools, blood, and genital secretions submitted for study.

One important source of bacterial and fungal contamination may be fetal bovine serum (FBS) (Boone *et al.*, 1972). This substance (FBS) is used routinely in the media of most tissue culture laboratories and it may therefore be responsible for the bacterial and fungal contaminations seen in cell culture. In a study of eight commercial suppliers of FBS over a period of 1 year, Boone *et al.* (1972) showed that roughly 10% of the commercial lots of FBS were contaminated with bacteria and/or fungi. Twelve out of 125 lot samples, or 9.6%, were contaminated: 8 with bacteria alone, 2 with fungi alone, and 2 with bacteria and fungi. At least three different bacteriological culture media, including blood agar plates were needed to adequately test sterility of the FBS. Presumably this is the reason why the commercial suppliers did not detect the contaminations.

B. General Properties of Bacteria and Fungi

Bacteria are unicellular microscopic organisms which are devoid of chlorophyll. Each bacterial cell contains all the necessary components including both DNA and RNA to metabolize, to increase in size, and to reproduce by simple binary fission. Bacteria vary in size from 0.5 to 1 μ in width to 0.5 to 8 μ in length with shapes that can be spheres, rods, or spirals.

Fungi are plants devoid of chlorophyll that are usually multicellular but are not differentiated into roots, stems, and leaves. Fungi range in size and shape from single-celled oval or spherical microscopic yeasts to giant mushrooms and puffballs. They are generally larger than bacteria with relatively rigid cell walls. The molds grow as tubular branching filaments known as hyphae which become interwoven to form a network known as a mycelium. The molds reproduce by binary fission and by forming asexual or sexual spores of various types while the yeasts commonly reproduce by binary fission and by budding, but may also form sexual spores.

Both bacteria and fungi can be effectively eliminated from solutions using sterilizing filtration systems with a 0.2 or 0.22 μ pore size membrane.

C. Detection of Bacteria and Fungi in Cell Culture

All cell media, sera, and reagents should be tested in a battery of bacteriological and mycological media prior to use on all cultures. The media used in such tests by a number of laboratories, including one commercial supplier of cell culture reagents, are listed in Table VII. All cultures including primary, cell line, and established cell line cultures should also be tested for bacteria, yeasts, and molds once a week on a similar battery of sterility media. It has now been adequately demonstrated that a single sterility medium like thioglycollate broth used by many small laboratories and commercial suppliers is simply not adequate for the detection of many bacterial and fungal agents (McGarrity and Coriell, 1971; Boone et al., 1972).

The question of sample size and number in such sterility tests is always an important consideration. For fluid reagents including sera, at least 0.5 ml and preferably 1 ml of sample per 10 ml of sterility test broth should be tested. For solid agar media in petri dishes, the sample should be spread over the surface of the agar with a bent glass rod. When testing cell cultures, a 0.1 to 0.2 ml sample containing approximately 2 to

TABLE VII

STERILITY TESTS FOR BACTERIA AND FUNGI

Media	McGarrity and Coriell (1971)		J. E. Shannon (1972), personal communication		Boone *et al.* (1972)		GIBCO[a] (1972)	
	37°C	30°C	37°C	26°C	37°C	26°C	37°C	26°C
Aerobic Incubation								
Blood agar	X	X	X		X		X	
Trypticase soy broth	X	X	X	X			X	X
Tryptose phosphate broth	X	X					X	X
Nutrient broth	X	X						
Brain-heart infusion broth					X			
Brain-heart infusion agar			X	X				
GIBCO special test agar							X	X
Sabouraud dextrose broth	X	X		X	X	X		X
Sabouraud dextrose agar	X	X			X			X
Yeast and mold broth	X	X						
Anaerobic Incubation								
Blood agar			X				X	
Thioglycollate broth			X	X	X		X	X

[a] GIBCO, Grand Island Biological Company, 1972 catalog.

4×10^5 cells should be used per tube or petri dish. The number of samples of a given lot will vary depending on the lot size. In general 10 to 15% of the lot should be sterility tested. This is usually done by taking periodic samples of the medium or reagent at regular intervals throughout the membrane filtration process.

As shown in Table VII, both aerobic and anaerobic incubation of certain media, for example blood agar, is desirable. Also it is helpful to incubate one set of cultures at body temperature (37°C) and one set at room temperature (26°C to 30°C). Sabouraud dextrose broth and agar as well as yeast and mold agar should be incubated preferably at room temperature. Length of incubation should be 14 to 21 days to ensure detecting slow growing microorganisms.

Of course, in addition to the sterility test media discussed above, *Mycoplasma* broth and agar media without antibiotics should also be inoculated as discussed in Section II,F. It has been shown by Coriell (1962), Barile and Schimke (1963), and Barile (1972) that many bacterial contaminants of cell culture are detectable in these *Mycoplasma* media when they would be missed in ordinary bacteriological media. This is especially true when antibiotics are employed to prevent contamination and an inapparent infection is actually present. It should be reemphasized that the quickest way to detect the presence of bacterial and fungal contamination of cell cultures is to simply remove antibiotics from the culture media. A number of laboratories are now omitting antibiotics and following a rigid adherence to aseptic techniques with the gratifying results that they are experiencing a reduction not only in bacterial and fungal contamination but in *Mycoplasma* contaminations as well.

D. Prevention of Contamination with Bacteria and Fungi

1. TISSUE CONTAMINATION

Contamination from tissue can be avoided by collecting specimens aseptically since most internal tissues of animals are sterile. This requires sterile operating techniques and the proper aseptic precautions during the culturing procedures.

In the case of tissues which are exposed to the external surfaces like respiratory, alimentary, and genital tract tissues, or if there is a chance that the tissue may have been contaminated during excision, then it is best to wash such tissues in a solution containing a high concentration of antibiotics. Earle's or Hank's balanced salt solution containing 1000 units of penicillin, 0.5 mg of streptomycin, 0.5 mg of neomycin, and 50 μg of Fungizone per milliliter is effective.

Many tissue specimens could not be cultured unless treated in this manner. In many cases washing in high concentrations of antibiotics before explantation or trypsinization destroys all the microorganisms and it is preferable to cultivate the cells subsequently in the absence of antibiotics. However with heavily contaminated tissue, it is advisable to cultivate and subculture the cells in a medium containing a normal concentration of such antibiotics (100 units penicillin, 0.1 mg streptomycin, and 2.5 μg of Fungizone per milliliter) for several weeks.

As soon as the cells have been subcultured to duplicate cultures (cell lines) it is best to remove the antibiotics from one culture to see if it remains sterile. It has been shown by Coriell (1962) that the presence of antibiotics masks or makes latent bacterial, yeast, and molds and that penicillin exerts selective pressure for the formation of stable L forms. This suggests that most contamination is due to poor aseptic techniques in the tissue culture laboratory which apparently develop under the false impression that if one uses antibiotics, sloppy techniques are permissible. For this reason Coriell (1962) initiated the practice of removing all antibiotics from culture media and instituting a regime of rigid aseptic techniques. Coriell (1962) reasoned that this should: (1) enforce strict adherence to bacteriological principles of sterility and asepsis, (2) permit early detection of contamination when it does occur, (3) minimize the formation of stable L forms, and (4) reduce the possibility of cross-contamination of cell lines. Eliminating antibiotics from culture medium has reduced the incidence of contaminations in Coriell's laboratory (Coriell, 1962; McGarrity and Coriell, 1971) as well as two other laboratories of the Cell Culture Collection Committee.

2. Air Contamination

Adherence to a regime of rigid aseptic technique is of no value if the air in the room in which the culturing operations are done is contaminated. Whether or not a specially fitted room with filtered air flow ducts is required will depend on the experimental program. For short-term cultures in which cells are kept for only a few days, experiments can be done in an ordinary laboratory and additional cultures set up to compensate for the occasional losses due to contamination from the air. Even in such surroundings, the simple use of slanted racks to hold bottles and tubes at a 45° angle will minimize the number of lost cultures due to contamination.

When longer term experiments are necessary lasting 2 to 3 weeks or more, then an aseptic room or a laminar flow hood with an absolute or a high-efficiency particulate air (HEPA) filter is desirable. Such a hood designed by Coriell and McGarrity (1968, 1970) was initially installed in an animal room to show that sterile, tissue culture work could be done in a contaminated environment. The model now in use is a horizontal laminar flow hood manufactured by Edgegard Hood, Baker Co., Inc., Biddeford, Maine, in which air passes from back to front through a HEPA filter which removes 99.97% of particles with a diameter of 0.3 μ. An air curtain across the hood permits easy access but separates the worker from aerosols produced in the hood and protects material inside the hood from room air contamination. A recent modification includes a series of perforations around the open edge of the hood which eliminates any cones of turbulent air that develop from large objects in the hood (Coriell and McGarrity, 1970). The use of such hoods has virtually eliminated the airborne route as a major source of contamination.

Similarly, laminar air flow rooms utilizing a solid bank of absolute or HEPA filters covering the entire ceiling or wall are being used in more tissue culture laboratories for the same purpose. To complement the use of sterile filtered air, laboratory cleaning procedures of proven effectiveness should be employed. These procedures include mopping floors and washing table surfaces with detergent-disinfectant solutions (Chlorophen, Wescodyne). A set of rules or regulations to prevent or reduce contaminations in tissue culture laboratories can be found in articles by Coriell (1962) and Paul (1970).

3. TECHNICIAN CONTAMINATION

The single most important precaution to take to prevent contamination of cell cultures is the rigid adherence to bacteriological aseptic techniques. To this end, all tissue culture personnel should be thoroughly trained in strict aseptic procedures preferably by a formal course of 4 to 5 units in general microbiology at the college level including adequate laboratory sessions with active student participation.

A technician contaminates cultures either directly or indirectly. Contamination is caused by transfer of microorganisms from parts of the technicians body directly to the cultures or to the media components which subsequently contaminate cultures (indirect). This is usually caused by faulty techniques in handling sterile equipment, e.g., like pipettes touching nonsterile objects prior to being used for sterile cultures. Also, exhaling of microorganisms during breathing, talking, sneezing, or coughing can directly contaminate cultures. This can be prevented by using a face mask or working within a hood in which a glass plate separates the technician's face from his hands and the materials he is working with. Safety devices like hoods also prevent contamination of cultures with bacteria-laden dust from the technician's hair or body. If a hood is not available then a technician should wear a cap as well as a face mask and a clean laboratory coat when working with cell cultures. Dust is a serious problem in the culture room. As mentioned previously the floor and table tops should be wiped frequently with a detergent-disinfectant solution to keep dust at a minimum and, in addition, it is important to avoid all unnecessary movement in the room when culturing activities are in progress. Despite all the precautions it should be assumed that dust is in the air and will get into the cultures unless the flasks, bottles, and tubes are kept at a 45° angle in a slant rack and manipulations are performed as quickly as possible. The slant rack to hold culture containers has the added advantage that it provides a covered area to place stoppers or screw caps of culture vessels while manipulations are carried out. Also it permits the flaming of the lips and necks of all culture vessels in one easy operation.

4. SERUM CONTAMINATION

Sterilization of sera by membrane filtration is the method of choice to eliminate bacteria and fungi. Since Boone *et al.* (1972) have shown an approximate 10% rate of bacterial and fungal contamination of commercial fetal bovine sera, it is advisable that individual users refilter the serum in their own laboratories. Fogh and Fogh (1969) have found that when this is done, contamination with bacteria and fungi is reduced to zero. Membranes with a pore size of 0.20 or 0.22 μ will retain bacteria and fungi but will not remove viruses or *Mycoplasma* present in serum.

5. Cure of Bacterial and Fungal Infections

In general if cell cultures become infected with bacteria, yeasts, or molds it is best to discard the specimens, taking precautions not to open them in the laboratory and making certain that they are autoclaved before the glassware is washed (Coriell, 1962). If the cell culture is an extremely valuable one and cannot be replaced by a previously frozen one, then a cure of the contaminated culture may be attempted using a high concentration of one or more antibiotics. Treatment of fungi-contaminated cultures with Mycostatin or Fungizone is occasionally successful in eradicating yeast or mold infections. For example, Hsu uses Mycostatin at 1000 units/ml of medium for 24 hours to treat primary animal cultures apparently heavily contaminated with yeast and molds. The cure rate is over 90%, including those with spores (T. C. Hsu, 1972, personal communication).

V. Parasite Contamination of Cultured Cells

A. Introduction

Hartmannella species of amebae have been identified on several occasions as contaminants of tissue cultured cells. "Spontaneous" contamination was reported in freshly trypsinized monkey kidney cells (Jahnes *et al.*, 1957; Culbertson *et al.*, 1958), in Chang's liver cells (Chang, 1961), in canine lymphosarcoma cells, LS #30 (Moore and Hlinka, 1968), and in HeLa cells (Casemore, 1969). Amebae have also been detected in cell cultures inoculated from pharyngeal swabs or other experimental or clinical materials (Hull *et al.*, 1958; Chi *et al.*, 1959; Wang and Feldman, 1967). The efficacy of tissue culture as an excellent growth medium for these species of *Hartmannella* was demonstrated by Wang and Feldman (1967) who isolated them with ease from the human nasopharynx and successfully cultivated them in many types (primary, diploid, established cell lines) and in several species (human, monkey, canine, bovine, rabbit, and chick embryo) of cultured tissue cells. Though it occurs relatively infrequently, lack of awareness of this potential source of contamination has caused considerable difficulty because of the similarity in size and morphology of the trophozoite form to tissue cells and the development of the cystic form in tissue culture which resembles the cytopathogenic effect (CPE) on cells caused by viruses.

B. Identification of the Amebae

1. CYTOPATHOLOGY

In cells infected with *Hartmannella* species either the trophozoite or the encysted form may predominate. Jahnes *et al.* (1957) described the cyclic changes, which occurred over a period of 20 days in monkey kidney cell culture, of cysts to trophozoite and back to the cystic form. Depending on the size of the inoculum and the predominant form present, tissue cells become granular and gradually or rapidly progress to complete disintegration. A general correlation between the pathogenicity of the amebae for mice and their rate of growth in tissue culture was noted by Culbertson (1971).

The cysts, which vary in size from 12 to 31 μ in diameter (Kingston and Warhurst, 1969), have been mistaken for viral CPE when living cultures have been examined with the light microscope. The trophozoite form, on the other hand, resembles a macrophage. It varies in size from 10 to 30 μ in diameter when rounded (Kingston and Warhurst, 1969) and has a granular and vacuolated cytoplasm. The features distinguishing the amebae from tissue cells are sluggish motility, a small spherical well-defined nucleus (3 μ) containing a relatively large eosinophilic karyosome measuring approximately one-half the diameter of the nucleus, and at times, a contractile vacuole with a periodicity of 1 to 2 minutes. Viewed with the electron microscope (Armstrong and Pereira, 1967), the cytopathology of infected cultures was degenerative in nature and revealed at the fine structural level swelling or rupture of the mitochondria and endoplasmic reticulum with eventual fragmentation of the cells. In late stages, only trophozoites could be found among the cellular debris. Nuclear changes suggestive of viral infection (altered nucleoli, eosinophilic inclusions, and clumped chromatin) were also found. The alteration in the nucleoli was described as nucleolar "capping," a segregation of nucleolar components. Jezequel *et al.* (1967) described a similar morphological change in the nucleoli caused by mycoplasma contamination.

2. PROPERTIES COMMON TO VIRUSES AND AMEBAE

Viruses and amebae have certain characteristics in common and these have led some investigators to the mistaken conclusion that a "bizarre" virus was the etiological agent: "Ryan virus" (Pereira *et al.*, 1966), "virus

(Chavis)" (Wang and Feldman, 1967), "lipovirus" (Dunnebacke and Williams, 1967). They resemble each other in cytopathology, in serial transmission in tissue culture, in sensitivity to ether, in resistance to acid at pH 4, in inhibition by 5-fluorodeoxyuridine, in failure to grow in bacteriological culture medium, in resistance to antibiotics, and in animal pathogenicity (Pereira *et al.*, 1966). The eosinophilic karyosome in the nucleus of the ameba has been mistaken for a viral inclusion. On the other hand, the amebae are retained by bacteriological filters and so differ from viruses in filterability.

C. Sources of Contamination

Most of the isolates from tissue culture appear to be *Hartmannella* sp. They belong to a relatively large group of amebae, found free-living in the soil, where they exist in the cystic and vegetative form feeding on bacteria, yeast, fungi, and other organic material. The first species belonging to the *Hartmannella* group was discovered by Castellani (1930), described by Douglas (1930), and classified as *Hartmannella castellanii* Douglas (1930). In the literature, one finds the taxonomic status of the free-living amebae to be somewhat confusing and unsettled (Culbertson, 1971). Two genera *Hartmannella* and *Acanthamoeba*, are listed by some authorities (Kudo, 1966; Chang, 1971). Kudo lists them in the family Amoebidae Bronn, and Chang places them in the family Hartmannellidae Singh (Singh, 1952). Kingston and Warhurst (1969) classify the species in the family Hartmannellidae, some in the genus *Hartmannella*, others in the genus *Hartmannella*, subgenus *Acanthamoeba*. Since many of the studies in this group are comparatively recent, the classification of the free-living amebae is still somewhat fluid.

Kingston and Warhurst (1969) demonstrated the presence of cysts in the air by isolating 36 strains of amebae from samples of the inside and outside air of a hospital obtained through a slit sampler or on "settle" plates. Among these were 15 strains belonging to the genus *Hartmannella* and a few species of *Naegleria*. The rate of settling (0.026 amebae per ft^2 or 0.28 per meter^2 per minute) suggested that settling during laboratory manipulation would be infrequent. The particle diameters were found to have high values (33 to 37 μ) compared with the diameter of the cysts (10 to 31 μ). This suggested to the authors that the cysts were associated with particles of soil which as airborne particles could be impacted in the nose. Large numbers of *Hartmannella* amebae were also recovered by Casemore (1969) from the air and soil. He demonstrated that cysts maintained at low temperature (4°C to -70°C) vegetated readily when placed on suitable culture medium. Thus, airborne contamination is believed to be a prime source in cell cultures.

Several investigators (Chi *et al.*, 1959; Pereira *et al.*, 1966; Wang and Feldman, 1967) have isolated free-living species of *Amoeba* from the human nasopharynx. The findings of Kingston and Warhurst (1969) suggest that the source of the human infection may be the soil-associated airborne cysts. Therefore, a second and occasional source of contamination in cultured cells may result by contact or droplet infection from the human nasopharynx.

A third source of cultured cell contamination was indicated by Culbertson *et al.* (1958) who found amebae in sections of the glomerular capillaries of the kidney obtained from infected monkeys. This suggested that contamination found in monkey kidney tissue cell cultures may have, in some instances, derived from the tissue of origin, Culbertson (1961, 1968) and others (Cerva *et al.*, 1968) have demonstrated the pathogenicity of these free-living amebae for mice, monkeys, and rabbits, and for humans.

D. Methods of Detection, Prevention, and Cure

1. MICROSCOPIC

Since at present this type of cell culture contamination appears to be relatively rare, the point to be stressed is the importance of alerting the investigator to potential contamination with amebae. Microscopic observation with the light and phase microscope at both low and high magnification should assist in making a tentative identification of either the cysts or the trophozoite form of the amebae. When present, the cysts show a superficial resemblance to shriveled cells. They have a double wall showing various degrees of wrinkling. The trophozoite form of the ameba may be identified by its sluggish motility. Filiform pseudopodia are formed from the forward extension of the clear cytoplasm. Motility can be enhanced by maintaining the preparation at 37°C or by suspension in 5% hog gastric mucin (Culbertson, 1961). The most distinguishing features are the morphology of the nucleus with its large round karyosome and the contractile vacuole. Tissue culture preparations on cover slips can be fixed with formalin or Bouin's solution and stained with hematoxylin and eosin for histological study of the trophozoite form of the amebae. Fixation with methyl alcohol followed by staining with either Wright's Giemsa, or May-Grunwald Giemsa stains can be used to bring out the eosinophilic karyosome in the nucleus. The cysts stain with greater difficulty. According to Jahnes *et al.* (1957) the Bauer reaction

(Lillie, 1954) stained the entire cyst red. Culbertson *et al.* (1966) list several other stains suitable for staining cysts.

2. CULTURE METHODS

The amebae grow on 1.5% agar containing 0 to 0.4% NaCl which has been mixed or spread with a lawn of living "edible" bacteria (*coli-aerogenes* group) (Singh, 1960; Culbertson, 1968, 1971) spread over the surface of the agar. *Hartmannella* species tolerate 0.85% NaCl well whereas *Naegleria* species do not grow well in agar containing more than 0.4% NaCl. On agar containing no NaCl, the flagellate form which is characteristic for the *Naegleria* species can be demonstrated (Culbertson, 1971). Some species of *Hartmannella* will grow in nutrient broth without bacteria and some require bacteria for growth. Similarly, some can grow in tissue culture medium in the absence of tissue cells (*H. glebae* and *H. rhysodes*), while others require the presence of tissue cells (*H. agricola*) (H. A. Feldman, 1971, personal communication). For growth in tissue cell cultures, 0.1 ml of a suspension of either cysts or amebae is inoculated onto 4- to 7-day-old cell sheets. Although the predominant isolates from tissue cultures have been identified as species of *Hartmannella,* the possibility of contamination with other genera of *Amoeba* or other parasites must also be considered.

3. PREVENTION AND CURE

Data from a limited number of studies suggest that the amebae are relatively insensitive to antibiotics. However, Culbertson *et al.* (1965) found that addition of sulfadiazine to the diet protected mice infected with *Hartmannella* amebae and large doses of amphotericin B combined with large doses of sulfadiazine in the diet showed some activity against experimental infection of mice with a *Naegleria* sp. Casemore (1969) tried Metronidazole, widely used in the treatment of trichomoniasis, and found that 10 μg/ml added to the culture medium to be used in uninoculated cell cultures, was nontoxic and prevented amebic contamination in a laboratory environment favoring contamination with amebae. Thorough cleaning of the laboratory did not in itself prevent further contamination. Conversely, Pereira *et al.* (1966) found that neither kanamycin nor neomycin at a concentration of 100 μg/ml had any effect on a *Hartmannella* sp. Similarly the therapeutic use of the sulfonamides, penicillin, tetracycline, oxytetracycline, chlortetracycline, oleandomycin, or erythromycin failed in the therapy of 16 human cases of amebic meningoencephalitis (Cerva *et al.,* 1968).

For prevention, alertness to the possibility of cell culture contamination with free-living amebae and other parasites is important in avoiding the rare occurrence in which these organisms cause problems. Cleanliness of the environment where tissue culture is performed to prevent airborne contamination combined with aseptic technique should prevent amebic contamination with the possible exception of the presence of amebae in the tissue of origin. Where airborne contamination is difficult to control, the use of Metronidazole in the culture medium to be used with uninoculated cell cultures may assist in suppressing the growth of the amebae. The lack of agents capable of curing a cell line of amebic infection stresses the importance of maintaining healthy stocks of cells preserved in the frozen state. A word of caution may be appropriate even for frozen cultures, for Ensminger and Culbertson (Culbertson, 1971) found that the *Hartmannella* amebae could be preserved in liquid nitrogen using dimethyl sulfoxide.

VI. Cell Contamination by Cultured Cells

A. Incidence and Sources of Contamination by Cells

Contamination of cell lines from one species with the cells from another species was initially suggested by Coriell *et al.* (1958) on the basis of similarity in cell antigens using the complement fixation (CF) test. Subsequently Rothfels *et al.* (1959) confirmed the similarity in cell antigens by CF and then using chromosome analysis showed that a culture of monkey kidney cells was contaminated by mouse L cells. Final proof that cross-contamination of established cell lines could not only happen but was often a frequent occurrence came from a number of laboratories using several immunological techniques (Brand and Syverton, 1960; Coombs *et al.*, 1961; Stulberg *et al.*, 1961; Franks *et al.*, 1963; Simpson and Stulberg, 1963; Greene *et al.*, 1964, 1966; Franks and Gurner, 1965) and cytogenetic procedures (Defendi *et al.*, 1960).

The problems related to such interspecies contamination of cell cultures have been known for some time and several immunological and karyological techniques have been widely used to detect such mishaps. That such cross-contaminations of cells with cells still occur today is well illustrated in a recent publication (Herrick *et al.*, 1970) in which a cell line designated guinea pig spleen (GPS) originated at Fort Detrick and had been disseminated to a number of laboratories. A combination of serological and karyological data revealed that the line actually consisted of L-M mouse cells rather than GPS cells.

For a long time many investigators suspected similar intraspecies contaminations among human cell culture. But intraspecies identification until recently has not been as convincing (Merchant et al., 1966; Franks, 1966; Brand and Chiu, 1966) as interspecies identification except for euploid cell lines containing chromosome markers and biochemical deficiencies (Harris, 1964; Krooth, 1965; Hayflick and Moorhead, 1961).

In 1966, tools became available to study the extent of such intraspecies contamination. Gartler (1967) reported that 18 established human cell lines derived independently from Caucasian and Negro subjects contained the A isoenzyme of glucose-6-phosphate dehydrogenase (G-6-PD) and type 1 isoenzyme of phosphoglucomutase (PGM). The A isoenzyme (fast moving) of G-6-PD is found in 30% of the American Negro population, but not in Caucasians who have the B isoenzyme (slow moving) while the type 1 PGM isoenzyme is found in 65% of both Negro and Caucasian populations. Since HeLa was the first human epithelial-like cell line established, and since it was derived from a Negro, Gartler concluded that most likely the other 17 established cell lines were HeLa cell contaminants.

In a follow-up study (Gartler, 1968), identical findings were reported for 20 heteroploid human cell lines established between 1952 and 1957. Conversely, recently isolated diploid or near diploid human cell lines, as well as 100 primary tumor cultures, contained isoenzyme patterns consistent with their racial phenotype. Furthermore, Mycoplasma-contaminated, SV40-inoculated, X-irradiated, or other treated cultures did not undergo alterations in their G-6-PD or PGM isoenzyme patterns.

These findings of Gartler have been confirmed and extended by three of the participating laboratories of the Animal Cell Culture Collection (Peterson et al., 1968; De Oca et al., 1969; Coriell, 1968; Stulberg et al., 1970). These three laboratories independently studied cells in the collection as well as other cell populations and found that the A and B types of G-6-PD occurred in cultured human diploid cell populations in accordance with the racial origin of the donor, while the heteroploid cell lines established during the 1950's possessed A type G-6-PD irrespective of donor race.

As has been pointed out (Peterson et al., 1968), one should not conclude that because heteroploid cell lines were contaminated by HeLa cells, all observed spontaneous transformations of human cells in vitro arose from cross-contamination with HeLa cells. Such HeLa cell contaminations may have taken place after the spontaneous transformation of human cells had occurred. This is one possible explanation for the latest reported results (Fraley and Ecker, 1970) on a 4-year-old established cell line MA 160 which originated from a spontaneous transformation of adult human prostatic epithelium. It was found (De Oca et al.,

1969) that the MA 160 had the rapidly migrating A component of G-6-PD, even though it arose from a Caucasian male and still retained the Y chromosome. Other possible explanations for the discrepancy were that the patient had Negro ancestry or the A band of G-6-PD may eventually be found in all cells carried for long-term culture (Fraley and Ecker, 1970). The latter argument implies an instability in the G-6-PD genetic marker in which the type B isoenzyme changes to type A possibly as a result of transformation from diploid to heteroploid karotype. However the recently reported studies of Auersperg and Gartler (1970) and of Peterson et al. (1971) on the establishment of two heteroploid cell lines from malignant tissues with the type B isoenzyme and its retention over 41 months (C-12R) and 36 months (Det 562) cultivation in vitro appears to counter the instability argument.

Conversely the development of type B heteroploid cell lines does not rule out the possibility that mutations affecting G-6-PD isoenzyme could have occurred in the past under cultivation techniques that are different from those being used today. Differences in medium, in serum, and in culture techniques all have been known to cause differences in cell expression. In this respect, it may be significant that the Det 562 was cultured in a manner and medium different from that in which earlier lines originated (Peterson et al., 1971). Yet, the stability of the G-6-PD type B isoenzyme to irradiation (Auersperg and Gartler, 1970) suggests that the sensitivity to mutation of the G-6-PD allele should be considerably greater than that brought about by changes in medium, serum, or techniques (Peterson et al., 1971).

Perhaps a more convincing argument that the controversy about a mutational change from G-6-PD type B to type A may still not be completely resolved is the fact that both the C-12R and the Det 562 established cell lines arose from malignant tissues and thus, the diploid to heteroploid transformation took place in vivo prior to cultivation in vitro and may therefore not be strictly comparable to transformation occurring in vitro (Auersperg and Gartler, 1970). Work now in progress in several laboratories (Stulberg et al., 1970) using these newer techniques of enzymatic analysis to monitor cells over extended periods of culture should satisfactorily answer this question in the near future.

Speculation about the source of cross-contamination of cell cultures by cells has not led to any adequate explanations. Human errors of incorrect labeling or mixing pipettes, or using the same medium for more than one cell culture type are the obvious ways such contamination can occur. Most workers, however, reject such possibilities occurring in their laboratories and feel that human errors per se do not convincingly explain the frequent occurrence of contamination in the past. Coriell (1962) demonstrated droplet infection by forcefully blowing HeLa or Detroit 6 from a pipette and catching the cells in a nearby open petri dish of culture medium containing monkey kidney cells. After three subcultures the monkey kidney cells were almost completely replaced with HeLa or Detroit 6 cells. In other tests Coriell (1962) used an atomizer to nebulize a HeLa culture into small droplet nuclei which were collected in open petri dishes held 1 foot away. Viable HeLa cells were obtained in this manner but Coriell (1962) was unsuccessful in recovering nebulized HeLa cells from room air by drawing small samples of air through culture media. There is, therefore, no evidence that aerosolization is the cause of such cell contaminations. We are left with the conclusion that human errors including mislabeling and haphazard cell culture techniques are probably the main cause of cross-contamination of cells.

B.   Detection of Contamination by Cells

1.   Serological

Several immunological methods are used to identify the species of origin of cell cultures.

a. Immunofluorescence. Antisera against known primary cultures, cell lines, and established cell lines of several animal species are prepared by immunizing guinea pigs or rabbits with washed cells mechanically removed from monolayer cultures. The gamma globulins obtained from such antisera by salt precipitation are then conjugated with fluorescein isothiocyanate and purified by gel filtration.

In practice, the test cells grown as monolayers are dispersed with 0.25% trypsin solution, washed three times with buffered saline at pH 7.0, and brought to 2% concentration by volume in buffered saline at pH 7.4. One-tenth milliliter of cell suspension is mixed with 0.1 ml of labeled antibody from the different species in wells of nonwettable plastic slide plates. Plates are incubated at room temperature in a humidity chamber or a rotary shaker and agitated for 30 minutes. The cell suspensions are then washed three times with buffered saline and one drop of the suspension is placed on a slide, covered with #1 cover slip, gently pressed with blotting paper, and sealed with clear fingernail lacquer. Preparations are examined within 2 hours by fluorescence microscopy using a BG-12 pass filter and an OG-1 barrier filter (Stulberg *et al.*, 1961). A surface fluorescence can be visually observed at the periphery of reacting cells. Proper positive and negative controls must be used to ensure a valid test. The advantage of this procedure is that in addition to determining species identity of cell cultures, it is also valuable in detecting the presence of only a few individual cells from a different species in a mixed population (low level contamination) (Stulberg and Simpson, 1973).

b. *Cytotoxic Antibody Test.* Antisera against known primary cultures, cell lines, and established cell lines of various animals, fish, and amphibian species are prepared by immunizing rabbits with living cells washed two to five times in Hank's balanced salt solution (Greene *et al.*, 1964, 1966). All sera are inactivated at 56°C for ½ hour before use.

Viable test cells in 1 ml of appropriate media (Eagle's or other) are placed in Wasserman tubes with 0.1 ml of each of the antisera. The cell-antisera mixture is incubated for 20 minutes at 37°C. Then 0.1 ml of guinea pig complement is added to each tube and the mixtures reincubated for 1 to 3 hours. A sample of 0.5 ml is removed and mixed with 0.1 ml of 0.25% trypan blue and the relative percentage of viable (nonstained) and nonviable (stained) cells is determined by counting duplicate samples in a hemocytometer. Cell suspensions are not used unless controls contain more than 80% viable cells. Death of 50% or more of the cells, as shown by trypan blue uptake, is considered evidence of specific immune cytotoxicity. This is a relatively simple, rapid test to determine the species of a given cell culture which most laboratories can carry out providing the different species antisera are available.

c. *Immunodiffusion Test.* The standard agar gel microimmunodiffusion method is employed in which melted 2% agar containing 0.85% sodium chloride is permitted to harden on slides (Crowle, 1958). Then, wells or depressions are cut out of the agar with the proper instrument. Extracts of cell cultures being tested are placed in the center well, and antisera (prepared as in Section VI,B,2) of the different animal species are placed in peripheral wells. The cell extracts are prepared by growing cells in monolayers, harvesting, washing in saline three times, and centrifuging the cells. Then, the cell pellets are frozen and thawed three times and used as antigens (Greene *et al.*, 1972). White lines of precipitate appear at the zones of optimal concentration between diffusing cell antigens and specific antibodies that have mutual reactivity. The agar gel immunodiffusion is an effective test for species identification of various poikilothermic and insect cell lines (Greene *et al.*, 1969, 1972; Green and Charney, 1973).

d. *Indirect Hemagglutination Test.* Antiserum against the test cells is prepared by immunizing a guinea pig. The serum obtained from bleeding the guinea pig is heated to 56°C for ½ hour. Two-tenths milliliter of the serial twofold diluted serum is mixed with 0.5-ml amounts of 0.3 to 0.5% red cell suspensions of various animal species in wells of plastic trays. After 4 to 6 hours incubation at 37°C with gentle agitation at half-hour intervals the wells or tubes are examined for hemagglutination. The titer is read as the highest serum dilution giving macroscopically visible aggregates with corresponding homologous erythocytes. Proper controls of antisera prepared against cell cultures of various known animal species must be used in each assay. In some cases, sera must be appropriately absorbed to remove nonspecific hemagglutinating factors. This system for identifying species or origin of cell cultures is an effective one and was the first immunological technique to unequivocally prove the interspecies contamination of cell cultures (Brand and Syverton, 1959, 1960). Seventeen cell lines from a number of laboratories were found to be incorrectly labeled. Cell lines from monkey, rabbit, swine, calf, hamster, or duck tissues were either mouse or human in hemagglutination properties (Brand and Syverton, 1962). However the time-consuming procedure of requiring the preparation of an antiserum against each test culture as well as the need for red cells of several animal species has prevented the use of this method as a routine procedure. Other immunological procedures have been more useful for routine assays since previously prepared antisera can be employed.

   e. Mixed Agglutination Test. This is another immunological test like the indirect hemagglutination test which depends on the presence of common species-specific antigens on the surface of red blood cells and tissue cells. In practice, antisera are prepared against red cells from several species usually in rabbits and are subsequently suitably absorbed to render the antisera selective in action. Two drops of the test cells washed twice and properly diluted with diluent (0.02% crystalline bovine plasma albumin in 0.9 sodium chloride) so that there are approximately 20 cells/ $40\times$ microscopic field are placed in siliconized 6×60 mm tubes. Two drops of anti-erythrocyte serum from each species diluted 1:10 with diluent are added to the cells and incubated for 1 hour at $28°C$ with frequent agitation. Then the tubes are centrifuged at $150g$ and the cells washed three times with diluent. Cells are resuspended in 1 drop of diluent, and 1 drop of 0.5% red cell suspension prepared in Alsever's solution from various species is added. Tubes are centrifuged for 2 minutes at $150g$ and a drop of each cell mixture is examined microscopically using a $10\times$ ocular and $40\times$ objective (preferably phase contrast) for evidence of mixed agglutination. A cluster of red cells surrounding each tissue cell is a positive reaction indicating common antigens between the red cell and tissue cell and therefore indicating that both cell types arose from the same species (Coombs *et al.*, 1961; Franks *et al.*, 1963). Proper positive and negative controls are necessary to ensure a valid test. This mixed agglutination test has other applications besides identifying species of origin of cell cultures. For example it is being used to study various blood group antigens on tissue cell surfaces. In detecting species of origin it has the drawback that fresh red cells from various species are needed as well as previously prepared antisera against those various types of red cells.

2. CHROMOSOMAL ANALYSIS

The use of chromosomal analysis to identify the species origin of a given cell culture was the first method used (Rothfels *et al.*, 1959) and is still a useful method in the hands of a competent cytologist. There are a number of techniques used to prepare chromosome spreads of cell cultures. In Rothfels and Siminovitch's (1958) procedure, cells are grown on cover slips for 24 to 48 hours and colchicine (1/40,000) or Colcemid (0.6 μg/ml) solution is added 6 to 8 hours before harvesting to accumulate cells at metaphase and condense the chromosomes. The cover slips are transferred to warm hyptotonic saline solution (1 part Hank's BSS to 9 parts distilled water) for 15 minutes for fibroblastic-like and 45 minutes for epithelial-like cells (swells the cells and prevents spindle formation). The cells are then fixed using acetic-alcohol (1 part glacial acetic acid to 3 parts ethanol) for 10 minutes. The cover slips are then left to air dry at room temperature. The chromosomes can be stained by various techniques including Feulgen method, aceto-orcein, Giesma, or Wright. The aceto-orcein method is usually used for chromosome studies. Stain is prepared by dissolving 2 gm natural orcein (G. T. Gurr, London) into 100 ml of the 50% glacial acetic acid. A drop or two of this stain is added to a slide and the cover slip immediately mounted over the stain. Sealing the edge of the cover slip with petroleum jelly prevents evaporation and preserves the preparation for a week or so. For enumeration of chromosomes, the use of Wright stain provides a simple and permanent preparation (P. P. Ludovici, unpublished data). The air-dried slide is flooded with Wright stain which is allowed to stand for 3 minutes. An equal amount of Wright stain buffer is added (KH_2PO_4 and Na_2HPO_4), the slide is blown gently to mix, allowed to stand for 4 minutes, and then washed quickly and completely with distilled water. It is then permitted to dry by standing on end and mounted with Permount.

Some individuals prefer to work with cells in suspension rather than as a monolayer on cover slips because the fixatives and other fluids penetrate the cells more evenly and rapidly. In this procedure cells are harvested after treatment with colchicine using trypsin or mechanical scrapings and are sedimented by centrifuge. Then they are resuspended in hypotonic saline for the times previously mentioned, recentrifuged, and resuspended in acetic-alcohol fixative solution. A drop of the cell suspension is allowed to air dry on a slide and is stained by one of the methods previously described. The preparations are examined with light microscopy first using a 10× objective to select cells with well-spread chromosomes and then using oil immersion for enumeration and analysis.

A certain amount of experience with the procedures is required before reliable results can be obtained. Adequate control material which has been previously characterized should be used to check techniques. The distribution of metacentrics, telocentrics, minutes, etc., between known control and test cultures should serve to characterize unknown cell populations.

3. Isoenzymatic Analysis

Isoenzymes are variant forms of enzymes which differ in their electrophoretic mobilities yet remain unchanged in their activities and substrate specificities. Many of these variant forms of enzymes are genetically determined and apparently the incidence of isoenzyme polymorphisms in several species is quite high. Such isoenzyme polymorphisms are well expressed in cell culture populations and therefore, as first shown by Gartler (1967), can be used to identify the origin of cells which arise from different individuals within the species.

a. Preparation of Cell Extracts. Harvested test cells are washed three to four times with 0.9% NaCl solution or saline solution containing $6.6 \times 10^{-4} M$ ethylenediamine tetraacetate and suspended in the same solution to 1×10^7 cells/ml. The cells are frozen and thawed three times using either liquid nitrogen or a methanol-dry-ice bath. The crude cell extract is centrifuged at 20,000g for one-half hour at 4°C. The supernatant containing the enzyme activity is removed and tested or stored frozen at $-65°C$ or lower until used.

b. Electrophoresis. The vertical starch gel technique described by Kirkman and Henderickson (1963) is usually used (Gartler, 1967; De Oca et al., 1969). Peterson et al. (1968) used cellulose acetate strips, gelatinized cellulose acetate strips, or sucrose-agar gel in barbitol buffer at pH 8.8 (G. K. Turner, Agarose Universal Electrophoresis Film). The sucrose-agar gel procedure provided higher resolution of the A and B bands of glucose-6-phosphate dehydrogenase (G-6-PD). Coriell (1968) used acrylimide gel with equal success for this enzyme. Electrophoresis is usually conducted in a chamber with a power supply delivering 6 mA for 90 minutes.

The development or staining procedure is generally the one used by Rattazzi et al. (1967). It is conducted in a high humidity chamber at 37°C for 15 minutes to 1 hour depending on the activity of the extracts.

Gels are stained by inverting them into a staining dish containing the staining reagent. For G-6-PD the reagent consists of 2 mg of phenazine methosulfate and 5 ml each of nitroblue tetrazolium (2 mg/ml), $0.025 M$ glucose-6-phosphate, $0.005 M$ nicotinamide adenine dinucleotide phosphate and $0.1 M$ $MgCl_2$. The reaction is stopped and background staining removed by placing the gels in a 1:40 solution of formaldehyde (Peterson et al., 1968). Gels are dried by incubation for 1 hour at 60°C and can be stored without color loss for approximately 6 months (Peterson et al., 1968).

Other isoenzyme polymorphisms besides G-6-PD that have been studied for the characterization and identification of cell cultures include phosphoglucomutase (PGM) (Gartler, 1967, 1968) and lactate dehydrogenase (LDH) (De Oca et al., 1969). No doubt others are now being investigated and eventually such isoenzyme polymorphisms should permit a concise "fingerprinting" of each new and old cell culture.

C. Prevention of Contamination by Cells

The adherence to rigid aseptic procedures would probably eliminate most of the incidences of cross-contamination of cells by cells of the same or other species. Extreme care should be exercised in the use of proper labeling techniques in the laboratory so that errors in labeling do not occur. This is especially true in laboratories where more than one cell line or established cell line is being cultivated. Also, under such circumstances it would be advisable to use separate batches of medium as well as other reagents including trypsin for each cell type and to wipe off the laboratory bench with alcohol or disinfectant when switching from one cell type to another. Another precaution to take is to refilter (0.20- or 0.22-μ membrane) all commercial reagents including sera.

Slow freezing of the cell line or established cell line should be initiated as early as possible to provide preserved seed stocks of cultures in case later contamination problems should arise. Storage at both $-65°C$ or lower (freezer) and at $-200°C$ (liquid nitrogen) should be used. Periodic checks of cultures in use should be carried out by immunological, karyological, and/or isoenzymatic analysis to routinely monitor for cross-contamination by cells.

VII.   Summary

The properties of mycoplasma, viruses, bacteria, fungi, parasites, and tissue cells are considered in relation to the contamination problems they pose for workers who use cell cultures. All of these agents are found as contaminants in cell cultures and in most cases their presence remains undetected unless special efforts are made to unmask them. The simplest method of unmasking bacteria and fungi is to remove antibiotics from the cell culture medium. Indirectly this also aids in preventing contamination with mycoplasma, parasites, and tissue cells because omission of antibiotics forces workers to follow a regime of strict aseptic techniques.

The methods used to detect each type of cell contaminant are reviewed and recommendations are made on the procedures to follow to prevent and where possible eradicate such contaminations. Attention is drawn to the possible misinterpretation of results if one uses such contaminated cell cultures.

Investigators are advised to initiate their work with reference cultures that have been certified by the Cell Culture Committee. These are obtainable from the Repository at the American Type Culture Collection, 12301 Parklawn Drive, Rockville, Maryland 20852. The reference cultures are preserved by slow freezing at controlled rates and stored in liquid nitrogen. Such cultures are tested and shown to be free of mycoplasmas, bacteria, fungi, protozoa, and cytopathic viruses (Stulberg et al., 1970). The presence or absence of virus-like particles by electron microscopy are also recorded when appropriate and the species of origin are verified by testing with specific antisera using one or more tests including: mixed agglutination, fluorescent antibody reaction, indirect hemagglutination, cytotoxic-antibody dye exclusion, and agar gel immunodiffusion. Karyotype and isoenzyme analyses are also used in conjunction with the immunological reactions (Stulberg et al., 1970).

Ideally, if investigators begin their research with contaminant-free reference cultures of this type, slow freeze a batch of seed stock cultures early in the project, routinely test their cultures for the presence of these contaminants throughout the research, pretest all reagents and sera for such contaminants before use, and follow a regime of rigid aseptic procedures, then such contaminations should be few in number and, when they occur, they can be handled with little loss in time to the research program. Finally it is recommended that workers include in all publications using cell cultures, the methods used to check for contaminations and the results obtained.

REFERENCES

Ackerman, W. W., and Kurtz, H. (1955). *J. Exp. Med.* **102**, 555.

Armstrong, J. A., and Pereira, M. S. (1967). *Brit. Med. J.* **1**, 212.

Auersperg, N., and Gartler, S. M. (1970). *Exp. Cell Res.* **61**, 465.

Aurelian, L., Strandberg, J. D., Melendez, L. V., and Johnson, L. A. (1972). *Science* **174**, 704.

Bang, F. B., and Gey, G. O. (1951). *Trans. N. Y. Acad. Sci.* [2] **13**, 324.

Barile, M. F. (1962). *Nat. Cancer Inst., Monogr.* **7**, 50.

Barile, M. F. (1966). *Nat. Cancer Inst., Monogr.* **29**, 201.

Barile, M. F. (1967). *Ann. N. Y. Acad. Sci.* **143**, 557.

Barile, M. F. (1968). *Nat. Cancer Inst., Monogr.* **29**, 201.

Barile, M. F. (1973). *In* "Cell Culture Contamination" (J. Fogh, ed.). Academic Press, New York (in press).

Barile, M. F., and Del Giudice, R. A. (1972). *Pathogenic Mycoplasma, Ciba Found. Symp.*

Barile, M. F., and Kern, J. (1971). *Proc. Soc. Exp. Biol. Med.* **138**, 432.

Barile, M. F., and Schimke, R. T. (1963). *Proc. Soc. Exp. Biol. Med.* **114**, 676.

Barile, M. F., Malizia, W. F., and Riggs, D. B. (1962). *J. Bacteriol.* **84**, 130.

Barile, M. F., Del Giudice, R. A., Carski, T. R., Yamashiroya, H. M., and Verna, J. A. (1970). *Proc. Soc. Exp. Biol. Med.* **134**, 146.

Blacklow, N. R., Hoggan, M. D., and Rowe, W. P. (1967). *Proc. Nat. Acad. Sci. U. S.* **58**, 1410.

Boone, C. W., Mantel, N., Caruso, T. D., Kazam, E., and Stevenson, R. E. (1972). *In Vitro* **7**, 174.

Brand, K. G., and Chiu, S. Y. (1966). *Nature (London)* **212**, 44.

Brand, K. G., and Syverton, J. T. (1959). *Proc. Amer. Ass. Cancer Res.* **3**, 8.

Brand, K. G., and Syverton, J. T. (1960). *J. Nat. Cancer Inst.* **24**, 1007.

Brand, K. G., and Syverton, J. T. (1962). *J. Nat. Cancer Inst.* **28**, 147.

Brown, A., and Officer, J. E. (1968). *In* "Methods in Virology" (K. Maramorosch and H. Koprowski, eds.), Vol. 4, pp. 531–564. Academic Press, New York.

Brown, A. L. (1968). *Nat. Cancer Inst., Monogr.* **29**, 523.

Brown, T. M., Swift, F., and Watson, R. F. (1940). *J. Bacteriol.* **40**, 857.

Brownstein, B., and Graham, A. F. (1961). *Virology* **14**, 303.

Burki, F., and Germann, E. (1964). *Berlin. Muenchen. Tieraeztzl. Wochenschr.* **16**, 324; **17**, 333.

Casemore, D. P. (1969). *J. Clin. Pathol.* **22**, 254.

Castellani, A. (1930). *J. Trop. Med.* **33**, 188.

Casto, B. C., Atchison, R. W., and Hammon, W. McD. (1967). *Virology* **32**, 52.

Cerva, L., Novak, K., and Culbertson, C. G. (1968). *Amer. J. Epidemiol.* **88**, 436.

Chang, R. S. (1961). *Proc. Soc. Exp. Biol. Med.* **107**, 135.

Chang, S. L. (1971). *Curr. Top. Pathobiol.* **1**, 241.

Chi, L., Vogel, J. E., and Shelokov, A. (1959). *Science* **130**, 1763.

Christian, R. T., Ludovici, P. P., Miller, N. F., and Riley, G. M. (1965). *Amer. J. Obstet. Gynecol.* **91**, 430.

Clark, H. W., Bailey, J. S., Fowler, R. C., and Brown, T. McP. (1963). *J. Bacteriol.* **85**, 111.

Clyde, W. A., Jr. (1964). *J. Immunol.* **92**, 958.

Coombs, R. A. A., Daniel, M. R., Gurner, B. W., and Kelus, A. (1961). *Immunology* **4**, 55.

Coons, A. J. (1958). *Gen. Cytochem. Methods* **1**, 399.

Coriell, L. L. (1962). *Nat. Cancer Inst., Monogr.* **7**, 33.

Coriell, L. L. (1968). *Nat. Cancer Inst., Monogr.* **29**, 179.

Coriell, L. L., and McGarrity, G. J. (1968). *Appl. Microbiol.* **16**, 1895.

Coriell, L. L., and McGarrity, G. J. (1970). *Appl. Microbiol.* **20**, 474.

Coriell, L. L., Tall, M., and Gaskill, H. (1958). *Science* **128**, 198.

Crandell, R. A., and Despeaux, E. W. (1959). *Proc. Soc. Exp. Biol. Med.* **101**, 494.

Crowle, A. J. (1958). *J. Lab. Clin. Med.* **52**, 784.

Culbertson, C. G. (1961). *Amer. J. Clin. Pathol.* **35**, 195.

Culbertson, C. G. (1968). *J. Protozool.* **15**, 353.

Culbertson, C. G. (1971). *Annu. Rev. Microbiol.* **25**, 231.

Culbertson, C. G., Smith, J. W., and Minner, J. R. (1958). *Science* **127**, 1506.

Culbertson, C. G., Holmes, D. H., and Overton, W. M. (1965). *Amer. J. Clin. Pathol.* **43**, 361.

Culbertson, C. G., Ensminger, P. W., and Overton, W. M. (1966). *Amer. J. Clin. Pathol.* **46**, 305.

Defendi, V., Billingham, R. E., Silvers, W. K., and Moorhead, P. F. (1960). *J. Nat. Cancer Inst.* **25**, 359.

Del Giudice, R. A., Robillard, N. F., and Carski, T. R. (1967). *J. Bacteriol.* **93**, 1205.

De Oca, F. M., Macy, M. L., and Shannon, J. E. (1969). *Proc. Soc. Exp. Biol. Med.* **132**, 462.

Douglas, M. (1930). *J. Trop. Med.* **33**, 258.

Duff, J. T. (1968). *Nat. Cancer Inst., Monogr.* **29**, 146.

Dunnebacke, T. H., and Williams, R. C. (1967). *Proc. Nat. Acad. Sci. U. S.* **57**, 1363.

Edward, D. G. ff., and Fitzgerald, W. A. (1954). *J. Pathol. Bacteriol.* **68**, 23.

Edward, D. G. ff., and Freundt, E. A. (1969). *In* "The Mycoplasmatales and the L-Phase of Bacteria" (L. Hayflick, ed.), pp. 147–200. Appleton, New York.

Edward, D. G. ff., and Freundt, E. A. (1970). *J. Gen. Microbiol.* **62**, 1.

Edward, G. A., and Fogh, J. (1960). *J. Bacteriol.* **79**, 267–276.

Fedoroff, S., Evans, V. J., Hopps, H. E., Sanford, K. K., and Boone, C. W. (1972). *In Vitro* **7**, 161.

Fogh, J. (1969). *Cancer Res.* **29**, 1721.

Fogh, J., ed. (1973). "Cell Culture Contamination." Academic Press, New York (in press).

Fogh, J., and Fogh, H. (1964). *Proc. Soc. Exp. Biol. Med.* **117**, 899.

Fogh, J., and Fogh, H. (1965). *Proc. Soc. Exp. Biol. Med.* **119**, 233.

Fogh, J., and Fogh, H. (1967a). *Proc. Soc. Exp. Biol. Med.* **125**, 423.

Fogh, J., and Fogh, H. (1967b). *Proc. Soc. Exp. Biol. Med.* **126**, 67.

Fogh, J., and Fogh, H. (1969). *Ann. N. Y. Acad. Sci.* **172**, 15.

Fogh, J., Holmgren, N. B., and Ludovici, P. P. (1971). *In Vitro* **7**, 26.

Fraley, E. E., and Ecker, S. (1970). *Science* **170**, 540.

Franks, D. (1966). *In Vitro* **2**, 74.

Franks, D., and Gurner, B. W. (1965). *Exp. Cell Res.* **37**, 693.

Franks, D., Gurner, B. W., Coombs, R. R. A., and Stevenson, R. E. (1963). *Exp. Cell Res.* **28**, 608.

Freed, J. J., and Schatz, S. A. (1969). *Exp. Cell Res.* **55**, 393.

Freeman, J. M., Magoffin, R. L., Lennette, E. H., and Herndon, R. M. (1967). *Lancet* **2**, 129.

Gartler, S. M. (1967). *Nat. Cancer Inst., Monogr.* **26**, 167.

Gartler, S. M. (1968). *Nature (London)* **217**, 750.

Gey, G. O., and Bang, F. B. (1951). *Trans. N. Y. Acad. Sci.* [2] **14**, 15.

Gillespie, J. H. (1968). *Nat. Cancer Inst., Monogr.* **29**, 133.

Girardi, A. J., Hamparian, V. V., Somerson, N. L., and Hayflick, L. (1965a). *Proc. Soc. Exp. Biol. Med.* **120**, 760–771.

Girardi, A. J., Hayflick, L., Lewis, A. M., and Somerson, N. L. (1965b). *Nature (London)* **205**, 188.

Gori, G. B., and Lee, D. Y. (1964). *Proc. Soc. Exp. Biol. Med.* **117**, 918.

Greene, A. E., and Charney, J. (1973). *In* "Methods and Applications of Tissue Culture" (P. Kruse and M. D. Patterson, eds.). Academic Press, New York (in press).

Greene, A. E., Coriell, L. L., and Charney, J. (1964). *J. Nat. Cancer Inst.* **32**, 779.

Greene, A. E., Goldner, H., and Coriell, L. L. (1966). *Growth* **30**, 305.

Greene, A. E., Coriell, L. L., and Charney, J. (1969). *Recent Results Cancer Res.,* *Spec. Suppl.* pp. 112–120.

Greene, A. E., Charney, J., Nichols, W. W., and Coriell, L. L. (1972). *In Vitro* **7**, 313.

Hackett, A. J. (1973). *In* "Methods and Applications of Tissue Culture" (P. Kruse and M. K. Patterson, Jr., eds.). Academic Press, New York (in press).

Hackett, A. J., and Manning, J. S. (1971). *J. Amer. Vet. Med. Ass.* **158**, 948.

Hackett, A. J., Pfiester, A., and Arnstein, P. (1970). *Proc. Soc. Exp. Biol. Med.* **135**, 899.

Hakala, M. T., Holland, J. F., and Horoszewicz, J. S. (1963). *Biochem. Biophys.* *Res. Commun.* **11**, 466.

Hampton, R. O. (1972). *Annu. Rev. Plant Physiol.* **23**, 389.

Hargreaves, F. D., and Leach, R. H. (1970). *J. Med. Microbiol.* **3**, 259.

Harris, M. (1964). *In* "Cell Culture and Somatic Variation," p. 547. Holt, New York.

Hayflick, L. (1960). *Nature (London)* **185**, 783.

Hayflick, L. (1965). *Tex. Rep. Biol. Med.* **23**, 285.

Hayflick, L. (1968). *Nat. Cancer Inst., Monogr.* **29**, 83.

Hayflick, L. (1972). *Pathogenic Mycoplasma, Ciba Found. Symp.* pp. 1–14.

Hayflick, L., and Moorhead, P. F. (1961). *Exp. Cell Res.* **25**, 585.

Hayflick, L., and Stanbridge, E. J. (1967). *Ann. N. Y. Acad. Sci.* **143**, 608.

Hederscheê, D., Ruys, A. C., and Van Rhijn, G. R. (1963). *Antonie van Leeuwenhoek; J. Microbiol. Serol.* **29**, 368.

Henle, G., and Henle, W. (1965). *J. Bacteriol.* **89**, 252.

Hennessen, W. (1968). *Nat. Cancer Inst., Monogr.* **29**, 161.

Herrick, P. R., Baumann, G. W., Merchant, D. J., Shearer, M. C., Shipman, C., Jr., and Brackett, R. G. (1970). *In Vitro* **6**, 143.

Hoggan, M. D., Blacklow, N. R., and Rowe, W. P. (1966). *Proc. Nat. Acad. Sci.* *U. S.* **55**, 1467.

Holland, J. F., Korn, R., O'Malley, J., Minnemeyer, H. J., and Tieckelmann, H. (1967). *Cancer Res.* **27**, 1867.

Hsiung, G. D. (1968a). *Nat. Cancer Inst., Monogr.* **29**, 351.

Hsiung, G. D. (1968b). *Bacteriol. Rev.* **32**, 185.

Hsiung, G. D., Fischman, H. R., Fong, C. K. Y., and Green, R. H. (1969). *Proc.* *Soc. Exp. Biol. Med.* **130**, 80.

Huebner, R. J. (1967). *Proc. Nat. Acad. Sci. U. S.* **58**, 835.

Hull, R. N. (1968). *Nat. Cancer Inst., Monogr.* **29**, 471 and 503.

Hull, R. N., Minner, J. R., and Smith, J. W. (1956). *Amer. J. Hyg.* **63**, 204.

Hull, R. N., Minner, J. R., and Mascoli, C. C. (1958) *Amer. J. Hyg.* **68**, 31.

Hull, R. N., Johnson, I. S., Culbertson, C. G., Reimer, C. B., and Wright, H. F. (1965). *Science* **150**, 1044.

Jacobson, W., and Webb, M. (1952). *Exp. Cell Res.* **3**, 163.

Jahnes, W. G., Fullmer, H. M., and Li, C. P. (1957). *Proc. Soc. Exp. Biol. Med.* **96**, 484.

Jezequel, A., Shreeve, M., and Steiner, J. W. (1967). *Lab. Invest.* **16**, 287.

Kaklamanis, E., Thomas, L., Stavropoulos, K., Borman, I., and Boshwitz, C. (1969). *Nature (London)* **221**, 860.

Kalter, S., and Heberling, R. (1968). *Nat. Cancer Inst., Monogr.* **29**, 149.

Kelton, W. H. (1962). *J. Bacteriol.* **83**, 948.

Kenny, G. E., and Pollock, M. E. (1963). *J. Infec. Dis.* **112**, 7.

Kingston, D., and Warhurst, D. C. (1969). *J. Med. Microbiol.* **2**, 27.

Kirkman, H. N., and Henderickson, E. M. (1963). *Amer. J. Hum. Genet.* **15**, 241.

Kniazeff, A. J. (1968). *Nat. Cancer Inst., Monogr.* **29**, 123.

Kniazeff, A. J., Rimer, V., and Gaeta, L. (1967). *Nature (London)* **214**, 805.

Krooth, R. S. (1965). *In* "Birth Defects Original Article Series: New Directions in Human Genetics" (D. Bergsma, ed.), Vol. 1, pp. 21–56. National Foundation of the March of Dimes.

Kruse, P., and Patterson, M. K., Jr. (1973). "Methods and Applications of Tissue Culture" (P. Kruse and M. K. Patterson, Jr., eds.). Academic Press, New York (in press).

Kudo, R. R. (1966). "Protozoology," 5th ed. Thomas, Springfield, Illinois.

Kundsin, R. B. (1968). *Appl. Microbiol.* **16**, 143.

Levashov, V. S., and Tsilinskii, Y. Y. (1964). *Tr. Inst. Poliomielita Virusn Entsetalitov.* **5**, 150; for translation, see *Ref. Zh., Biol.* No. 20B70 (1965).

Levine, E. M., Burleigh, I. G., Boone, C. W., and Eagle, H. (1967). *Proc. Nat. Acad. Sci. U. S.* **57**, 431.

Levine, E. M., Thomas, L., McGregor, D., Hayflick, L., and Eagle, H. (1968). *Proc. Nat. Acad. Sci. U. S.* **60**, 583.

Lille, R. D. (1954). "Histopathologic Technic and Practical Histochemistry," 2nd ed. McGraw-Hill (Blakiston), New York.

Ludovici, P. P., and Christian, R. T. (1969). *Proc. Soc. Exp. Biol. Med.* **131**, 301.

Luginbuhl, R. E. (1968). *Nat. Cancer Inst., Monogr.* **29**, 109.

McGarrity, G. J., and Coriell, L. L. (1971). *In Vitro* **6**, 257.

McGee, Z. A., Rogul, M., and Wittler, R. G. (1967). *Ann. N. Y. Acad. Sci.* **143**, 21.

Macpherson, I. (1966). *J. Cell Sci.* **1**, 145.

Macpherson, I., and Russell, W. (1966). *Nature (London)* **210**, 1343.

Malizia, W. F., Barile, M. F., and Riggs, D. B. (1961). *Nature (London)* **191**, 190.

Malmquist, W. A. (1968). *J. Amer. Vet. Med. Ass.* **152**, 763.

Maniloff, J., and Morowitz, H. J. (1967). *Ann. N. Y. Acad. Sci.* **143**, 59.

Markov, G. G., Bradvarova, I., Mintcheva, A., Petrov, P., Shishkov, N., and Tsanev, R. G. (1969). *Exp. Cell Res.* **57**, 374.

Marmion, B. P., and Goodburn, G. M. (1961). *Nature (London)* **189**,, 247.

Melnick, J. L. (1968). *Nat. Cancer Inst., Monogr.* **29**, 337.

Merchant, D. J. (1968). *Nat. Cancer Inst., Monogr.* **29**, 1–589.

Merchant, D. J., Gangal, S. J., and Holmgren, N. B. (1966). *In Vitro* **2**, 97.

Molander, C. W., Paley, A., Boone, C. W., Kniazeff, A., and Imagawa, D. T. (1968). *In Vitro* **4**, 148.

Molander, C. W., Kniazeff, A., Paley, A., and Imagawa, D. T. (1969). *20th Annu. Meet. Tissue Cult. Ass.* p. 9.

Molander, C. W., Kniazeff, A. J., Boone, C. W., Paley, A., and Imagawa, D. T. (1972). *In Vitro* **7**, 168.

Moore, A. E., and Hlinka, J. (1968). *J. Nat. Cancer Inst.* **40**, 569.

Morowitz, H. J., and Maniloff, J. (1966). *J. Bacteriol.* **91**, 1638.

Morowitz, H. J., Tourtellotte, M. E., and Pollock, M. E. (1963). *J. Bacteriol.* **85**, 134.

Nardone, R. M., Todd, J., Gonzales, P., and Gaffney, E. V. (1965). *Science* **149**, 1100.

Negroni, G. (1964). *Brit. Med. J.* **1**, 927.

Neimark, H. (1964). *Nature (London)* **203**, 549.

Neimark, H. (1967). *Ann. N. Y. Acad. Sci.* **143**, 31.

Neimark, H. (1970). *J. Gen. Microbiol.* **63**, 249.

O'Connell, R. C., Wittler, R. G., and Faber, J. E. (1964). *Appl. Microbiol.* **12**, 337.

Ogata, M., and Koshimizu, K. (1967). *Jap. J. Microbiol.* **11**, 289.

Parks, W. P., Melnick, J. L., Rongey, R., and Mayor, H. D. (1967). *J. Virol.* **1**, 171.

Paton, G. R., and Allison, A. C. (1970). *Nature (London)* **227**, 707–708.

Paul, J. (1970). *In* "Cell and Tissue Culture," 4th ed., pp. 152–158. Williams & Wilkins, Baltimore, Maryland.

Pereira, M. S., Marsden, H. B., Corbitt, G., and Tobin, J. O'H. (1966). *Brit. Med. J.* **1**, 130.

Perlman, D., Rahman, S. B., and Semar, J. B. (1967). *Appl. Microbiol.* **15**, 82.

Peterson, W. D., Jr., Stulberg, C. S., Swanborg, N. K., and Robinson, A. R. (1968). *Proc. Soc. Exp. Biol. Med.* **128**, 772.

Peterson, W. D., Jr., Stulberg, C. S., and Simpson, W. F. (1971). *Proc. Soc. Exp. Biol. Med.* **136**, 1187.

Pollock, M. E., and Kenny, G. E. (1963). *Proc. Soc. Exp. Biol. Med.* **112**, 176.

Pollock, M. E., Kenny, G. E., and Syverton, J. T. (1960). *Proc. Soc. Exp. Biol. Med.* **105**, 10.

Ponten, G., and Macpherson, I. (1966). *Ann. Med. Exp. Biol. Fenn.* **44**, 2260.

Powelson, D. M. (1961). *J. Bacteriol.* **82**, 288.

Purcell, R. H., Chanock, R. M., and Taylor-Robinson, D. (1969). *In* "The Mycoplasmatales and the L-Phase of Bacteria" (L. Hayflick, ed.), pp. 237–241. Appleton, New York.

Rakavskaya, I. V. (1965). *Vop. Virus.* **10**, 233.

Randall, C. C., Gafford, L. G., Gentry, G. A., and Lawson, L. A. (1965). *Science* **149**, 1098.

Rapp, F. (1968). *Nat. Cancer Inst., Monogr.* **29**, 405.

Rattazzi, M. C., Bernini, L. F., Fiorelli, G., and Mannucci, P. M. (1967). *Nature (London)* **213**, 79.

Rawls, W. E., and Melnick, J. L. (1966). *J. Exp. Med.* **123**, 795.

Rivers, T. M., Haagen, E., and Muckenfuss, R. S. (1929). *J. Exp. Med.* **50**, 673.

Robinson, L. B., Wichlhausen, R. H., and Roizman, B. (1956). *Science* **124**, 1147.

Romváry, J. (1965). *Acta Vet. (Budapest)* **15**, 341.

Rothblat, G. H., and Morton, H. E. (1959). *Proc. Soc. Exp. Biol. Med.* **100**, 87.

Rothfels, K. H., and Siminovitch, L. (1958). *Stain Technol.* **33**, 73.

Rothfels, K. H., Axelrad, A. A., Siminovitch, L., McCulloch, E. A., and Parker, R. C. (1959). *Proc. Can. Cancer Res. Conf.* **3**, 189.

Rouse, H. C., and Schlesinger, R. W. (1967). *Virology* **33**, 513.

Rouse, H. C., Bonifas, V. H., and Schlesinger, R. W. (1963). *Virology* **20**, 357.

Rowe, W. P. (1953). *Proc. Soc. Exp. Biol. Med.* **84**, 570.

Rowe, W. P. (1968). *Nat. Cancer Inst., Monogr.* **29**, 369.

Russell, W. C. (1966). *Nature (London)* **212**, 1537.

Russell, W. C., Niven, J., and Berman, L. (1968). *Int. J. Cancer* **3**, 191.

Schimke, R. T. (1967). *Ann. N. Y. Acad. Sci.* **143**, 573.

Shaw, C-M., Buchan, G. C., and Carlson, C. B. (1967). *N. Engl. J. Med.* **277**, 511.

Shepard, M. C. (1969). *In* "The Mycoplasmatales and the L-Phase of Bacteria" (L. Hayflick, ed.), pp. 49–65. Appleton, New York.

Simpson, W. F., and Stulberg, C. S. (1963). *Nature* (*London*) **199**, 616.

Singh, B. N. (1952). *Phil. Trans. Roy. Soc. London, Ser. B* **236**, 405.

Singh, B. N. (1960). *Proc. Indian Sci. Congr., 47th Sess.*, Part 2, p. 145.

Smith, K. O. (1968). *Nat. Cancer Inst., Monogr.* **29**, 371.

Smith, P. F. (1971). "The Biology of Mycoplasma," pp. 99–103, 162–171, and 225–229. Academic Press, New York.

Sobeslavsky, O., Prescott, B., and Chanock, R. M. (1968). *J. Bacteriol.* **96**, 695.

Somerson, N. L., and Cook, M. K. (1965). *J. Bacteriol.* **90**, 534.

Stanbridge, E. J. (1971). *Bacteriol. Rev.* **35**, 206–227.

Stanbridge, E. J., and Hayflick, L. (1967). *J. Bacteriol.* **93**, 1392.

Stanbridge, E. J., Onen, M., Perkins, F. T., and Hayflick, L. (1969). *Exp. Cell. Res.* **57**, 397.

Stanbridge, E. J., Hayflick, L., and Perkins, F. T. (1971). *Nature* (*London*) *New Biol.* **232**, 242.

Stock, D. A., and Gentry, G. A. (1969). *J. Virol.* **3**, 313.

Stulberg, C. S. (1973). *In* "Cell Culture Contamination" (J. Fogh, ed.). Academic Press, New York (in press).

Stulberg, C. S., and Simpson, W. F. (1973). *In* "Methods and Applications of Tissue Culture" (P. Kruse and M. K. Patterson, eds.). Academic Press, New York (in press).

Stulberg, C. S., Simpson, W. F., and Berman, L. (1961). *Proc. Soc. Exp. Biol. Med.* **108**, 434.

Stulberg, C. S., Coriell, L. L., Kniazeff, A. J., and Shannon, J. E. (1970). *In Vitro* **5**, 1.

Thomas, L. (1968). *Yale J. Biol. Med.* **40**, 444.

Thomas, L. (1969). *In* "Cellular Recognition" (R. T. Smith and R. A. Good, eds.), pp. 139–141. Appleton, New York.

Todaro, G. J., Zere, V., and Aaronson, S. A. (1970). *Nature* (*London*) **226**, 1077.

Todaro, G. J., Aaronson, S. A., and Rands, E. (1971). *Exp. Cell Res.* **65**, 257.

van Herick, W., and Eaton, M. D. (1945). *J. Bacteriol.* **50**, 47.

Wang, S. S., and Feldman, H. A. (1967). *N. Engl. J. Med.* **277**, 1174.

Wexler, S., Ludovici, P. P., and Moore, M. L. (1970). *Appl. Microbiol.* **20**, 309.

Williams, C. O., Wittler, R. G., and Burris, C. (1969). *J. Bacteriol.* **99**, 341.

Wilner, B. I. (1968). *Nat. Cancer Inst., Monogr.* **29**, 144.

Witter, R. L. (1968). *Nat. Cancer Inst., Monogr.* **29**, 119.

Zgorniak-Nowosielska, I., Sedwick, W. D., Hummeler, K., and Koprowski, H. (1967). *J. Virol.* **1**, 1227.

Chapter 7

Isolation of Mutants of Cultured Mammalian Cells

LARRY H. THOMPSON AND RAYMOND M. BAKER

The Ontario Cancer Institute and
Department of Medical Biophysics, University of Toronto,
Toronto, Canada

I. Introduction

The progress of molecular genetics has been firmly rooted in the technology for isolating mutations affecting diverse biological functions and for manipulating the mutant genes in order to study consequences of the changes. At present little is known about the molecular organization and modes of expression of the genetic material in mammalian cells. However, with the base of information provided by studies of microbial systems the vigorous application of existing genetic concepts to the cells of higher organisms should be rewarding. Isolation of mutant cells combined with the development of new genetic techniques should be invaluable in building an understanding of the fundamental aspects of eukaryotic cell growth and differentiation.

A *priori*, the probable redundancy of genetic information in cells of diploid origin implies that the observation of mutant phenotypes will be infrequent compared to the occurrence of mutant genotypes. One should be able to detect only mutations that are dominant or codominant, or recessive mutations that are observable due to inactivation by point mutation, deletion, or repression at the homologous locus (loci). This consideration and the experimental difficulty of manipulating mammalian cells in culture as compared with microorganisms have in the past undoubtedly discouraged attempts to isolate somatic cell mutants. In view of the increasing variety of lines of altered phenotype which are now being isolated, however, it is no longer clear that the problem of ploidy is a major obstacle to obtaining mutations affecting the whole range of cellular functions.

A. Criteria for Classification of Phenotype as Mutant

In the conventional sense of the word a "mutation" is a permanent hereditary alteration in the chromosomal genetic material (DNA), if we exclude from consideration extranuclear genes. Historically, the term "mutant" has often been used to describe cell lines of altered phenotype which have been isolated in culture or which have been induced *in vivo*, e.g., drug-resistant lines of tumor cells. However, this use of the term has usually been based solely on the criterion that the altered characteristic of such lines is a heritable property which appears to be stable. The actual nature of the heritable alterations has remained largely undefined because of the lack of appropriate methods for genetic analysis.

For somatic cells we apply the following criteria to define "mutant":

(1) The altered phenotype breeds true in the sense that it is stably transmitted through consecutive generations.

(2) The frequency of occurrence of the phenotype can be enhanced by the application of mutagenic agents.

(3) The phenotype can ordinarily be associated with an altered gene product (usually a protein).

(4) The phenotype can be attributed to a specific region of the genome, i.e., it can be mapped in a linkage group which behaves in a Mendelian manner.

Undoubtedly evidence of the latter type, concerning linkage, would be most unequivocal and thus most desirable. However, because adequate systems of recombination and segregation are not readily available in somatic cells, one must usually rely on the more indirect criteria. Due to the indirect nature of the evidence some doubt has been expressed [e.g., *Nature (London), New Biol.*, **234**, 226 (1971)] that heritable variation observed *in vitro* is indeed due to "classical mutation." First, it is often believed that altered phenotypes observed in somatic cells are unduly unstable; mutation rates have been a cause of suspicion because they are usually higher than those obtained with microorganisms (cf. Sections V and VI). Second, the mutational origin of markers derived *in vitro* has been questioned in two recent reports in which the frequency of cells resistant to certain agents did not vary as expected with ploidy and treatment with mutagen (Harris, 1971; Mezger-Freed, 1972). What is the evidence then that altered phenotypes derived from cultured somatic cells meet the requirements stated above?

It is important first to emphasize that although somatic cell variants have had a reputation, at least among some workers, of being "notoriously unstable" [*Nature (London), New Biol.* **237**, 98 (1972)], the evidence available on this point does not appear to warrant such a generalized assessment. Most types of drug-resistant lines which have been isolated have maintained their characteristic resistance when cultured for months or years in the absence of the selecting agent (Szybalski *et al.*, 1962; Littlefield, 1964b; Chu *et al.*, 1969), or when sometimes challenged in an attempt to alter their phenotype (Littlefield, 1963). A number of auxotrophic lines of Chinese hamster cells have maintained stable phenotypes for long periods and under various conditions (Kao and Puck, 1968, 1972; Puck *et al.*, 1971).

There are several explanations which might account for past confusion concerning the stability of somatic cell phenotypes. First, a failure to design and to validate selection conditions adequately can result in observations that suggest phenotypic instability. For example, the mere development of a colony of cells during prolonged incubation under selective conditions is obviously not *per se* proof that those cells possess an altered phenotype[1] (Harris, 1967; Fox, 1971); further characterization is essential to rule out artifacts (cf. Section III,A). Second, apparent instability (Morrow, 1970) can reflect selection pressures present in a population which may cause a variant phenotype to be overgrown by revertant cells during continuous subculture (Morrow, 1972). Third, gene-dosage modulation of the basic phenotype may occur in some cell lines due to genome heterogeneity within the population, and in some instances this could lead to second-order instability of a marker (cf. Section VI). Thus heteroploid[2] cell lines such as mouse L cells might be expected to have limited suitability for genetic studies. Finally, it is of course possible that some of the relatively unstable variant lines which have been isolated, for example, certain puromycin-resistant frog cells (Mezger-Freed, 1971), may indeed originate by a nongenetic mechanism.

Even when a marker phenotype does appear stable, one cannot be certain only on this basis that it indicates a true genetic mutation. Stable phenotypes in culture frequently involve differentiated traits, and the loss of such traits (Coffino *et al.*, 1972; Levisohn and Thompson, 1972; Aviv and Thompson, 1972) might not involve gene mutation. Stable phenotypes due to epigenetic changes have been well documented in other systems (Nanney, 1968), and some phenotypic states even appear to have a nonchromosomal basis (Sonneborn, 1970).

With regard to the criterion of response to mutagen for ascribing a mutational basis to an altered phenotype, the frequencies of most somatic cell markers which have been tested are enhanced by mutagen treatment. The use of mutagens substantially increases the frequency of both drug-resistance markers (Chu and Malling, 1968; Till *et al.*, 1973; Albertini and DeMars, 1973) and auxotrophic markers (Kao and Puck, 1969), although Szybalski (1964) observed no effect in his pioneering studies on purine-analog resistance (see Section V,B).

Isolation and characterization of an altered protein as evidence for a mutant gene has been lacking for mutant lines induced in culture. Several recent reports, however, indicate that altered gene products can be identified in drug-resistant lines (Albrecht *et al.*, 1972; Gillin *et al.*, 1972; Chan *et al.*, 1972).

---

[1] For purposes of clarity, our use of the term "phenotype" shall exclude "transient" phenotypes that are not reproducible upon subsequent testing (and might have been due initially to statistically atypical or cell-density dependent interactions of a clone with its selective environment—see Section III,A,2 and 3).

[2] We use the term here as it has been used by Kraemer *et al.* (1972), to denote "populations with abnormal variability of chromosome number and structure."

There is no methodology for recombination and segregation analysis which can be generally applied to presumptive somatic cell mutants, but the technology for mapping markers on chromosomes is currently being developed at the level of determining crude linkage relationships (Ruddle et al., 1971; Ruddle, 1972). By analyzing the consequences of chromosome loss from somatic cell hybrids, several mutant markers which were derived in culture (e.g., TK⁻ and HGPRT⁻) have been associated with particular human chromosomes (Miller et al., 1971; Ruddle, 1972).

In summary, much of the skepticism expressed about the nature and usefulness of somatic cell mutations seems inconsistent with the results now being obtained with a variety of phenotypes. Many of the past difficulties can probably be attributed to the choice of cellular material or to various pitfalls in the execution of experimental procedure and interpretation of results. Further evaluation of the evidence for a mutational origin of somatic cell phenotypes induced in culture will be given in Section VI following detailed consideration of selection methodologies and properties of the various classes of markers. In discussing most markers we shall now use the term "mutant" in the broader sense, with the understanding that in some instances it may be less appropriate than "variant."

B.   Mutant Types Isolated in Culture

As a class of mutants, drug-resistance phenotypes have received the most attention, undoubtedly because of the relative ease of the selection procedures. Beginning with Szybalski and Smith (1959), a number of investigators (Littlefield, 1963; Chu and Malling, 1968; Gillin et al., 1972) have readily isolated spontaneous mutant clones resistant to 8-azaguanine (or related purine analogs) by using single-step selection procedures. These workers have shown that resistance commonly arises from a loss of activity of the enzyme hypoxanthine-guanine phosphoribosyltransferase (HGPRTase). Similarly, the pyrimidine analog 5-bromodeoxyuridine (BrdU) has been used to select resistant lines which have lost activity for the enzyme thymidine kinase (Kit et al., 1963; Littlefield, 1965).

The drug puromycin has also been used by several workers as the selective agent. Lieberman and Ove (1959a,b) demonstrated two levels of resistance to puromycin in L cells and calculated the corresponding mutation rates. The validity of attempting to use puromycin resistance as a discrete genetic marker was later questioned by Harris (1967) because of evidence that one factor affecting resistance to the drug is nonhereditary and dependent on population density in the culture (Cass,

1972). Recently, Mezger-Freed (1971) has obtained evidence to indicate that the puromycin-resistance phenotype which can be observed in frog cells may be nongenetic in nature. In order to explain her data, she has suggested that resistance might arise from altered forms of self-replicating membrane subunits.

A wide variety of other drugs have also been used to select resistant phenotypes of somatic cells in culture. They include actinomycin D (Bosmann, 1971), ara-C (Bach, 1969; Smith and Chu, 1972), 2-fluoroadenine (Bennett et al., 1966), 2,6-diaminopurine (Atkins and Gartler, 1968; Kusano et al., 1971), tritiated thymidine (Breslow and Goldsby, 1969), 2-deoxyglucose (Barban, 1962), colchicine (Till et al., 1973; see Section III,E), ouabain (Baker and Till, 1971; see Section III,D), concanavalin A (Wright, 1973; Till et al., 1973), α-amanitin (Chan et al., 1972), and antifolate drugs such as aminopterin (Orkin and Littlefield, 1971), and amethopterin (Albrecht et al., 1972). Reference to many of the earlier papers concerning drug resistance, induced both in vitro and in vivo, can be found in the review by Gartler and Pious (1966).

Recently evidence has been presented for multiple genetic loci underlying drug-resistance phenotypes. Using baby hamster kidney (BHK) cells, Orkin and Littlefield (1971) have found that clones isolated for resistance to aminopterin after treatment with chemical mutagens display a different mechanism of resistance from lines occurring spontaneously. Gillin et al. (1972) using V79 Chinese hamster cells and DeMars and Held (1972) using diploid human fibroblasts have demonstrated the existence of different classes of 8-azaguanine-resistance phenotypes. These findings are important because they emphasize that caution should be exercised in attempting to generalize the interpretation of any phenotype from studies of only a limited number of independent isolates.

A second class of mutants, nutritional auxotrophs, have so far been isolated by only a few laboratories. Glutamine-requiring lines have been selected with both HeLa cells (DeMars and Hooper, 1960) and V79 cells (Chu et al., 1969), and a variety of auxotrophic lines representing different gene loci have been obtained with Chinese hamster ovary (CHO) cells (Kao and Puck, 1968, 1969, 1972; Kao et al., 1969a,b; Taylor et al., 1971). The auxotrophs of CHO cells have been regarded as single gene mutants (Kao and Puck, 1968) because of their stable properties and the observation that their frequency of occurrence is increased by the application of well-known mutagens (Kao and Puck, 1969).

The isolation of temperature-sensitive conditional lethal mutants of mammalian cells was first reported by Naha (1969) using monkey kidney cells. A number of temperature-sensitive (ts) lines obtained from mouse L cells in our laboratory have been described (Thompson et al., 1970, 1971). More recently we have isolated ts mutants with the near-diploid

CHO line (see Section IV,C), and one of these lines has been shown to be markedly defective in protein synthesis (Thompson and Stanners, 1972). In BHK cells, several complementation groups of *ts* mutants have been reported (Meiss and Basilico, 1972), and a line has been described which exhibits a lesion in cytokinesis at the nonpermissive temperature (Smith and Wigglesworth, 1972). The limited number of isolations reported to date for this potentially very large class of mutants can probably be attributed primarily to the lack of well-developed, efficient selection procedures.

Mutations of great potential importance in understanding development and differentiation are those which produce alterations in differentiated functions that can be expressed in culture. Coffino *et al.* (1972) have developed methods for detecting quantitatively and isolating variant sublines of mouse myeloma cells which are defective in the production of either heavy chains or both heavy and light chains of immunoglobulins. Levisohn and Thompson (1972) have described a variant of hepatoma cells in which the enzyme tyrosine aminotransferase fails to respond to the inducing steroid dexamethasone. Undoubtedly mutant markers will be isolated from a variety of cell types displaying tissue-specific functions as such lines become better characterized.

In addition to the mutant phenotypes which can be induced and isolated *in vitro*, mention should be made of the use of cultures established from tissue material known to possess specific mutations associated with genetic diseases (Krooth and Sell, 1970). A promising approach for obtaining mutant markers in human cells for somatic cell genetic studies would appear to be through the establishment of permanent lines of lymphocytoid cells (Moore and McLimans, 1968). Such a line derived from a patient having the Lesch-Nyhan syndrome was reported by Choi and Bloom (1970) and carried a mutation which causes loss of activity of the enzyme HGPRTase.

C.   Uses of Somatic Cell Mutants

The applications which can be made of mutants isolated in culture are as numerous and varied as the facets of genetics itself. We shall only mention briefly some of the major ways in which somatic cell mutants can now be applied to increase our understanding of mammalian cell function.

1.   DIRECT STUDY OF MUTANTS

Study of the mutant phenotypes themselves can be very useful in understanding the physiology of the normal cell. This has been clearly demonstrated with bacteria and other microorganisms in which metabolic pathways were elucidated mainly through the use of mutants. The principal reason why mutants provide so powerful an approach to understanding cell function is that they allow one to compare abnormal states with the normal state under conditions of an isogenic background. By allowing one to examine the metabolic consequences of altering one particular component of the cell, a mutant serves as a tool for probing functional relationships. Thus, mutants may be used in a way analogous to inhibitory drugs, but with the advantage that specificity of action is much more certain.

Auxotrophic mutations are well-suited to mapping synthetic pathways, and drug-resistance markers provide a means to study the properties of a target molecule in terms of its role in normal cell function. Conditional lethal mutations are undoubtedly the most useful, particularly for the study of regulation. When the lethality is temperature-sensitive, there is the potential for correlating the effects of temperature shift on a cellular property with changes in the functional or structural integrity of an isolated molecular component. Thus, ts mutants provide a means of evaluating the biological importance of enzymes or other proteins which can be isolated and characterized in vitro. Eventually, conditional lethal mutants may prove to be invaluable in the study of differentiation in mammalian tissue, in ways analogous to their use in elucidating morphogenesis in bacterial phages such as T4 and λ and to their current application in the study of development in *Drosophila* (Suzuki, 1970).

2. MUTATIONS AS GENETIC MARKERS

Progress toward developing a sophisticated methodology for recombination and segregation of genes in somatic cells will depend on the availability of the appropriate multiply marked lines for detecting these events. Current efforts to assign markers to specific chromosomes and to establish linkage relationships in the human genome make use of the technique of cell hybridization and the loss of human chromosomes which usually occurs in interspecific hybrids (Green, 1969; Miller *et al.*, 1971; Ruddle *et al.*, 1971; Kusano *et al.*, 1971). Thus, a mutant marker which can be isolated in culture and characterized biochemically in cells of nonhuman origin should be useful in the appropriate hybrids for identifying the human chromosome providing that particular function.

In the isolation of cell hybrids mutant markers which can be selected against are necessary for the effective elimination of nonhybrid cells. The selective system described by Littlefield (1964c) which makes use of the markers 5-bromodeoxyuridine-resistance and 8-azaguanine-resistance for recovery of hybrids in HAT medium (cf. Section III,B) has been used by many laboratories. Similarly, the selection of hybrids can be accomplished by employing other drug-resistance or auxotrophic markers in the appropriate selective medium or by using conditional-lethality markers under the nonpermissive conditions (cf. Section II,I).

For the study of mutagenesis in mammalian cells *in vitro* (see Section V,C), it is desirable to have well-characterized genetic markers which can be reliably and efficiently detected. Promising assay systems based on the 8-azaguanine-resistance marker have recently been developed (Chu, 1971a; Albertini and DeMars, 1973).

3. STUDY OF GENE EXPRESSION

For study of the regulation of gene expression, mutant markers should be useful in much the same way as differentiated functions and other nonmutant markers have already been employed (see review by Davidson, 1971; Davidson, 1972; Fougère *et al.*, 1972). By combining different genomes within a single viable hybrid cell one can study the interaction of the markers by analyzing dominance versus recessivity relationships and gene dosage effects.

II. Culture Requirements

A. Cell Lines

While no available cellular material appears ideally suited to genetic studies of mammalian cells in culture, several existing lines have proved suitable for the isolation of mutants of different classes as indicated above. Consideration of the karyotype is an important aspect of selecting a cell type from which to obtain mutants. Stable haploid cell lines would appear to be most appropriate from a theoretical viewpoint, but no such material is yet available from mammalian tissue. Although it has been possible to establish haploid lines from frog embryos (Freed and Mezger-Freed, 1970), it has proven to be difficult to isolate single-gene mutations in these cells (Mezger-Freed, 1971, 1972). Nor have any cell lines been well characterized with respect to monosomy for particular chromosomal regions. Aside from the ploidy, another important consideration is the normality and stability of the karyotype of the experimental material. Primary cultures of fibroblasts (from man and other animals) which show a quite uniform diploid karyotype have been readily obtainable and might be appropriate for selecting mutants were it not for the facts that: (1) such cells have a finite life span in culture (Hayflick, 1965) which will limit their growth to only two or three successive steps of cloning followed by outgrowth; and (2) they generally cannot be handled with the ease of established cell lines. Their plating efficiency tends to be low, and it is not possible to grow them in suspension culture.

In order to circumvent these latter limitations, one turns to the "established" cell lines, cultures which appear to have acquired immortality, perhaps at the expense of changes in karyotype. With respect to their karyotypes, established lines vary widely in the extent of their aneuploidy. Some lines like HeLa and mouse L cells, designated heteroploid, exhibit variable and often excessive numbers of chromosomes as well as rearrangements. Other lines, particularly Chinese hamster lines such as V79 (Ford and Yerganian, 1958; Yu and Sinclair, 1964), Don C (Stubblefield, 1966), and CHO (Puck et al., 1958; Kao and Puck, 1967; Deaven and Petersen, 1972), possess relatively stable karyotypes and usually have approximately the diploid number of chromosomes. These lines are often referred to as "near-diploid," but they do contain chromosomal rearrangements and possibly also deletions or reduplications. In recent years, permanent lines (Moore and McLimans, 1968) of lymphocytoid (or lymphoblast) cells which derive from human peripheral blood cultures have become available (Foley et al., 1965; Moore et al., 1967; Glade et al., 1968), and these lines often appear to have normal or nearly normal diploid karyotypes (Steel, 1971; Sato et al., 1972). The application to these lymphocytoid cells of recently developed methods for banding chromosomes should help to determine with improved resolution the degree to which their karyotypes are normal. The most apparent disadvantage of these latter lines is that they grow only in suspension and do not attach to glass or plastic substrates. Therefore, cloning cannot be readily accomplished without the use of a semisolid matrix such as agarose to anchor the cells (Sato et al., 1972) or by limiting dilution into Linbro-type dishes.

In this laboratory we turned to the CHO line, which had already proved useful in isolating nutritional auxotrophs (Kao and Puck, 1968); it appears to have a number of favorable characteristics for genetic studies:

Compared to many other established cell lines, CHO cells exhibit a relatively stable karyotype. The modal chromosome number is 21, or 20 in the case of one subclone isolated by Kao and Puck (1968), compared with a diploid number of 22 for the Chinese hamster. Although CHO cells carry many chromosomes which differ from those of the diploid Chinese hamster karyotype (Kao and Puck, 1969), the modal chromosome number of a clone is usually constant over many months or years. The low chromosome number and relative stability greatly facilitate cytogenetic analysis.

CHO cells grow well in either suspension or monolayer culture with a relatively short doubling time (approximately 15 hours at 37°C in suspension), and have a high plating efficiency. The time required for formation of macroscopic colonies is short compared to many cell lines, e.g., 8 days for CHO cells versus 12 days for L cells at 37°C.

CHO cells have specific advantages for use in somatic cell hybridization experiments. When interspecific crosses are carried out between CHO and human cells (Kao and Puck, 1970) most of the human chromosomes are quickly lost from the hybrid cells. Such rapid and specific reduction in the representation of one set of parental chromosomes greatly facilitates the assignment of marker genes to specific chromosomes or linkage groups in the human karyotype.

Finally, CHO cells respond particularly well to synchronization methods based on mitotic detachment (Petersen et al., 1968) either with or without Colcemid as a blocking agent.

As will be discussed presently, CHO cells have proved to be quite suitable in our hands for selecting a wide variety of mutant types, including drug resistance and conditional lethality, as well as the previously reported nutritional auxotrophy (Kao and Puck, 1969, 1972).

B. Monolayer versus Suspension Culture

Suspension culture allows considerable technical flexibility when the cells can also be grown under monolayer conditions. Some of the comparative advantages of suspension and monolayer culture are as follows: (1) The subculture of cells in suspension is simpler, and the growth rate can be monitored rapidly and precisely with an electronic Coulter-type counter. In the case of CHO cells in suspension the concentration is usually maintained between 2×10^4 and 4×10^5 cells/ml. With monolayer cultures it is often possible to dilute more heavily so that less frequent subculturing is required. (2) Suspension culture eliminates prolonged cell-cell contact and perhaps results in greater homogeneity with respect to the growth rate of individual cells than in monolayer culture, where local gradations in cell density occur. This consideration becomes important in designing a selection procedure for mutants under conditions where the action of a cytocidal agent requires cell proliferation (cf. Section IV,A), and the presence of a fraction of cells which are not rapidly cycling would impair the efficiency of the selection. (3) Under conditions where precise temperature regulation is important, as in the study of temperature-sensitive mutants, suspension culture is often preferable because of the relative ease of regulating culture temperature in water baths as opposed to incubators (see Section II,D). (4) Monolayer culture is often more suitable than suspension for certain kinds of experimental operations, e.g., for cloning or for the outgrowth of relatively small numbers of surviving cells after a selective treatment has been imposed on a population.

Heteroploid lines such as mouse L and HeLa are often grown in suspension. Presumably other near-diploid lines in addition to CHO (Puck et al., 1964) might also be adapted from monolayer to suspension culture. Our experience indicates that there is no difficulty in transferring cultures of CHO back and forth between the suspension and monolayer states; virtually no adaptation period is necessary.

C.   Media and Sera

With the availability in recent years of commercially prepared powdered media, the inconvenience and complications of preparing tissue culture media have been greatly reduced. We have routinely used a medium designated MEM-α, which was developed by Dr. C. P. Stanners at the Ontario Cancer Institute (Stanners et al., 1971). This medium has a base of Eagle's minimal essential medium (MEM) with certain additions, resulting in a general tissue culture medium which has been used for a wide variety of cell types including CHO, HeLa, KB, mouse L, mouse embryo, 3T3, RAG, hamster embryo, human bone marrow, mouse bone marrow, etc. Currently, we purchase powdered α medium from Flow Laboratories (Rockville, Maryland) in large lots that are pretested. In batches of α medium custom-blended in the laboratory, a number of the ingredients can be omitted such that one can select for auxotrophic mutants of CHO cells (see Section IV,B,1).

Many of the technical difficulties which arise in culturing cells can be ascribed to the sera which are routinely used as supplements for stimulating cell growth. Since the particular growth factors involved have not been well characterized, the quality of a particular lot of serum must be gauged empirically by its ability to support cell growth. Thus, our preferred procedure in buying serum is to test a sample in advance prior to purchasing the lot. The acquisition of large lots, stored at −20°C, helps to ensure experimental uniformity.

Many experiments involving the selection or the characterization of mutant lines may require a medium that is completely defined for small molecules such as amino acids and nucleosides. In these cases whole serum, e.g., fetal bovine serum, is unsatisfactory, and it is necessary to utilize serum from which small molecules have been selectively removed by separation on a column such as Sephadex (Kao and Puck, 1967) or by dialysis. We routinely modify the serum by dialysis,[3] a procedure that does not appreciably affect the ability of the serum to support growth under nonselective conditions.

D.   Temperature Regulation

Studies on mutants which are temperature-sensitive for growth or for some other property such as drug resistance involve a special technical problem not ordinarily encountered, namely that of precise and reliable temperature regulation. The greatest difficulties arise when mutant cells are being incubated at the nonpermissive temperature in monolayer culture since fluctuations in temperature of 0.5°C and even less can sometimes greatly affect the extent of growth observed. One needs an incubator having not only a precise temperature regulator with long-term stability but also sufficient air circulation to provide rapid recovery when the door is opened and to reduce the possibility of stable gradients within the compartment. Although no exceptionally good incubators appear to be available commercially, we have successfully used the Forma Unitrol Model 3000 Unit (Forma Scientific, Marietta, Ohio). This unit generally regulates with a precision of approximately ±0.1°C, and has internal gradients which are small and give a total spatial variation of no more than approximately 0.5°C when the unit is loaded. When culture containers can be sealed so that continuous flushing with CO_2 is unnecessary, a possible alternative to the incubator unit is a warm room equipped with a sensitive thermostat and very efficient air circulation.

The problem of temperature regulation in suspension culture is less acute since water baths can be used and a relatively inexpensive regulator (such as General Electric's Zero Voltage Switching AC Power Controller #S200 A-1 fitted with the appropriate thermistor and heater) will readily provide temperature uniformity with variations not exceeding ±0.1°C.

E.   Cold Storage of Mutant Lines

In a program in which mutant lines are being isolated it is essential to have an effective means of preserving populations of cells for extended or indefinite periods. Once a mutant line has been isolated, it is obviously desirable to protect it against loss due to contamination and to store it when there is no immediate demand for it. In addition, preservation of

---

[3] A 2-liter quantity of serum, sealed in dialysis tubing under sterile conditions, is dialyzed at 4°C for 24 hours against 40 liters of precooled phosphate-buffered saline agitated with a magnetic stirring bar. The cycle is repeated three times with fresh, cold saline, and then the serum is sterilized by filtration.

cells allows one to minimize the complications which can arise when a series of experiments are carried out on a culture whose properties may change over a prolonged period of serial cultivation.

For indefinite, long-term storage of cells, most workers appear to prefer immersion of samples in liquid nitrogen ($-196°C$). This is probably unnecessary in many instances and has the disadvantages of the cost and inconvenience of maintenance, and of preparing the cells for storage in sealed ampoules. A low-temperature mechanical freezer ($-76°C$ or colder) can provide a more versatile method of storage with minimal cost beyond the purchase price. For cell storage at either $-196°C$ or $-76°C$, we have found that dimethyl sulfoxide (DMSO) is quite effective as a preservative for a variety of cell types when used at a concentration of 10% in medium containing 10% serum.

Our procedure for frozen storage involves resuspending the cells (usually $> 10^6$) in about 2 ml of medium containing DMSO. For storage at $-76°C$ the sample is placed in a small plastic tube (12×75 mm; Falcon Plastics, Oxnard, California), tightly capped, and placed directly at $-76°C$. For recovery, the sample is removed from the freezer and thawed rapidly in a $37°C$ water bath, then transferred to a bottle or flask containing 20 to 40 ml of fresh medium. After 2 or 3 hours incubation at the desired temperature, the viable cells will have attached so that the medium containing DMSO can be poured off and replaced with fresh medium. Alternatively, one can remove the DMSO from the cells by centrifugation immediately after thawing so that a subsequent medium change is unnecessary. Under these conditions the recovery as measured by the percent of cells which attach is routinely 75% or better for many clones of CHO cells as well as various mouse lines like RAG and 3T3. Tests on CHO cells indicate no reduction in viability during storage periods of at least six months at $-76°C$.

For storage in liquid nitrogen, our procedure involves prefreezing the sample (enclosed in a flame-sealed glass ampoule) at $-20°C$ for 1–2 hours before immersion into nitrogen. Storage at $-196°C$ under these conditions has sometimes resulted in recovery that is somewhat lower and more variable than at $-76°$. The causes of cell killing are as yet poorly understood (Mazur et al., 1972), and the optimal conditions of cooling rate, DMSO concentration, etc., may vary for different cell lines.

Another method of cold storage which we have found useful on a short-term basis is simply to place the culture at $4°C$ directly from the incubator. This procedure works well for most CHO clones, but other lines such as mouse L cells show more rapid loss of viability. The method must thus be tested for the particular line involved. CHO cells can often be rescued from a mass culture stored under these conditions for up to a month or longer, but for reasonable recovery it is advisable to limit the storage time to 1 or 2 weeks. A necessary precaution is to ensure that the pH be kept in the normal range by making the culture vessel airtight.

F.   Cloning and Replica Plating

In genetic studies especially it is important to have accurate, efficient methods of clone isolation. There are several possible reasons why a colony isolated under normal plating conditions might not represent a pure clone: (1) The cells may be somewhat clumped at the time of plating so that a colony does not arise from a single progenitor cell. (2) During colony formation cells may migrate from one colony to another. (3) The process of removing the colony from its substrate may result in cross-contamination with cells from another colony. These problems can be dealt with as follows: (1) Plating a single-cell suspension at limiting dilution will help ensure that only one clone appears in a given culture vessel. Suitable for this purpose is a tray (Linbro, Chemical Co., #ISFB-96TC) containing 96 wells of ≃0.2 ml capacity or a Microtest Plate (Falcon Plastics). The concentration of cells inoculated can be adjusted so that the likelihood of more than one colony per well is small. (2) The population to be plated should be dispersed well and checked for the presence of clumping; after plating, wells containing a single cell can be identified visually if desired. (3) If colonies must be isolated from plates containing other colonies, it is advisable to rinse the plate before isolation to remove loosely attached or floating cells. For maximum certainty of purity recloning is desirable.

When the appropriate colonies have been located for isolation, they can be removed by trypsinization or by some mechanical method. We prefer the latter, utilizing Pasteur pipettes prepared by fire-polishing the tip to reduce roughness and making a right-angle bend in the tip about 1 cm from the end before autoclaving. Actual removal of the colony is accomplished by simultaneously scraping the colony with the pipette tip and applying a slight suction. The cells usually detach as clumps which can then be dispersed by pipetting. For the cell types we have used (CHO, V79, mouse L, 3T3, HeLa, and KB) the transfer efficiency is satisfactory. This method can be applied to colonies growing on petri dishes or in bottles as well as in cloning trays, subject to the limitations described above. An alternative method for isolating colonies from petri dishes is to surround a colony before trypsinization with a metal or glass cylinder coated on one end with silicone grease. For a detailed discussion of the methodology for cloning mammalian cells the reader should consult Ham (1972).

Replica plating techniques analogous to those which have greatly facilitated the isolation of mutants of bacteria and viruses have not been well developed for mammalian cells. One mechanical replicating device which makes use of the 96-well Linbro trays mentioned above has been described (Goldsby and Zipser, 1969), and a simplified version of this approach utilizing a hand replicator has also been reported (Suzuki *et al.*, 1971). The fidelity of replication reported for these devices was 93.5% to 97.1% in the first case and 97.8% in the latter. Such devices may prove to be useful, particularly if one requires a number of replicates of a given set of clones. Where the emphasis is upon testing a large number of clones, a main limitation of these replicators is that they must be cleaned of cells and kept sterile between master trays containing different clones. Moreover, if the colonies are randomly distributed, many wells would not be useful because they would not contain single colonies.

In testing for auxotrophic and temperature-sensitive mutants, we have chosen to prepare a few replicates of the clones manually rather than with a mechanical device. When the clones have been isolated into small plastic tubes containing several milliliters of medium, they are divided among 35-mm tissue culture dishes (Falcon Plastics) or 24-well Linbro trays (FB16-24TC) for testing. This method assures the inoculation of a similar number of cells into each replicate, and the fidelity of replication approaches 100%.

G. Maintaining Culture Uniformity

There is ample evidence that the properties of cultures grown for extended periods may change, often unpredictably. Such variation can confound the interpretation of experiments performed with cells maintained by serial subculture over a long term. There are essentially two approaches which can be utilized to circumvent the problem.

Probably the most reliable policy is to replace a culture periodically with material from frozen reference stock. When a clone has been isolated it should be grown to a large population so that a number of samples can be frozen to serve as a source for reestablishing the original culture as necessary. Because the pertinent cellular properties which are subject to change are not always defined and measurable, an advisable policy is to utilize serial subcultures from a given stock for only a limited period of time, unless the long-term stability of a particular phenotype is itself being tested. As a rule of thumb we consider the useful life of a culture to lie in the range of 2 to 4 months.

The alternative method for maintaining population homogeneity is to subclone periodically and select for the desired phenotype. A significant disadvantage of this approach is the possibility that a subclone line will differ from the parental line in some additional property besides the one of primary interest. A modification of the subcloning approach would be simply to dilute the culture heavily. One can sometimes adequately remove an undesirable subpopulation by regrowing the culture from a small number (i.e., 10^2–10^3) of cells. It should be emphasized that diluting to improve population homogeneity will be successful only on a statistical basis; the operation can be expected occasionally to increase the heterogeneity of a culture. Thus, the procedure should be confirmed by testing whenever possible.

In a laboratory in which more than one cell line is being maintained, the danger of cross-contamination of cell cultures is a very real one (cf. Gartler and Pious, 1966). The problem can be particularly severe in genetic studies since many different lines are being handled, and population purity is often critically important for experimental interpretation. Cross-contamination can occur, for example, when the same container is used to dispense medium to different cultures; cells splashed into the medium stock might find their way into other cultures. As far as possible, the design of routine procedures should eliminate the possibility of cross-contamination. This may be accomplished in several ways: (1) either a given bottle of medium must be used only for one particular cell type; or (2) the fresh medium must always be added to the new culture container before an inoculum of cells is introduced; or (3) a different sterile intermediate vessel must be used between the medium and each culture.

H.   Independent Mutant Isolates

It is often desirable to ensure that different mutant isolates are independent in origin and not derived from progeny of the same clone. This is particularly important when temperature-sensitive (*ts*) or auxotrophic mutants are isolated by the use of multiple selection cycles because of the opportunity for preferential enrichment of faster growing mutant types during the course of the selection.

Independent isolates are obtained from separate cultures in which the mutants have arisen independently. In the course of initiating such separate cultures from a parental population, any preexisting mutants must be eliminated. The simplest and most foolproof way of accomplishing this is to establish cultures from different subclones of the parental lines. A slightly less rigorous alternative is to set up a series of subcultures of the parental line from small numbers of cells in order to dilute out the preexisting mutants. In the case of *ts* conditional lethal or auxotrophic phenotypes, a cell population can be purified of preexisting mutants by growth under the nonpermissive conditions for a period of time. The selection can then be performed on a series of separate cultures initiated from the purified population. If the object is to isolate spontaneous mutants, it will generally be advisable to maintain the separate purified cultures for an extended period before selection in order to allow mutants to accumulate.

If a mutagen is being used, the mutagenesis can of course be carried out independently on each separate subpopulation to be selected. In cases where it is known that the efficiency of mutagenesis is high and the frequency of spontaneous mutants in a population is low, the protocol for obtaining independent mutants can be simplified. The population as a whole can be treated with mutagen and then split immediately, before significant cell division can occur, into separate cultures for selection. Furthermore, because such a mutagenized population will contain a large number of independent mutants, mutant colonies obtained by plating in a single-step selection soon after mutagenesis will themselves very likely be of different origin (cf. Section V,C,2).

I. Selective Markers and Cell Hybridization

The technique of cell-cell hybridization constitutes a uniquely powerful tool for the somatic cell geneticist because it is the only method available for combining genetic material from different cells in order to study gene interaction. Since the original discovery that cells in culture could fuse to form a viable hybrid cell containing the genomes of each parent (Barski et al., 1960), and the subsequent observations that the fusion process could be appreciably enhanced by the use of inactivated Sendai virus (Harris and Watkins, 1965), cell hybridization has become the most widely used genetic technique in cell biology. The methodology for fusing cells and for selectively recovering hybrids has been recently reviewed by Rao and Johnson (1972). In the context of our present discussion concerning mutants isolated in culture, it is pertinent to mention two aspects of cell hybridization—the use of mutations as selective markers in isolating hybrid cells and the use of the hybridization technique to study the properties of mutants themselves.

In the first instance, mutant lines with selective markers are necessary for the efficient recovery of viable hybrid clones. Although the most commonly used selective system has been that employing 8-azaguanine resistance and BrdU-resistance markers in combination with HAT medium (Littlefield, 1964b,c; cf. Section III,B and C), auxotrophic markers have also been shown to be very useful (Kao and Puck, 1970; Puck et al., 1971), and other markers such as ts conditional lethality will probably prove effective as well.

Second, the hybridization technique is an indispensable tool for the genetic characterization of isolated mutations. Two basic questions which immediately arise in regard to the genetic properties of mutant lines are: (1) Are the various mutant lines different in the sense that they involve different genetic loci? (2) Are the mutations dominant, recessive, or codominant in expression? Before mutant isolates of similar phenotype (e.g., temperature sensitivity) can be analyzed for uniqueness on the basis of their ability to complement, it will generally be necessary to establish first that each mutation behaves as recessive in the hybrid configuration.

The presence of a second marker on a mutant line can be an important advantage in the characterization of the mutation. For example, in testing a mutation for dominance, codominance, or recessivity or for complementation in a hybrid cell, it is preferable to isolate the hybrids to be examined without utilizing the mutation under study as a selective marker. Thus, it is advantageous to ensure that mutants isolated with the intent of performing genetic analysis be selected from cell lines already carrying a mutation which can serve as a selective marker. In our own mutant isolations with CHO cells, for example, we have utilized auxotrophic markers as second markers on *ts* mutants.

An important point to recognize is that because the dominance-recessivity test can at present be readily performed only in a hybrid cell having double the normal complement of chromosomes, interpretation of dominance versus recessivity of an allele in the classical Mendelian sense is complicated by the abnormal gene dosage. It has already been shown that the expression of a differentiation trait in hybrid cells may depend upon the relative number of genomes contributed by each parental cell (Davidson, 1972; Fougère et al., 1972). Refinements on the above method of genetic analysis must await the development of techniques for tailoring the chromosome constitution of hybrid cells, for example, through selective removal or addition of particular chromosomes.

III.   Isolation of Drug-Resistant Mutants

A.   General Considerations in Direct Selection for Drug Resistance

In principle the direct selection of drug-resistant somatic cell mutants is straightforward. One exposes a mass culture of wild-type cells to a cytotoxic drug and isolates the surviving clones of drug-resistant phenotype.

For the purpose of discussion in this article, we define a "drug-resistant" cell line as one that retains a normal plating efficiency, but not necessarily a normal growth rate, in the presence of a drug concentration that substantially reduces the plating efficiency of wild-type cells (i.e., a dose $\geq$ D_S in Fig. 1, Section III,A,2 below). Before examining the selections and characteristics of several specific types of drug-resistant mutants, we shall first survey some general topics relevant to successful application of this approach; these include choice of an appropriate selective drug, dose response of the cells to the drug, and several factors that often hinder isolation of a desired phenotype.

1. Selective Drug

From the geneticist's point of view, the most suitable drugs for use as selective agents are those whose action is well understood and for which it seems likely *a priori* that simple molecular alterations might confer resistance. Such a drug may be an agent that binds preferentially to a particular enzyme or structural component or an antimetabolite whose toxicity is dependent upon a specific sequence of reactions which lead to cell killing. Ouabain and colchicine, discussed below, are examples of drugs in the former category; 8-azaguanine and 5-bromodeoxyuridine (BrdU) are examples of the latter type of drug.

From a biochemical viewpoint, resistant mutants may be desired for use as a tool in characterizing an agent's mechanism of action. The suitability of a particular compound as a selective drug can then be determined only empirically, according to the considerations outlined below.

2. Dose Response

The dose response of wild-type cells to a potential selective drug could be measured either in terms of growth rate or plating efficiency in the presence of the drug. In most cases a plating assay is preferable because it permits an evaluation of population heterogeneity throughout the dose range and is also a more sensitive indicator of the minimum doses necessary to kill the wild-type cells and to discern any unusually resistant clones.

The type of response generally observed in a plating assay with increasing concentration of drug in the medium can be illustrated with the hypothetical survival curve shown in Fig. 1. For drug concentrations above a threshold dose D_T, there is a rapid decline in plating efficiency over a rather narrow dose interval up to a dose D_S, which may be two- or threefold greater than D_T. For doses higher than D_S, if the curve tails off as indicated by the dashed lines "A" and "B," discrete surviving colonies may reflect the presence of a subpopulation of resistant cells. This

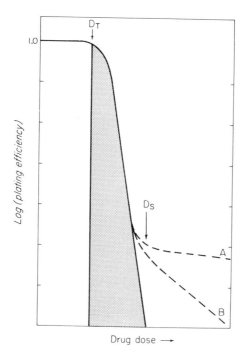

FIG. 1. An idealized dose-response curve, depicting the typical effect of increasing concentration of a cytotoxic drug in the culture medium upon the plating efficiency of wild-type cells. See text for explanation.

must of course be substantiated by isolating surviving colonies and testing their drug response in comparison to wild-type cells. Compared to curve "B," a curve like that labeled "A" would imply that the subpopulation of resistant cells is less affected by the drug at higher concentrations (and is thus, perhaps, less heterogeneous).

Surviving colonies obtained at doses corresponding to the steep region of the dose-response curve (indicated by the shading in Fig. 1) are usually small and heterogeneous in size. They will not ordinarily represent resistant cells; rather, they probably reflect the statistical nature of colony formation under partially selective conditions. An effect of the drug in this dose range may be to reduce the average probability of division, P, for any individual cell. At the macroscopic level the effect of this phenomenon, which will be referred to as the P-effect. (Whitmore and Till, 1964), would be to broaden the spectrum of colony size. Thus, simply on the basis of chance alone it would be possible for some relatively large colonies to develop. As the value of P becomes small with further increases in drug dose, nonresistant colonies rapidly disappear. For doses $\lesssim D_s$, however, there is the possibility of confusing P-effect colonies with those that are actually resistant.

3. POSSIBLE COMPLICATING FACTORS

A complication in selections for most types of drug resistance is that the results of the plating assay for mutant phenotypes depend upon cell density and vary with the number and the dispersion of cells on a culture dish. Such phenomena can arise from (a) metabolic cooperation between wild type and mutant cells (Subak-Sharpe *et al.*, 1969; Cox *et al.*, 1972), (b) a requirement for extensive cell growth in order that the drug produce its toxic effect, and (c) depletion of the medium by drug-sensitive cells prior to their metabolic death. (Examples of these problems are furnished below in the discussions of selections for 8-azaguanine resistance, 5-bromodeoxyuridine resistance, and colchicine resistance.) In a given situation these factors are not necessarily independent of each other or of the P-effect discussed directly above. For example, in our experience the frequency of "P-effect" colonies is usually dependent upon the cell density.

In order to obtain cell lines resistant to high concentrations of a drug, it may be desirable to extend the single-step isolation procedure to a multistep or a continuous selection. This situation may be encountered when there are two or more possible mechanisms for increasing resistance or when the level of resistance depends upon the dosage of multiple alleles of a gene. Successive increments in drug resistance might be obtained simply by submitting the cell population to consecutive single-step selections with graded increases in drug concentration (e.g., colchicine; Section III,E). On the other hand it may be impossible to reconcile a requirement of low cell density for an effective selection against sensitive cells with a requirement for high cell numbers in order to obtain rare resistant mutants (e.g., BrdU; Section III,C). It is then necessary to enrich for the desired mutants by maintaining a mass population under continuous selection pressure, culturing it in gradually increasing drug concentrations, until cells with a satisfactory degree of resistance overgrow the culture or are frequent enough to isolate at low cell density. A disadvantage of this tactic is the difficulty of evaluating the number of mutant alleles or loci that might be necessary to produce the selected phenotype, since the frequency of the mutant phenotype in the wild-type population cannot be measured directly.

For drugs that act intracellularly, one must distinguish between resistance as a consequence of altered permeability to the selective agent and resistance due to changes in intracellular structure or function. Permeability changes have been found to be a major mechanism of resistance to actinomycin D (Bosmann, 1971) and colchicine (Section III,E) and have also been encountered in cells resistant to 8-azaguanine (Szybalski *et al.*, 1962), 5-bromodeoxyuridine (Mezger-Freed, 1972), and α-amanitin (Chan *et al.*, 1972). In order to determine whether a resistant phenotype can be explained by more than one mechanism it is necessary to characterize a number of independently derived mutant lines.

B. 8-Azaguanine Resistance

The selection systems most widely utilized to date for isolation and characterization of drug-resistant somatic cells are those for resistance to the purine analog 8-azaguanine (AG) or closely related compounds. The combination of advantageous characteristics offered by these markers is at present essentially unique: 8-azaguanine-resistant (AG^R) cells can generally be obtained in a single-step selection; a reduction in enzyme activity that commonly (but not always) accompanies the AG^R phenotype has been identified and appears analogous to the defect in a human genetic disease (the Lesch-Nyhan syndrome); and a back selection is available for assaying revertants of this class of AG^R mutants. Potential limitations on the usefulness of the AG^R marker stem from certain technical requirements for a reliable selection procedure and the possibility of confusing phenotypically different classes of AG^R mutants.

In addition to 8-azaguanine, the analogs 6-thioguanine, 6-mercaptopurine, 8-azaguanosine, and 8-azahypoxanthine have been used as selective agents to isolate cell lines with enhanced resistance to 8-azaguanine. Isolates selected for resistance to one or the other of these drugs have now been obtained from a variety of somatic cells, most notably heteroploid human D98 cells (Szybalski and Smith, 1959; Szybalski *et al.*, 1962) and heteroploid mouse cells (Lieberman and Ove, 1959a; Littlefield, 1963), near-diploid Chinese hamster cells (Chu *et al.*, 1969; Chu, 1971a; Gillin *et al.*, 1972), human lymphoblasts (Sato *et al.*, 1972), and diploid human fibroblasts (Albertini and DeMars, 1970, 1973; DeMars and Held, 1972). Because of the extensive work performed in this area, we will survey the pertinent properties of mutants studied to date before explicitly discussing the methodology for mutant isolation.

1. PROPERTIES OF AG-RESISTANT CELL LINES

It has generally been possible to isolate AG-resistant mutants in a single selective step from somatic cell lines. Under appropriate assay conditions (see Section III,B,2 below), the curves describing the dose response of wild-type cells to 8-azaguanine are analogous to those labeled "A" or "B" in Fig. 1. From colonies that form in selecting concentrations of the drug, clones typically resistant to doses 10- to 100-fold greater than those toxic to wild type have been obtained (Szybalski and Smith, 1959; Littlefield, 1963; Chu et al., 1969; Sato et al., 1972; DeMars and Held, 1972).

The AG^R phenotype has generally proved to be highly stable in the absence of selective pressure, although Morrow (1970) has reported unstable phenotypes for a low level of resistance in a heteroploid mouse line. By Luria-Delbruck fluctuation analyses (see Section V,A,2), the random origin of AG^R phenotypes of D98 cells (Szybalski, 1959), V79 Chinese hamster cells (Chu et al., 1969; Harris, 1971), and diploid human fibroblasts (DeMars and Held, 1972) has been established and the mutations rates estimated as, respectively, 5×10^{-4}, 2×10^{-5} to 1.6×10^{-8}, and 1×10^{-6} per cell per generation. Induction of AG^R Chinese hamster cells and human diploid fibroblasts and lymphoblasts by a variety of chemical and physical mutagens has now been demonstrated (Chu and Malling, 1968; Bridges and Huckle, 1970; Albertini and DeMars, 1973; Sato et al., 1972) (cf. Section V,B).

Cell lines with different levels of resistance to 8-azaguanine can be obtained by selecting for mutants at different drug doses (Littlefield, 1963). The degree of resistance exhibited by AG^R cells isolated by single-step selection can usually be augmented by further manipulation, either multistep selection at increasing drug concentration (Littlefield, 1963) or sequential selection with different analogs (Szybalski et al., 1962). For example, from an AG^R population of D98 cells Szybalski and collaborators were able to isolate cells with greatly increased resistance to 8-azaguanosine or to 8-azahypoxanthine, which displayed enhanced cross-resistance to 8-azaguanine.

An alteration in activity of the intracellular enzyme hypoxanthine-guanine phosphoribosyl transferase (HGPRTase, E.C. 2.4.2.8)[4] is often associated with the AG^R phenotype (Szybalski et al., 1962; Littlefield, 1963; Gillin et al., 1972; DeMars and Held, 1972). In these cases the drug resistance no doubt entails an inability of the cells to utilize exogenous guanine, hypoxanthine, or their cytotoxic analogs (Brockman, 1963). Littlefield (1963, 1964a,b) found a direct correlation in L cells between levels of AG-resistance and loss of HGPRTase activity assayed in vitro.

The phenotype of HGPRT$^-$ AG^R cells isolated in culture is comparable to that of fibroblasts from humans with the X-linked genetic dis-

---

[4] Often referred to in the early literature as inosinic acid pyrophosphorylase.

ease known as the Lesch-Nyhan syndrome (Seegmiller et al., 1967; Nyhan, 1968; DeMars, 1971; DeMars and Held, 1972). This similarity suggests that the alteration in phenotype observed in culture is due to a true genetic mutation. Direct evidence that the HGPRT⁻ phenotypes derived in culture involve a structurally altered enzyme is now becoming available. Beaudet (cited in Gillin et al., 1972) has demonstrated immunologically active but enzymatically inactive HGPRTase in extracts from some V79 hamster cell mutants. DeMars and Held (1972) have observed qualitative differences from wild type in the residual HGPRTase activity of an AG^R clone isolated in culture from human diploid fibroblasts.

Effective back selection is available for detecting revertants of HGPRT⁻ AG^R cells. If the medium contains hypoxanthine and a folic acid antagonist (aminopterin or amethopterin) to poison endogenous purine synthesis, only those cells that have regained HGPRTase activity can proliferate (Szybalski et al., 1962). Since the inhibitor also blocks endogenous thymidine and glycine syntheses, these compounds must also be included in the selective medium (hence the term "HAT" or "THAG" medium).

Like the forward selection for AG^R mutants, the back selection for HAT-resistant (HAT^R) cells can be accomplished in a single-step plating experiment. The frequencies of revertants assayed in different mutant lines vary widely, ranging from 10^{-4} to $<10^{-7}$ (Chu, 1971a; Gillin et al., 1972). It has been shown that the HGPRT⁻ phenotype behaves as a recessive character in somatic cell hybrids (Littlefield, 1964b,c; Chu et al., 1969), thus permitting selection for HAT^R hybrid cells (Section II,I).

Cells selected from HGPRT⁻ lines for reversion from HAT-sensitivity to HAT-resistance are characterized by at least a partial recovery of HGPRTase activity (Szybalski et al., 1962; Littlefield, 1963, 1964a; Chu et al., 1969). In some cases the regained activity is expressed only when endogenous purine synthesis is inhibited, so that the HAT^R cell still exhibits an AG^R phenotype in the absence of aminopterin (Chu et al., 1969; Gillin et al., 1972).

Evidence is now accumulating that forward mutation to AG-resistance is often not accompanied by acquisition of HAT-sensitivity and loss of HGPRTase activity. Indeed, the D98 cells initially isolated for resistance to 8-azaguanine and 8-azaguanosine by Szybalski et al. (1962) showed little change in HGPRTase activity although the net conversion of guanine to guanylic acid by whole cells was much less efficient than in wild-type cells. A glimpse of the possible complexity of AG^R phenotypes is provided by a recent study by Gillin et al. (1972) of 35 different

EMS-induced and MNNG-induced AG[R] clonal lines of V79 hamster cells. In addition to being AG-resistant, each line was *either* thioguanine-resistant and HAT-sensitive, *or* thioguanine-sensitive and HAT-resistant, *or* thioguanine-resistant and HAT-resistant. On the basis of *in vitro* assays for HGPRTase activity and *in vivo* assays for uptake of labeled hypoxanthine in the presence of aminopterin, the phenotypes of 21 of the mutants (all from the first two categories listed above) were explicable on the basis of altered levels of HGPRTase activity. The authors suggest that the behavior of the remaining 14 mutants, which take up exogenous purines poorly, might be explained on the basis either of partial transport defects or of altered regulation of endogenous purine biosynthesis.

Comparable findings have been reported from studies of human diploid fibroblasts. DeMars and Held (1972) found that only one of ten spontaneous mutants to AG-resistance from normal fibroblasts had a HAT-sensitive, HGPRT⁻ phenotype comparable to that of Lesch-Nyhan fibroblasts. Eight of the spontaneous mutants had somewhat reduced HGPRTase activity but were HAT-resistant, i.e., capable of utilizing exogenous hypoxanthine in the presence of aminopterin. The remaining mutant clone, also HAT-resistant, displayed quantitatively normal HGPRTase activity and closely resembled fibroblasts from a patient with an X-linked gout associated with qualitative alteration in HGPRT activity (Benke and Herrick, 1972). In an evaluation of 51 X-ray-induced mutants to AG-resistance in normal fibroblasts, Albertini and DeMars (1973) observed that 80% of the mutant clones had at least partially reduced HGPRTase activities but *all* of them proved to be resistant to HAT medium.

In biochemical terms, at least five mechanisms can be envisioned to account for acquisition of resistance to guanine or hypoxanthine analogs (Murray, 1971; Gillin *et al.*, 1972; G. A. LePage, personal communication): (a) decreased activity (or altered substrate specificity) of HGPRTase, as discussed above; (b) decreased transport of analog into the cell (Szybalski *et al.*, 1962); (c) decreased feedback inhibition of *de novo* purine biosynthesis (Nagy, 1970); (d) increased degradation of nucleotides by alkaline phosphohydrolases (Wolpert *et al.*, 1971); and (e) altered substrate specificity of nucleotide phosphorylase or ribonucleoside diphosphate reductase (Miech *et al.*, 1969; Peery and LePage, 1969). Although only the first of these mechanisms has yet been conclusively demonstrated for drug-resistant somatic cells isolated in culture, the observations of Gillin *et al.* (1972) and DeMars and Held (1972) cited above certainly support the view that various AG-resistant phenotypes can arise from mutations at different genetic loci. An obvious implication of the phenotypic heterogeneity apparent among AG-resistant cells is that it cannot be assumed without quantitative documentation that a particular mutant line is HGPRT-deficient, much less that it is HAT-sensitive and that the character is recessive.

The relative frequencies at which different phenotypes occur may well vary with the *cell type* (compare Szybalski *et al.*, 1962, and Gillin *et al.*, 1972) and also with the particular *analog* employed as the selective agent (Szybalski *et al.*, 1962). One might expect 6-thioguanine to be preferable for selecting HGPRT⁻ human and Chinese hamster cells, because HGPRTases from these cells are reported to have significantly greater substrate specificity for 6-thioguanine than for 8-azaguanine or 8-azahypoxanthine (Krenitsky *et al.*, 1969; Gillin *et al.*, 1972). The frequency of the various phenotypes may also vary with the *mutagen*, if one is used, and with the selecting *drug dose*. In selections of AG-resistant human diploid fibroblasts the HAT-resistant phenotype was quite frequent at low selecting dose, but the lone HAT-sensitive isolate was resistant to considerably higher AG concentrations than were the HAT-resistant cells (DeMars and Held, 1972).

2. Special Problems Associated with the Selection

Several technical problems have been encountered in attempts to establish quantitatively reliable single-step plating assays for AG^R mutants. These pitfalls in the selection may take the form of spurious colony formation by drug-sensitive cells on the one hand, or of suppression of the mutant phenotype on the other hand. The principal difficulties involve interference with the selection by components of normal serum and the dependence of the selection upon cell density.

Ingredients, presumably purines, antagonistic to the inhibition of cell growth by 8-azaguanine or 8-azahypoxanthine are often present in fetal bovine serum (but apparently not in calf serum) (Felix and DeMars, 1971; Vaughan and Steele, 1971). In our experiments employing CHO cells and fetal calf serum, we have found it essential that the serum be dialyzed in order to achieve a clean and reproducible selection assay. Even in the presence of a high concentration of 8-azaguanine (40 $\mu g/ml$) with undialyzed serum considerable growth occurs on plates inoculated with 10^4 or fewer cells. As the size of the inoculum is increased to 10^6 cells per dish, a better discrimination is obtained between presumptive AG^R colonies and those apparently reflecting *P*-effect residual growth of wild-type cells, perhaps because the large number of metabolizing cells initially present depletes the medium of the serum factors antagonistic to the selection. However, the gross nonlinearity and leakiness of this assay render it difficult to score and of very doubtful reproducibility. In our experience, while a high drug concentration in the selective medium does not eliminate the phenomenon, the problem is circumvented when dialyzed serum is used in the selection. Photographs illustrating a comparison of assay plates prepared with undialyzed and dialyzed serum are shown in Fig. 2.

Even when dialyzed serum is used in the selective medium, the yield of colonies in a selection for AG-resistance may not be a strictly linear

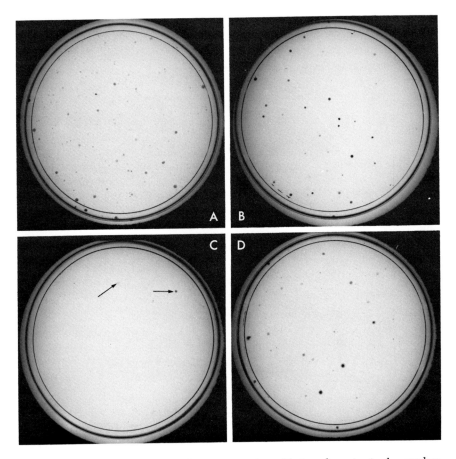

FIG. 2. Representative plates showing variation with inoculum size in the number of CHO colonies "phenotypically resistant" to 8-azaguanine in dialyzed and undialyzed fetal calf serum. For selective medium containing 20 μg/ml azaguanine plus 10% *whole* serum: (A) 10^5 cells inoculated yielded >100 heterogeneous colonies; (B) 10^6 cells inoculated yielded about 40 more discrete colonies. For medium containing 10 μg/ml azaguanine plus 10% *dialyzed* serum: (C) 2 colonies (arrows) from 10^5 cells inoculated; (D) about 30 colonies from 10^6 cells plated. The dishes (100 mm) were incubated for 11 days at 37°C and stained with methylene blue.

function of the number of wild-type cells inoculated. At least two factors affect this relationship.

First, it has been shown in many cell culture systems that metabolic cooperation can occur between AG-resistant and AG-sensitive cells (Subak-Sharpe et al., 1969; Dancis et al., 1969). Most evidence is consistent with a material transfer mediated by cell to cell contact (van Zeeland et al., 1972; Ashkenazi and Gartler, 1971; Cox et al., 1972). In practice, this phenomenon has the effect of suppressing the expression of the mutant phenotype at high cell densities. In selecting for the AG[R] phenotype in such a system, a balance must therefore be struck between plating enough cells to ensure the presence of a significant number of mutant cells and minimizing the cell density to retard any metabolic cooperation. Reduced recovery of AG[R] mutants of V79 hamster cells has been observed at high cell densities (Chu and Malling, 1968; Bridges and Huckle, 1970). In selection experiments with CHO cells, however, we have observed a small increase rather than a decrease in the frequency of colonies at high cell densities. This suggests that the factor discussed next predominates in the CHO system.

A second major factor which can affect the yield of colonies with varying cell density is the variation of effective azaguanine dose with time in cultures inoculated with different numbers of cells. Several authors have stressed the importance of refeeding the selection cultures periodically with fresh medium in order to maintain the effective drug dose (Littlefield, 1963; Chu, 1971a). Metabolites released into the medium by the initial breakdown of some cells may antagonize the further cytotoxic action of the drug on remaining sensitive cells. One would expect this effect to be more pronounced at higher initial cell concentration and thus to favor an increase in the frequency of colonies with increasing size of the inoculum if the cultures are not refed. We have observed a small effect of this nature in selections of CHO cells. On the other hand, for a given inoculum, the yield of AG[R] colonies is significantly reduced by refeeding.

To establish an assay for detecting mutagen-induced increases in the frequency of AG[R] colonies, we have adopted a procedure involving a fixed inoculum of moderate size (10^5 cells/18–21 ml of medium in a 100-mm dish), an intermediate dose (10–12 μg/ml), and have foregone refeeding in order to avoid the risk of initiating satellite colonies. As described below, this assay detects AG[R] colonies that breed true, but they are usually fully resistant (i.e., exhibit 100% plating efficiency) only at a dose somewhat lower than that present in the selective medium. Where the main objective is simply to isolate AG[R] mutants with high resistance, it may be preferable to change the medium periodically or to alter the inoculum size or initial drug dose, since quantitation and re-producibility will be of less concern.

3. Procedure for Isolation of Mutants

For descriptions of selective procedures employed with human lymphoblasts and human diploid fibroblasts, the reader is referred to the recent articles by Sato et al. (1972) and DeMars and Held (1972), respectively. Many other systems currently in use for selection of AG^R cells are based on the protocol introduced by Chu and Malling (1968; Chu, 1971a) for the near-diploid V79 hamster line. In this laboratory, we have satisfactorily utilized the following procedure for assay and single-step isolation of AG^R CHO cells.

The selective medium is prepared by dissolving 8-azaguanine (Sigma Chemical Co.) at the desired concentration, usually 12 $\mu g/ml$, directly in α medium lacking nucleosides. Solubilization of the drug is aided by temporarily adjusting the medium to pH 10. The medium is then supplemented with 10% dialyzed fetal bovine serum (see Section II,C).

The population to be selected is taken from an exponentially growing suspension culture and washed and resuspended with medium lacking nucleosides; the cell concentration is adjusted to $10^5/ml$. For selection, 100-mm petri dishes (Falcon 3003) containing 17 to 20 ml of the selective medium with 12 $\mu g/ml$ 8-azaguanine are inoculated with 10^5 cells and agitated to promote uniform dispersion of the cells. Alternatively, cells suspended in the selective medium at a concentration of about $5 \times 10^3/ml$ may be dispensed in the wells of Linbro trays, from which single resistant colonies may be isolated with minimal likelihood of cross-contamination (see Section II,F). The cells should be incubated at the desired temperature for at least 2 or 3 days longer than the standard time for colony formation.

This procedure will typically yield 1 to 10 colonies of AG^R cells per 10^5 cells assayed. The fidelity of the selection is illustrated by the following control experiment. Fifteen clones, each containing 50 or more cells after 11 days incubation at 37°C in Linbro trays prepared as above, were picked, cultured in nonselective medium, and tested for AG-resistance about 1 month after isolation. Each of these clones showed normal plating efficiency (P.E.) in 3 $\mu g/ml$ azaguanine, a concentration about fivefold that which significantly reduces the P.E. of wild-type cells, and most showed significant growth (although usually a reduced P.E.) in 10 $\mu g/ml$ drug. Thirteen of the fifteen clones, including those least resistant to 8-azaguanine, were sensitive to HAT medium; two of the clones were HAT-resistant.

C. 5-Bromodeoxyuridine Resistance

Mutants resistant to the thymidine analog 5-bromodeoxyuridine (BrdU) have been isolated from many cell lines. Incorporation of BrdU into DNA by wild-type cells leads to cell death, evidently as a consequence of chromosome aberrations (Hsu and Somers, 1961) and mutations (cf. Freese, 1971). Utilization of exogenous BrdU is dependent upon activity of the enzyme ATP:thymidine-5′-phosphotransferase (E.C. 2.7.1.21—commonly referred to as thymidine kinase and abbreviated TK) (Morris and Fischer, 1960; Kit *et al.*, 1963). By culturing wild-type cells in the presence of BrdU, resistant sublines with decreased thymidine kinase activity have been obtained (Djordjevic and Szybalski, 1960; Hsu and Somers, 1962; Littlefield, 1965; Kit *et al.*, 1966). Such cells have been shown to be cross-resistant, as might be expected, to 5-iododeoxyuridine (IdU) and excessive concentrations of thymidine (Dubbs and Kit, 1964).

The HAT medium back-selective system (see Section III,B,1) can be applied to obtain revertants of BrdU-resistant cells lacking thymidine kinase activity (TK⁻ cells) in the same way as described above for selecting revertants of HGPRT⁻ AG-resistant cells. Since aminopterin (or amethopterin) poisons endogenous thymidine synthesis by inhibiting thymidylate synthetase, only those cells in a BrdU-resistant culture that have regained the capacity to utilize exogenously supplied thymidine are able to proliferate in HAT-supplemented medium (Littlefield, 1964b, 1965). Like the HGPRT⁻ marker, the TK⁻ marker is recessive and enables selection against a parental cell type in cell-cell hybridization experiments (Littlefield, 1964b,c).

Cell populations generally acquire resistance to BrdU in a stepwise manner. The likely relevance of multiple *genes* to the BrdU-resistance phenotype will be discussed shortly. The phenotype is evidently also affected by multiple, perhaps two to four, *alleles* for thymidine kinase. Heteroploid mouse L cell and human HeLa cell lines isolated for resistance to intermediate BrdU concentrations show partial reductions in both thymidine kinase activity and incorporation of BrdU into their DNA (Kit *et al.*, 1963, 1966; Littlefield, 1965). Such lines may show a reduced growth rate but a normal plating efficiency in HAT medium. Similarly, revertants selected in HAT medium from cells resistant to high BrdU concentrations typically exhibit only partial recovery of thymidine kinase activity (Littlefield, 1965; Clive *et al.*, 1972).

Mutants that are highly resistant to BrdU and sensitive to HAT medium are less frequent by orders of magnitude than the analogous

HAT-sensitive, AG-resistant mutants in L cells (cf. Littlefield, 1963, 1965) and CHO cells (R. M. Baker, unpublished observations); it is rarely possible to isolate them in a single step (Clive et al., 1972). From some cell lines they can be obtained by consecutive single-step selections using increasing drug doses (Littlefield and Basilico, 1966). However, generally their isolation is further hindered by technical difficulties encountered in conducting a selection.

Drug-sensitive cells exposed to selecting concentrations of BrdU in monolayer culture display extensive background growth and giant cell formation; they typically detach quite slowly from a surface and it is therefore seldom possible to discern discrete colonies of resistant cells. Following subculture, it becomes apparent that most of the cells carried through the selection step are either nonviable or in poor condition due to bromouracil substitution in the DNA. This problem is aggravated by metabolic cooperation (Subak-Sharpe et al., 1969), which apparently occurs through contact between TK^+ and TK^- cells and causes suppression of the mutant phenotype (Chu, 1971a; Morrow, 1972). For these reasons the selection for BrdU-resistance operates optimally only in very well dispersed cultures. However, with monolayer cultures it is usually impractical to conduct the selection in this manner because of the relatively low frequencies of resistant cells.

In view of such difficulties, two approaches may be taken to selecting BrdU-resistant cells. (a) One may carry out the selection in monolayer cultures by continuously exposing an initially sizable population of cells to the drug during consecutive subcultures and refeedings, until a resistant subpopulation is established (Hsu and Somers, 1962). In order to increase the level of resistance, subsequent selection steps can be instituted using progressively higher drug concentrations (Kit et al., 1966). This method has been successfully used for L cells and CHO cells in this laboratory (Kim, 1969; R. M. Baker, unpublished experiments). We employ cultures initially containing 10^6 cells in 16-oz glass prescription bottles and concentrations of BrdU in approximately twofold increments from 2 $\mu g/ml$ to 200 $\mu g/ml$. (b) An effective selection can be performed in long-term suspension cultures, inoculated with a large number of sensitive cells and maintained in the presence of the selective dose of BrdU (with replacement of medium as necessary) until overgrown by a drug-resistant subpopulation (Littlefield, 1964b, 1965). As for monolayer cultures, consecutive levels of resistance may be selected. Where it is experimentally feasible, we think this method of selection should be preferable. Compared to a monolayer selection, a selection in suspension can conveniently be performed on a larger initial cell population, more likely to include a desired mutant. Also the uniform dispersion of cells in a suspension should minimize cooperative effects due to cell-cell contact and provide for better outgrowth of the mutant phenotype.

Once obtained, the BrdU-resistance phenotype is stable in the absence of selection. Revertants resistant to HAT medium occur at frequencies $\lesssim 10^{-5}$ (Littlefield, 1965; R. M. Baker, unpublished results). In the case of the LM-TK⁻ cell line (Kit et al., 1963), apparently a deletion mutant, no revertants have been observed. By examination of chromosome loss patterns in hybrid cells, this marker has been mapped on the E-17 human chromosome (Miller et al., 1971), substantiating its genetic basis.

It has been reported that X-rays and EMS increase the frequency of mouse lymphoma cells that are stably resistant to IdU (Fox, 1971) or to BrdU (Clive et al., 1972).

Although much work on BrdU-resistance has been related to the TK⁻ phenotype, there is evidence that other genes may also be involved in determining the level of drug resistance. Breslow and Goldsby (1969) have described V79 hamster cells isolated on the basis of resistance to the cytotoxic effects of tritiated thymidine. These cells occurred spontaneously at quite high frequency and showed greatly reduced uptake of thymidine but normal thymidine kinase activity in vitro. These characteristics imply a defect in transport. Mezger-Freed (1972) has obtained stably BrdU-resistant lines of haploid frog cells that also have nearly normal thymidine kinase activity although their uptake of thymidine is but 2–3% that of wild-type cells.

D. Ouabain Resistance

Mutants of somatic cells resistant to the membrane-active drug ouabain (strophanthin G) have been isolated and studied in this laboratory (Baker and Till, 1971; Till et al., 1973; Baker et al., 1973). Ouabain is a specific inhibitor of the Na⁺-K⁺-activated ATPase of the plasma membrane, the enzyme responsible for the active transport of potassium into the cell and extrusion of sodium (Glynn, 1964; Albers et al., 1968). Thus the rationale for selecting ouabain-resistant mutants was that cells refractory to the cytotoxic action of this drug might have an altered ATPase with either reduced ability to bind the drug or a changed response to bound drug. Because the drug acts on the cell surface, the isolation and characterization of resistant mutants could not be complicated by a class of cells with altered permeability to the selective agent.

The ouabain concentration necessary to effect substantial killing in tissue culture varies with cell species. Drug concentrations of approximately 10^{-7} M are toxic for human KB or HeLa cells, but doses of 10^{-3} M are necessary for toxicity to mouse 3T3 or L cells or CHO cells. For historical reasons our work to date has focused on the latter cell lines, although the high drug doses required have been a distinct disadvantage. Because of the solubility limit of ouabain, it is not possible to utilize doses much greater than 3 mM. However, we have found that the sensitivity of wild-type cells to ouabain can be increased about tenfold by lowering the potassium concentration in the medium from 5 mM to 0.5 mM (Till et al., 1973).[5] This observation is consistent with reports that potassium is antagonistic to ouabain binding and inhibition of Na^+-K^+-activated ATPase activity (Baker and Willis, 1970) and supports the inference that cell killing by ouabain is due to potassium starvation (Mayhew and Levinson, 1968). This problem of dosimetry is the only technical complication that we have encountered in assays of ouabain resistance.

The dose response of wild-type cells to ouabain is well suited for the isolation of drug-resistant mutants in a single-step selection experiment. The plating efficiency of CHO or L cells drops sharply from $\simeq 1$ at 0.3–0.5 mM ouabain (5 mM K^+) to 10^{-4}–10^{-7} at 1.0 mM ouabain (Till et al., 1973). With further increases in dose, the rate of decline in plating efficiency is much reduced (as in curve A, Fig. 1). The colonies detected at these doses represent ouabain-resistant cells. For example, of eleven clones picked at random from a selection of CHO cells in 3 mM ouabain and subsequently grown to mass culture in the absence of the drug, each showed 100% relative plating efficiency at the selecting drug concentration.

After a normal period of incubation, most of the colonies that appear under selective conditions are in the normal size range, but some are relatively small. By isolating and testing cells from small colonies at low selective dose and from large colonies at high dose, lines with varying degrees of drug resistance can be obtained. In this way we have derived mutant clones that are from threefold to about fiftyfold more resistant than wild-type clones in terms of minimum toxic drug dose. These variations in resistance are consistent with the fact that there is a gradual decrease in frequency of mutants assayed at progressively higher selective doses. For moderately high selective doses (3 mM ouabain with ≤ 5 mM K^+) the number of mutant clones detected varies linearly with the num-

---

[5] In order to obtain low potassium concentrations, the α medium is prepared without KCl and then supplemented with whole fetal bovine serum of measured K^+ content (usually ~ 10 mM) and concentrated KCl as desired. In our experience, the plating efficiency of L cells and CHO cells is normal for concentrations of $K^+ \geq 0.2$ mM (2% serum in α lacking KCl).

ber of wild-type cells inoculated (up to well over 10^6 cells per 100-mm culture dish with 20 ml medium). Thus in practice the assay is not dependent on cell density.

The ouabain-resistant cell lines satisfy several genetic criteria necessary to their classification as mutants. In general the clones breed true with respect to the extent of drug resistance. The phenotype of every ouabain-resistant line examined has been stable in the absence of the drug over an extended period of time (up to 11 months) in serial culture. Furthermore, the frequency of ouabain-resistant cells in a wild-type culture can be markedly increased by the chemical mutagen EMS (Section V,B).

Ouabain-resistant L cells have been compared to the wild-type cells with respect to ouabain inhibition of ^{42}K uptake by whole cells and of Na^+-K^+–ATPase activity in isolated plasma membranes (Brunette, 1972; Till et al., 1973; Baker et al., 1973). With increasing concentrations of ouabain in the selective dose range, there is progressive inhibition of uptake of extracellular ^{42}K by wild-type cells and of Na^+-K^+-activated ATPase activity in plasma membranes isolated from wild-type cells. Over this dose range the ^{42}K uptake and ATPase activity of ouabain-resistant cells are less affected. At drug doses sufficiently high to inhibit the growth of resistant cells, they *also* exhibit reduced uptake and reduced ATPase activity. Thus the effects of ouabain upon ATPase activity, ^{42}K uptake, and cell viability are strongly correlated.

On the basis of the evidence summarized above, we think it almost certain that ouabain-resistant cells are genetic mutants with an altered ATPase response to ouabain. It is not yet clear whether the altered response is typically due to a change in the ATPase protein, or whether it might be induced by membrane cooperative effects (cf. Cereijido and Rotunno, 1970). Efforts to detect a physiological difference between the wild-type and mutant lines in addition to their response to ouabain have so far not been fruitful. The ATPase activities of the wild-type and of a ouabain-resistant L cell line examined in detail are indistinguishable with respect to kinetics and optimal requirements for various constituents of the transport reaction (Brunette and Till, 1972; Brunette, 1972). Feasible procedures for isolation and assay of the ATPase enzyme from solubilized membranes are yet to be developed (Dunham and Hoffman, 1970; Kyte, 1971). Although the probable site of the alteration to ouabain resistance is known and the altered response of mutant cell lines has been observed both *in vivo* and *in vitro*, demonstration of an altered gene product at the molecular level remains to be accomplished.

The principal disadvantage of the ouabain-resistant phenotype as a genetic marker is the lack of a direct back selection for assay of revertants. On the other hand its advantages include knowledge of the principal affected enzyme, the ease and efficiency of the single-step selection, and the absence of cell-density effects in the assays at high drug dose. These properties suggest that the ouabain-resistance marker will be particularly useful for assay of mutagen induction of forward point mutations (Section V,C). In addition ouabain-resistance is codominant in somatic cell hybrids (R. Mankovitz and R. M. Baker, unpublished data). Thus this mutation will be useful for studies of gene dosage effects on phenotypic expression and as a selective marker in hybridization experiments.

E. Colchicine Resistance

The drug colchicine would appear to be another appropriate selecting agent for mutants which have been altered in a defined cellular structure. Colchicine is thought to bind specifically to a protein subunit of cellular microtubules (Borisy and Taylor, 1967a,b). This hypothesis is consistent with the observations that *in vivo* the primary effect of the drug is to arrest the cell at metaphase (Taylor, 1965) by interfering with the formation of microtubules so that a normal mitotic spindle does not develop (Brinkley *et al.*, 1967). Thus, it might be expected that the microtubule protein of resistant mutants would have a reduced affinity for the drug. Alternatively, resistance might arise from reduced permeability of the cell to the drug, as for 8-azaguanine and 5-bromodeoxyuridine discussed above.

A number of colchicine-resistant lines of CHO have been obtained in our laboratory and characterized in terms of the nature of the alteration which confers resistance and their suitability for use as genetic markers (Ling and Thompson, 1973). The selection of lines resistant to colchicine is relatively straightforward and can be accomplished by performing single-step plating experiments. Wild-type cells undergo rapid killing with increasing concentration of drug at doses in the neighborhood of 0.05 μg/ml. The colonies observed at concentrations along the steep portion of the killing curve are small and heterogeneous in size, probably reflecting a P-effect (Section III,A). The tail of the curve is analogous to curve "B" in Fig. 1 above. At high drug concentrations, e.g., 10 μg/ml, no resistant colonies appear in a single-step plating even if the population has been mutagen treated.

The maximum inoculum size consistent with efficient recovery of colchicine-resistant clones of CHO cells is about 5×10^5 cells per 100-mm petri dish containing 30 ml of α medium. Giant-cell formation typically occurs and cell lysis is delayed, resulting in appreciable depletion of the medium by nonsurviving cells. The incubation interval required to produce the maximum number of macroscopic colonies is generally somewhat longer than normal, i.e., about 11 days or more at 37°C.

At a selecting concentration of 0.10 μg/ml colchicine, first-step selections have yielded a spectrum of clones with different levels of resistance. Some clones are fully resistant to the selecting concentration, whereas many others are not, but are more resistant than wild-type clones. In a recently cloned wild-type population the frequency of cells resistant to 0.10 μg/ml colchicine is of the order of 10^{-6} in the absence of mutagen treatment. Attempts to obtain lines of increased resistance by replating such resistant clonal populations at higher concentration of drug have yielded clones resistant to 5 μg/ml after a total of three selection steps in which mutagen was used (Till et al., 1973). These highly resistant clones are characterized by long population doubling times (about twofold that of wild-type cells), even when grown in the absence of the drug.

The genetic properties of the resistant colonies isolated are consistent with the criteria for true mutants. First, when the dose-response curve is determined for individual isolates early after growth to mass culture in the absence of drug, the lines invariably exhibit increased resistance to colchicine as compared to wild-type cells. The lines are phenotypically stable during passage in culture. Second, the frequency of resistant cells in populations of either wild-type or partially resistant cells can be markedly increased by treatment with a chemical mutagen, ethyl methanesulfonate (Section V,B).

A preliminary biochemical characterization of the alteration(s) which confers resistance in these lines can be performed utilizing two complementary assays. First, to test the hypothesis that a resistant phenotype derives from a structural alteration in the colchicine-binding protein, the ability of cell extracts from mutant lines to bind colchicine-^{3}H has been examined. A resistant line might bind the drug less effectively than wild-type cells. Second, the rate of uptake of colchicine-^{3}H by whole cells has been measured as an index of the permeability of the cell to colchicine. Most, if not all, lines which we have tested appear to derive their resistance through reduced permeability since the efficiency of colchicine binding in cytoplasmic extracts is normal and the rate of accumulation of the drug by growing cells is greatly decreased (Till et al., 1973). There also appears to be a positive correlation between the level of resistance of a mutant line and the ability of the cells to exclude colchicine. Although such results are disappointing with respect to the isolation of a mutation affecting microtubule protein, the mutants are interesting in themselves as another class of genetic change affecting permeability.

IV. Isolation of Auxotrophic and Conditional Lethal Mutants by Indirect Selection

A. Selection for Noncycling Cells

The isolation of cells carrying mutations to auxotrophy or to conditional lethality involves retrieval from a wild-type population of those cells that do not grow under nonpermissive conditions. This may be accomplished either by nonselective screening procedures or by using selection techniques, but the relatively low frequency of mutants in a population will generally necessitate the use of a selective isolation. In order to enrich for mutant cells in a selective isolation, one must subject a culture under the nonpermissive conditions to a treatment that kills the normally growing wild-type cells but allows noncycling mutant cells to survive. A basic limitation of this approach is that the mutant cells will usually also lose viability simply as a consequence of being exposed to the nonpermissive conditions, and thus such selections generally cannot yield quantitative recovery of mutants from a population.

Most of the selective isolation procedures which have been described to date have been designed to select for noncycling cells and have employed agents which act during the S phase of the cell cycle. Auxotrophic mutants have been isolated by Kao and Puck (1968) using a procedure which involves BrdU incorporation followed by irradiation with visible light (Puck and Kao, 1967). Temperature-sensitive (*ts*) conditional lethal mutations have been selected using tritiated-thymidine (dT-^{3}H) (Thompson *et al.*, 1970, 1971), 1-β-4-arabinofuranosylcytosine (ara-C) (Smith and Wigglesworth, 1972; Thompson *et al.*, 1971), and 5-fluoro-2'-deoxyuridine (Meiss and Basilico, 1972), as well as the BrdU-light treatment (Naha, 1969). In principle it should be possible to use other cytotoxic agents as well, for example, drugs such as colchicine and vinblastine sulfate which specifically kill cells passing into mitosis.

Our procedures employing dT-^{3}H suicide to select both auxotrophic and temperature-sensitive mutants are similar in principle, and apparently also in effectiveness, to the BrdU-light method described by Puck and Kao (1967). We chose the dT-^{3}H treatment because it seemed somewhat easier to manipulate technically. Both procedures suffer from the theoretical limitation that thymidine kinase activity and DNA synthesis are required for a lethal effect. Although we used both dT-^{3}H and ara-C, either singly or in combination, in our selections of L cells, we have not employed ara-C with CHO cells because dT-^{3}H alone appears to be sufficient to produce maximal killing (cf. Section IV,A,5). Moreover, under

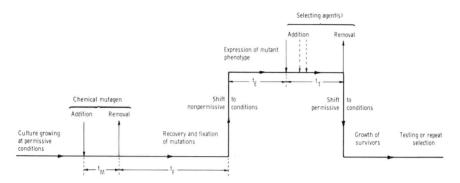

FIG. 3. Outline of selection procedure for nutritional auxotrophs or temperature-sensitive conditional lethal mutations. See text for experimental details.

some conditions ara-C may be less effective than dT-^{3}H as a killing agent (Graham and Whitmore, 1970). With L cells ara-C treatment resulted in accelerated lysis, while dT-^{3}H treatment caused formation of giant cells. Although rapid lysis might be an advantage if the selection were being done on monolayer cultures, in suspension culture this can be a limitation because of cell-concentration dependency of growth. A disadvantage of the use of dT-^{3}H is the problem of isotope disposal.

The conditions for selection of auxotrophic and of *ts* conditional lethal mutants will be discussed in general terms applicable to both before considering those aspects specific to either class of mutants. The selective techniques developed in our laboratory for the two types of mutants have a number of features in common, as indicated in the outline of the basic enrichment procedure for selecting noncycling cells given in Fig. 3. Permissive and nonpermissive conditions refer either to different temperatures, in the case of *ts* mutants, or to the appropriate complete and deficient media, respectively, in the case of auxotrophic mutants. The various parameters of the selection will now be discussed in some detail.

1. MUTAGEN TREATMENT

Selections in our laboratory have usually been performed on cultures grown in suspension in 200-ml spinner flasks (Johns Glass Company, Toronto, Ontario) at a concentration of about 10^5 cells/ml. In order to increase the frequency of mutants, the culture is first treated with a chemical mutagen, N-methyl-*n*′-nitro-N-nitrosoguanidine (MNNG or NTG) (Mann Research Laboratories) or ethyl methanesulfonate (EMS) (Sigma Chemical Co)., using conditions which result in a survival after mutagen treatment on the order of 10–50%. The appropriate dosage must be determined for each cell line; our experience indicates that even different strains of a given line can exhibit somewhat different sensitivities to a mutagen treatment. For wild-type CHO cells, an exposure to 200 μgm/ml EMS for 16 hours reduces the viability to about 50%; treatment with MNNG for 3 hours at a concentration of 0.1 μg/ml results in survival of about 10%.

2. FIXATION OF MUTATIONS

After removal of the mutagen from the culture by centrifugation and resuspension of the cells in fresh medium, an interval (t_F) is allowed for recovery of surviving cells from sublethal effects of the treatment and for fixation and expression of the mutant genes (see Chu, 1971a). We have generally allowed an increase in cell number after mutagen treatment of about tenfold. If the time interval, t_F, is greatly prolonged, the frequency of mutants in the culture may decline if they are at a selective disadvantage, as is the case for many ts lines and for some auxotrophs.

3. PHENOTYPIC EXPRESSION UNDER NONPERMISSIVE CONDITIONS

Following the interval t_F, the culture is placed under nonpermissive conditions. A phenotypic expression time t_E is allowed to pass so that the growth of mutants present in the population will be arrested and they will be refractory to the selecting agent. This interval may be varied but is usually on the order of 6 to 24 hours (see Sections IV,B,2 and C,2 below). It is clearly important that the expression time not be unnecessarily prolonged since this would contribute to increased loss of viability of arrested mutant cells.

4. ADDITION OF THE SELECTING AGENT

Since the efficiency of the selection will be determined largely by the fraction of the wild-type cell population which passes through S phase during the treatment interval, t_T, it is essential that the culture be growing well and that there be no limitation of growth due to medium depletion during the treatment. The starting cell concentration in suspension culture should be no greater than about 10^5 cells/ml. High specific activity (20–30 Ci/mmole) dT-^{3}H is added to the culture to kill by radiation suicide all cells passing through S phase (Whitmore and Gulyas, 1966). The amount of isotope needed will depend on the efficiency with which the cell line utilizes exogenous thymidine and the length of treatment, t_T. A dT-^{3}H concentration of 3 μCi/ml is suitable for a 24-hour treatment of wild-type CHO cells when 10% dialyzed fetal calf serum is used in α medium lacking thymidine. With L cells the greater efficiency of thymidine utilization allows use of whole serum (which generally contains thymidine) in combination with the above isotope concentration. The length of the treatment, t_T, ranged from 2 to 5 days in our previously reported L cell selections, but in more recent selections with CHO cells we have usually employed a treatment interval of 24 hours in order to minimize the total period under nonpermissive conditions. For $t_T > 24$ hours additional quantities of isotope are often required, as indicated in Fig. 3, to compensate for its depletion from the medium. The selective treatment is terminated either by centrifuging and resuspending the cells in fresh medium or by adding nonradioactive thymidine (10–100 μg/ml) to the medium.

5. Completion of the Selection

Under the above conditions, there is sufficient incorporation of isotopic precursor to result in maximal killing of wild-type cells without the necessity of a cold storage interval for the accumulation of radiation damage. At this point the cells are transferred to monolayer culture at permissive conditions so that the surviving cells may multiply. Macroscopic colonies formed by the survivors may be picked and tested for mutant phenotype, or they may be pooled by trypsinization and returned to suspension culture for a repetition of the enrichment cycle. In practice we have sometimes found it necessary with the shorter treatment times (e.g., 24 hours) to perform the selection cycle two or more times on a population in order to increase the frequency of mutants to a readily detectable level, say 10%. The killing of wild-type cells appears to be limited in each selection cycle by the presence of a small fraction of slowly cycling cells which do not pass through S phase during t_T, but which later cycle normally. Tests have shown that in the case of CHO cells the nonmutant cells which survive the selection are sensitive to the killing effects of dT-^{3}H and are, therefore, not resistant phenotypes analogous to the dT-^{3}H and BrdU resistant cell types (Breslow and Goldsby, 1969; Dubbs and Kit, 1964).

A disadvantage of the enrichment procedure conducted with a single suspension as described above is that mutant clones which are isolated are not necessarily independent, i.e., they may be siblings. Because attempts to demonstrate that different clones actually have different mutations can be quite time consuming, it is desirable to ensure independence of origin whenever possible (Section II,H).

B.   Isolation of Nutritional Auxotrophs

Auxotrophic mutations are important for a variety of reasons. They can assist in the elucidation of the nature of mutations in animal cells and in the characterization of metabolic pathways. They can serve as genetic markers for a number of purposes, such as mutagenesis, selection of hybrid cells, and determination of linkage relationships. Moreover, information and experience gained from their selection may be useful in developing selections for *ts* conditional lethal mutations.

Aside from the glutamine-requiring lines isolated from HeLa cells (DeMars and Hooper, 1960) and V79 hamster cells (Chu *et al.*, 1969), essentially all of the reported isolations of auxotrophs have been performed using strains of the CHO line. The feasibility of isolating a variety of auxotrophic markers was first demonstrated by Kao and Puck (1968, 1969) with a selection procedure which was developed by using a naturally occurring line auxotrophic for the nonessential amino acid proline as the basis of a model selection (Puck and Kao, 1967). As originally described and characterized (Kao and Puck, 1967), the "wild-type" CHO cells were found to have a fortuitous growth requirement for proline. Thus, by mixing cells carrying this marker (Pro⁻) into populations of other hamster cells (CHL Pro⁺), Puck and Kao demonstrated by reconstruction experiments an effective selection procedure for enriching for cells of the Pro⁻ phenotype. To date mutations that fall into a total of eight different complementation groups, including four for glycine auxotrophy (Kao *et al.*, 1969b), have been selected in that laboratory (Kao and Puck, 1972). Similar auxotrophic phenotypes have been isolated in the same CHO line (subclone K1 containing 20 chromosomes) by Taylor *et al.* (1971), and by members of our laboratory (D. Ness and M. McBurney, personal communications) using a line of CHO with a modal chromosome number of 21.

1. Media Requirements

In order to select for nutritional auxotrophs one must formulate a permissive and a nonpermissive medium for the mutant types being sought. In theory this might be accomplished either by removing unnecessary components from the normal growth medium to produce a medium which will be permissive for wild-type cells but nonpermissive for the desired mutant, or by regarding the normal medium to be nonpermissive and making certain additions to it which might permit growth of mutants. Members of our laboratory chose to select for auxotrophic mutants of CHO cells comparable to those which had been reported by Kao and Puck (1968, 1972). In order to derive a nonpermissive medium from complete α medium (cf. Section II,C) we removed the same components that Kao and Puck had omitted from F12 medium to form F12D (Kao and Puck, 1968). The ingredients omitted were inositol, lipoic acid, vitamin B_{12}, aspartic acid, glutamic acid, alanine, glycine, and all eight nucleosides and deoxynucleosides, resulting in a medium we designated deficient α or $D\alpha$. (One might not expect to obtain single-gene mutations to auxotrophy for the nonessential amino acids alanine, glutamic acid, or aspartic acid, since each is a key intermediate in metabolism and a product of several different pathways. In fact, none of these phenotypes has yet been reported.) The Dα medium is always supplemented with *dialyzed* fetal calf serum (cf. Section II,C) at a concentration of 10%.

2. OUTLINE OF TYPICAL SELECTION

The following is an outline of a representative result (D. Ness, personal communication) based on a total of six experiments which have been carried out in our laboratory to isolate auxotrophs according to the dT-^{3}H enrichment procedure indicated in Fig. 3: (1) After treatment with EMS for 16 hours at 200 μg/ml (fractional survival $\simeq$ 0.20) and a mutation fixation interval of 6 days, the culture was rinsed twice with phosphate-buffered saline and resuspended after centrifugation in Dα medium containing 10% dialyzed fetal calf serum. (2) The culture was divided into two parts to allow phenotypic expression times of 6 or 24 hours before adding the selecting agent dT-^{3}H. (3) In each case a treatment time of 24 hours was used; the survivals were 3×10^{-3} and 5×10^{-4} for the cultures of 6- and 24-hour expression times, respectively. (4) The survivors were grown out in α medium and the selection cycle was repeated in the same way. (5) After outgrowth of the survivors a second time, the two cultures were plated for clone formation in 96-well Linbro trays (ISFB-96TC). (6) Eight-day-old clones were picked for testing, divided equally into 24-well Linbro trays (FB16-24TC), and incubated in α medium for 2 days to produce rapidly growing cultures. (7) The test for auxotrophy was performed by aspirating the medium from each well, rinsing thoroughly with PBS, and then adding α medium to one and Dα medium to the other of each pair of cultures. (8) After several days incubation, the wells containing Dα medium were examined under low-power, bright-field microscopy for the absence of growth and for disintegration of the cells. When this occurred, the corresponding culture was grown up for further testing. In this experiment the frequencies of mutant colonies corresponding to expression times of 6 and 24 hours were 36/237 and 14/236 after two selection cycles. In similar experiments the enrichment efficiency of the selection was measured; e.g., the frequency of mutant colonies increased appreciably between the first and second selection cycles from 0/97 to 22/138 when MNNG was used as the mutagen and a 24-hour expression time, t_E, was allowed.

3. PRELIMINARY CHARACTERIZATION PROCEDURES

In order to screen a number of mutant clones for their growth requirements, we have utilized 96-well Linbro trays (#96CV-TC) which hold about 1 ml of medium per culture well. The appropriate test media, and the desired number of cells (e.g., 10^3) in a very small volume of $D\alpha$ are added to each well. After incubation for 5–6 days at 37°C the trays are stained and examined for wells with no growth. This method is rapid and allows a semiquantitative assessment of the growth requirements of each mutant. In our experiments the media for the preliminary tests are usually formulated such that there is a single omission from α medium. The results will then indicate whether a given mutant has one or several absolute requirements for growth. Alternative requirements (e.g., adenosine or deoxyadenosine) must be determined by testing in nonpermissive medium containing only certain additions.

When the growth requirements are known, the lines are plated at various cell concentrations in $D\alpha$ and α medium to characterize more fully their phenotypes in terms of normality of growth, reversion frequencies, etc. When plated at low cell concentration into α medium, the plating efficiency of the mutant lines is generally 20 to 80%; colony size may be normal or below normal. Some mutant lines grow better when other, nonessential growth factors are present. All the lines are essentially stable in phenotype when cultured over a period of weeks or months. The plating efficiencies in $D\alpha$, assayed shortly after isolation, range from $\simeq 10^{-5}$ to $< 10^{-7}$ in different mutant lines. In general, the auxotrophs isolated in this laboratory appear to be similar in their properties to those which have been previously reported by Kao and Puck (1968, 1972) and by Taylor *et al.* (1971).

4. FACTORS AFFECTING SELECTION YIELD

Although a variety of auxotrophic mutations can be isolated from the near-diploid CHO lines, the different possible kinds of auxotrophs are obtained with quite variable frequencies. For example, mutations in four of the eight different complementation groups identified by Kao and Puck (1972) lead to glycine auxotrophy (Kao *et al.*, 1969b), but in our selections only one glycine mutant was obtained. Of six types of auxotrophs isolated here, one phenotype (dT, A, or dA)$^-$ appears to represent a complementation group not previously reported for CHO cells (D. Ness, personal communication). Moreover, in one selection (M. McBurney, personal communication) the phenotypes of all of the mutant clones tested after two cycles were either (gly, dT, A)$^-$ or (gly, T)$^-$, which appeared to be noncomplementing. Some phenotypes such as auxotrophy for vitamin B_{12} have not been recovered at all. Among the factors which may significantly influence the ease of selecting a particular mutant phenotype are the following.

a. Karyotype. Success in isolating auxotrophs may be related to the occurrence of monosomic chromosomal regions in the wild-type cell lines, as suggested by Kao and Puck (1968) (see also Section VI). They utilized a subclone of CHO containing only 20 chromosomes (as opposed to 21 chromosomes in the parental line and the 22-chromosome karyotype *in vivo*). Our parental line of CHO cells has a mode of 21 chromosomes and appears to differ slightly from the parental line of Kao and Puck (1969) in one or more of the smaller chromosomes. Such karyotypic variations may involve differences in the extent of hemizygosity, which could be reflected by variations in mutant yield. In addition, the selections themselves may enrich for subpopulations with particular monosomic regions because of the significant degree of chromosomal damage and possibly rearrangement which occurs with mutagens like EMS and MNNG (Kao and Puck, 1969). Kao and Puck (1968) have reported that their mutants usually possess the same number of chromosomes as the parental line. Most of our auxotrophic lines (17 out of 22 lines examined) also had the modal chromosome number of the parental line (21). (The remaining clones had either 23, 22, or 20, or a mixture of 21–22 or 20–21 chromosomes.) Certainly there was no tendency for the mutant lines to have fewer chromosomes than the wild type, but the extent of chromosomal rearrangements they may have undergone is not clear.

b. Viability of Mutants under Nonpermissive Conditions. Undoubtedly a very important parameter affecting the relative recovery of mutants is the degree to which they remain viable under conditions of arrest during the selection. Measurements on the rate of loss of viability in deficient medium for various mutant lines indicate sizable differences in the rate of killing (Kao and Puck, 1969). For mutants of phenotypes (gly, dT, A)$^-$ and gly$^-$ isolated in this laboratory, for example, the survival after 48 hours in nonpermissive medium was approximately 10% and 50%, respectively. In a selection one could expect differences in the survival both of different types of auxotrophs and of independent mutants of the same type.

c. Degree of Phenotypic Expression at Time of Selection. In order to survive the selection, a mutant cell must not only have retained its viability at the end of the interval, t_T (Fig. 3), but it must have expressed its mutant phenotype at the beginning of t_T. The efficiency of selecting different mutants will vary with the degree of phenotypic expression of each type at the end of t_E. Different mutant lines show considerable variability when examined for the rate of decrease in their incorporation of dT-^{3}H in nonpermissive medium. Under the actual conditions of selection the rate of expression of the mutant phenotype may be slowed or modified by the presence of the excess numbers of wild-type cells as suggested below.

d. Effects of Cross-Feeding on Mutant Phenotype. Since the expression of the auxotrophic phenotype is determined by the presence or absence of soluble factors in the medium it might be expected that nonmutant cells in deficient medium could sometimes release into the medium the nutrient(s) required by an auxotrophic mutant. The effect of such cross-feeding in a selection would be to prolong the time required for phenotypic expression or even to prevent the arrest of mutant cells.

Various observations on our auxotrophic lines indicate that cross-

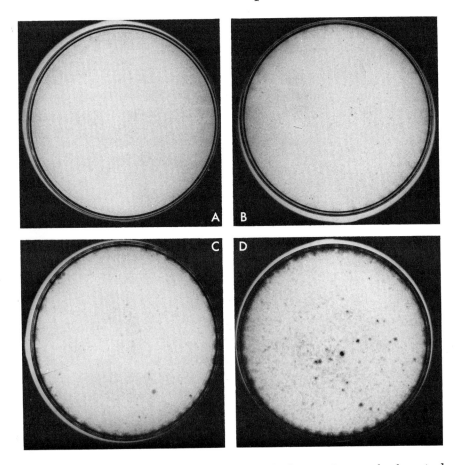

FIG. 4. Demonstration of the effect of cross-feeding on the growth of a mixed population of two auxotrophic lines of CHO cells. The mutants, designated *AUXE5* and *AUXE29*, were isolated by D. Ness in this laboratory; each has a growth requirement satisfied by adenosine and belongs to a different complementation group (Section II,I). Each 60-mm petri dish shown was inoculated with a total of 4×10^5 cells and incubated at 37°C for 8 days in Dα medium (see text Section IV,B,1). (A) *AUXE5* cells, no medium change; (B) *AUXE29* cells, no medium change; (C) *AUXE5* + *AUXE29* cells, medium changed every 48 hours; (D) *AUXE5* + *AUXE29* cells, no medium change.

feeding does occur and probably affects the spectrum of mutant types recovered from a selection. Figure 4 illustrates this phenomenon. The stained dishes show that when two different auxotrophic lines are incubated in Dα medium separately there is little or no growth (dishes A and B), but when the two are incubated as a mixture there is considerable growth (dish D), which can be reduced by frequently changing the medium (dish C). In instances where cross-feeding appears to pose a real difficulty in selecting a particular kind of mutant, it might be advisable to conduct the selection at a lower cell concentration than normally used or to change the medium frequently.

C. Isolation of Conditional Lethal Temperature-Sensitive Mutants

Although temperature-sensitive mutations comprise the broadest and potentially the most useful class of mutations, only a few laboratories have reported the isolation of temperature-sensitive (ts) somatic cell lines (Naha, 1969; Thompson et al., 1970, 1971; Meiss and Basilico, 1972; Smith and Wigglesworth, 1972; Scheffler and Buttin, 1973). Conditional lethality dependent on pH, ionic strength, the presence or absence of suppressor mutations, etc., has not yet been described for mammalian cells. The following properties of conditional-lethal ts mutants make them uniquely useful for studying the relationship between structure and function: (1) The conditionality of expression of the phenotype allows one to obtain and work with lethal mutations in characters which could not otherwise be isolated. (2) Knowledge of a particular function is not required for isolating a mutant that is defective in that function, as long as it is necessary for survival. (3) A culture of mutant cells at the permissive temperature will serve as an isogenic control for a culture under study at the nonpermissive temperature. (4) The ts defect will probably arise from a change in the thermal kinetics of denaturation or assembly of a particular protein.[6] Identification of the temperature-sensitive protein permits study of its molecular properties relevant to the defective cell function.

1. ESTABLISHING TEMPERATURES FOR THE SELECTION

Since cultured mammalian cells grow only in a temperature range from about 32°C to about 40°C, the maximum temperature differential is

---

[6] Temperature sensitivity is sometimes attributable to changes in tRNA (Yamamoto et al., 1972). It is conceivable that ts phenotype could also arise in some instances from a phase-transition in membrane lipids at the nonpermissive temperature (see Overath et al., 1970, 1971).

smaller than that employed with bacteria and other organisms in which *ts* mutations have been isolated. In theory, it would seem desirable to use permissive and nonpermissive temperatures spanning as wide a range as possible in order to optimize the likelihood of obtaining mutants. Our experience suggests to us that it is preferable to be slightly conservative in choosing the temperature limits for a particular cell line so that near-normal growth of wild-type cells occurs at both temperatures. If the permissive temperature is so low that the population doubling time is substantially prolonged, an impractical time interval is required for the selection. Furthermore, our analysis of the growth of several mutant lines of L cells and CHO cells as a function of temperature suggests that the temperature sensitivity is usually expressed in the upper portion of the available temperature range, e.g., above 37°C (see Thompson *et al.*, 1970). If the nonpermissive temperature is chosen so high that wild-type cells begin to show abnormal growth properties, the efficiency of the selection may be noticeably impaired. For L cells and for CHO cells appropriate selection temperatures are 34° and 38.5°C although CHO cells continue to grow well up to approximately 39.5°C. At these temperatures the plating efficiencies and doubling times compare well with the values obtained at 37°C.

In some circumstances it may be advantageous to investigate the behavior of mutant isolates at temperatures other than those used in the selection. In particular, if the biochemical basis of the phenotype is being studied, it may be possible to accelerate the expression of the temperature-sensitive lesion by employing temperatures slightly higher than those used in the selection.

2. Enrichment Procedures Based on dT-³H Treatment

The selection procedures for conditional lethal *ts* mutants that we have employed have been based primarily on the use of dT-³H and the basic selection protocol as outlined in Fig. 3 and discussed above. Many of the preceding comments on the isolation of auxotrophic mutations are relevant to the selection of *ts* mutants; we shall now consider in detail several of the selection parameters with specific reference to the isolation of *ts* mutants, based on our combined experience with L cells initially (Thompson *et al.*, 1970) and more recently with CHO cells. The number of selection cycles and the length of treatment in each cycle (t_T) will be discussed together because they have been connected variables in our experiments. The phenotypic expression time, t_E, will then be considered as another major variable. Because of the length of time and the work required to complete a selection experiment, it has not been feasible to test the selection parameters in a highly systematic way and to optimize rigorously the variables which affect mutant frequency and/or mutant phenotype. Thus, the primary aim in this section is to indicate a range of selective conditions which appear to be productive, to serve as a guide to the investigator who may wish to modify the details of the procedures to suit his own objectives.

As indicated above (Section IV,A), the design of a selection hinges upon establishing a treatment time long enough to kill substantially wild-type cells but not so long that all mutants lose their viability under the nonpermissive conditions. The precise length of treatment, t_T, in a particular situation will depend on the phenotype of the mutation being sought. If t_T is relatively short, e.g., 24 hours, it will likely be necessary to repeat the dT-^{3}H selection cycle one or more times in order to raise the frequency of mutants in the culture to a level at which they can be easily isolated by cloning and testing.

In combination with a fixed expression time of 24 hours, we have tried schemes ranging from single treatments of 1 to 5 days duration, to multiple (two to five) treatments, each of 1 to 2 days duration. In each case, the selective conditions have yielded ts lines although the frequency of recovery, the "quality" of the isolates, and probably also the degree of independence of the phenotypes, has varied considerably. Our original isolation with L cells (Thompson et al., 1970), which involved four 48-hour treatment cycles with dT-^{3}H followed by an extended treatment of colonies with ara-C, yielded a high percentage of mutants, but the procedure was too long to be routinely practical. Subsequent selection experiments with L cells (Thompson et al., 1971) demonstrated that the frequency of ts cells in a freshly mutagenized population is quite high (e.g., $\simeq 6 \times 10^{-3}$) and that mutants can readily be recovered by using single, prolonged selection treatments 5 days in length, during which time the medium and isotope are replenished at least once. These conditions result in very low survival of wild-type cells (10^{-5}–10^{-6}) and mutant frequencies as high as 40% among the survivors. The main limitation here is that most mutants in the culture do not survive the treatment; calculations indicate that more than 99% of the mutant cells are killed. Thus the retrieval of some phenotypes undoubtedly requires a shorter t_T and probably additional application of the dT-^{3}H cycle.

For CHO cells somewhat more information is available concerning selection design with respect to t_T and the effect of repetition of treatment. The kinetics of killing with dT-^{3}H was examined in a single-cycle selection experiment, using a 24-hour expression time and treatment times of 24, 48, or 72 hours. The rate of killing of wild-type cells was substantially reduced after the first 24 hours; between 24 and 72 hours the survival declined only from 3.3×10^{-4} to 3.6×10^{-5} even though

TABLE I

dT-³H SELECTIONS FOR *ts* MUTANTS WITH CHO LINE[a] *AUXB1*

Selection experiment	Mutagen and surviving fraction	Fixation time[b] (in days)	Expression time[b] (in hours)	Treatment time[b] (in hours)	Frequency of mutants after	
					Two cycles	Three cycles
I	MNNG, 0.002	11	6	18	2/90	—
J	EMS, 0.20	5	6	18	2/100	19/131
K	EMS, 0.20	5	24	24	9/130	—
L	MNNG, 0.03	8	9	18	0/103	4/92
M	MNNG, 0.03	8	24	24	1/99	9/96

[a] This line has a triple requirement for glycine, thymidine, and adenosine; in order to perform the dT-³H selections the thymidine requirement was fulfilled by adding Leucovorin (formyl tetrahydropteroylglutamic acid) at a concentration of 4 μg/ml.

[b] Defined in Fig. 3.

the medium and dT-³H were renewed at 48 hours. The surviving colonies, some of which were picked and tested, exhibited temperature sensitivity at a frequency of about 10% regardless of the length of treatment. (These isolates tended to be less temperature-sensitive, however, than lines obtained after multiple selection cycles using the altered protocol described in this article.)

Table I outlines data on multiple-cycle selections with CHO cells. The frequencies of *ts* clones, determined after two or three selection cycles, are based on tests of 12-day-old clones picked from Linbro trays after plating a mass population which grew up from selection survivors. The test was performed by inoculating the cells of each colony into duplicate 24-well Linbro trays and first incubating the trays at 34°C for 2 days to allow growth to be well established. After placing one tray of each duplicate pair at 38.5°C, the trays were further incubated for about 5 days. Identification of mutant phenotypes was based on visual examination of the cultures for gross differences in growth and was therefore somewhat arbitrary. Most of the clones considered to be *ts* according to this standard in the preliminary screen were shown by subsequent testing to differ by at least 100-fold in plating efficiency at the two temperatures. For cases where the increase in mutant frequency between cycles in the same selection experiment was measured, the data in Table I show an estimated enrichment of about ten-fold for each cycle. Since the surviving fraction at the end of a dT-³H treatment was typically $\leq 10^{-3}$, it is clear that these procedures are relatively inefficient in enriching for mutant subpopulations.

Let us now briefly consider the factors which affect choice of the expression time, t_E. The value of 24 hours for t_E which was used for the L cell selections and several CHO selections was chosen somewhat arbitrarily, but with the idea that its approximate equivalence to one generation time might be appropriate for selecting mutants blocked at a particular phase of the cell cycle. It would appear desirable to decrease the sum of $t_E + t_T$ in order to cope with the problem of rapid death of some mutants under nonpermissive conditions. The possibility of shortening t_E was emphasized to us by the properties of one CHO mutant (tsH1) which shows an unusually clean ts phenotype, including a very rapid response to the nonpermissive temperature (Thompson and Stanners, 1972). The fractional survival of this mutant after 22 hours at 38.5°C is only 0.004. The data in Table I imply that shorter expression times, 6 to 9 hours, even when accompanied by a somewhat abbreviated t_T, which did not greatly reduce the degree of killing of wild-type cells, do not result in higher frequencies of mutants than are obtained with a 24-hour expression.

Based on the above information, recommended parameters for the selection of ts mutants of unspecified phenotype from CHO cells are: $t_E \simeq 24$ hours and $t_T \simeq 24$ hours, with either one or two applications of the selection cycle prior to testing a hundred or more clones for temperature sensitivity.

3. ALTERNATIVE SELECTING AGENTS

The selections described above are often lengthy, and undoubtedly they produce a biased representation of the phenotypes which exist in the culture after mutagenesis. There is no apparent way of overcoming the limited efficiency of a selecting agent which is dependent on cell progression around the cycle for its lethal effect.

Certain kinds of mutations which arrest growth can probably be isolated more effectively by methods which do not involve selection on the basis of cell progression through any particular phase of the cell cycle. For example, mutations affecting protein synthesis can likely be selected directly on the basis of a reduction in the rate of protein synthesis. Either conditions of thymineless death or, particularly, treatment with tritiated amino acids might be effective for such a selection (Kaplan and Anderson, 1968; Tocchini-Valentini et al., 1969). Since all viable cells in a population can be expected to engage in protein synthesis, there should be no resistant subpopulation in this type of selection, and a single-step selection could be practical and efficient.

In an analogous way it might be possible to isolate ts mutations specifically affecting RNA synthesis by using tritiated uridine, as a selecting agent. Other approaches utilizing the appropriate radioactive precursors might be fruitful for selecting mutations affecting functions involved in the synthesis of cellular components such as phospholipid (Cronan *et al.*, 1970) or glycoprotein.

4. Selection of Cell Division Mutants on the Basis of Differential Attachment

Conditional lethal mutations affecting the biochemical events or structural components of cell division would be particularly useful. They could be critical tools in examining the molecular processes which underlie the profound structural changes involving the chromosomes, mitotic spindle, and nuclear membrane. In many cell lines, including CHO, the mitotic interval is unique in that it is a period of dramatic morphological alteration. When interphase cells growing attached to a surface in a spread configuration enter mitosis, they assume a spherical shape and become loosely attached or completely detached from their substrate. This property of mitotic cells has often been utilized to collect them as pure populations simply by shaking the culture gently so that they preferentially detach (Terasima and Tolmach, 1961; Petersen *et al.*, 1968).

This phenomenon of mitotic detachment suggests that *ts* mutants arrested in cell division might be obtained by selecting for cells which are unable to attach at the nonpermissive temperature. Mitotic cells would be collected at the nonpermissive temperature, replated, and incubated for a short time (e.g., 2–3 hours) at the same temperature to allow the nonmutant cells to attach, and then the unattached cells would be collected and returned to the permissive temperature. The enrichment cycle would be repeated as necessary to enrich for *ts* mutants that are reversibly blocked at cell division. There are at least two technical limitations of this approach: (1) The enrichment step will generally not be highly efficient because it is based on differential attachment of interphase as compared to mitotic cells. Interphase cells often exhibit some detachment in normal cultures. (2) The selection will not be completely specific for cell division mutants because it will select for any mutant which cannot attach at the nonpermissive temperature.

An attempt in our laboratory to select for cell-division mutants using the above approach (5 to 7 selection cycles) yielded several mutant lines which are temperature-sensitive. One of them (*ts*MS1-1) appears to have a defect in cytokinesis. At the nonpermissive temperature, nuclear division usually occurs and mitosis proceeds until telophase when cytoplasmic cleavage is initiated but not completed. The resulting abortive division often produces a binucleated cell. Thus, this mutant is defective but not actually *arrested* in division at the nonpermissive temperature; it appears to have been selected because of an altered morphology during inter-phase at the nonpermissive temperature. A mutant line having a some-what similar *ts* phenotype has been isolated with BHK cells by Smith and Wigglesworth (1972). Although the BHK mutant does not appear temperature-sensitive at high cell concentration, *ts*MS1-1 exhibits no ap-parent concentration dependency for its phenotypic expression.

5. Nonselective Isolation of Temperature-Sensitive Mutants

Most selection methods designed to enrich for *ts* mutants will yield a biased spectrum of phenotypes with respect to the unselected popula-tion. Phenotypes for which the arrest of growth is irreversibly lethal will not be represented. An alternative approach to isolating mutants is simply to screen clonal isolates in the absence of any selective agent; then any phenotypes created by a mutagenic treatment may be represented among the isolates. Such a nonselective isolation procedure is obviously feasible only in a situation where the frequency of mutants is sufficiently high that they can be detected in an experiment of reasonable size.

In our laboratory a rather large nonselective isolation experiment was performed using mouse L cells and methods described in Section II,F. That this method of isolating mutants might be possible was indicated to us by unpublished results kindly communicated by Smith and Chu who were working with V79 hamster cells at Oak Ridge National Labora-tory. Our finding, based on isolating and testing some 1300 clones after treatment of an L cell population with MNNG, was that *ts* clones could be detected at a level of $\simeq 6 \times 10^{-3}$ (Thompson *et al.*, 1971). This frequency is indeed high, but compared with frequencies of other mutant phenotypes (Sections V,B and VI) it is not unreasonable in view of the multiplicity of possible *ts* genotypes. The eight mutant lines obtained in this way appeared similar in their *ts* properties to those mutants isolated by the dT-³H procedures, but they tended to show less rapid inhibition of cell multiplication when shifted to the nonpermissive temperature. These results suggest that nonselective isolation methods may be useful, especially if technical innovations render the handling of large numbers of clones more feasible.

6. Approach to Characterization of *ts* Mutants

Mutants which have been isolated must be assessed in terms of their plating efficiencies, rate of response to the nonpermissive temperature, ability to complement other mutants, etc. Plating efficiency (P.E.) is usually the most important assay for evaluating the degree of temperature sensitivity of a clone in terms of its potential usefulness. Plating over a range of concentrations at the nonpermissive temperature is particularly important because it will indicate to what extent the phenotype of the line is cell-concentration dependent. This can be done by plating 10^3, 10^4, 10^5, and 10^6 cells per dish for incubation at the nonpermissive temperature and 10^2 and 10^3 cells per dish for the permissive temperature. We have found it convenient to use 60-mm plastic petri dishes for all inocula except 10^6 and sometimes 10^5 cells per dish, which are plated into 100-mm dishes (Falcon Plastics). It is appropriate to incubate the dishes at the two temperatures for unequal intervals if the time required for colony formation of wild-type cells at the two temperatures is different. Even at the permissive temperature, mutant lines must often be incubated longer than wild-type cells because of their reduced growth rates. In addition to plating, a determination of the growth curve of a mutant line after shift to the nonpermissive temperature is useful because it indicates more precisely the kinetics of residual growth and the time when cells begin to disintegrate.

The degrees of temperature sensitivity observed with the various L cell and CHO mutants selected as described in this section are qualitatively similar (Thompson et al., 1970, 1971). The lines show variations in phenotype ranging from absolutely no growth when plated at $\geq 10^6$ cells per dish at 38.5°C to cases of appreciable growth when plated at 10^5 per dish, and sometimes at 10^4 cells per dish. Some of the lines, especially at high cell concentration, exhibit growth at 38.5°C to an extent which will probably preclude their usefulness for most types of biochemical or genetic characterization. In addition, some isolates do not grow or plate well at the permissive temperature; they also will probably have limited usefulness.

In general, mutant growth at the nonpermissive temperature may be ascribed to one or a combination of three phenomena: (1) Many or all of the cells in the culture may grow at a reduced rate, which may be concentration dependent. This indicates that the temperature-sensitive defect is not absolute; the lesion is "leaky" in a biochemical sense although genetically it may be very stable. (2) Growth may result from a distinct subpopulation representing revertant or partially revertant phenotypes due to back mutations or suppressor mutations. Discrete colonies that appear under nonpermissive conditions can be checked for revertant phenotype by isolation and testing for temperature sensitivity. For example, of five colonies of varying size isolated from a culture of tsH1 plated at 38.5°C, four of the isolates had a P.E. ≥ 0.20 at 38.5°C and the P.E. of the other clone was $< 10^{-4}$. However, for three mutant lines of L cells, colonies occurring at $\simeq 10^{-5}$ were isolated, tested, and typically found to have P.E.'s $\simeq 10^{-2}$. Such an explicit determination of colony phenotype can be important because the presence of colonies under selective conditions does not necessarily indicate clones of altered phenotype. (3) Under some conditions the mutant culture may give rise to colonies at low frequency simply due to the stochastic nature of colony formation, for example as described by the division probability or P model (cf. Section III,A). This type of colony is most likely to arise in a situation where temperature sensitivity is modified by cell density effects. For example, a microcolony resulting from a few "lucky" divisions may continue to grow, once established, if the ts phenotype is partially reversed by some mechanism resulting from cell-cell interaction. Some of the ts lines which we have tested appear to behave in this manner, e.g., one mutant produced no evidence of growth when plated at 10^4 cells per dish, but at 10^5 cells per dish hundreds of small colonies arose.

To determine whether mutations in various isolates are unique and will complement, cell hybridization procedures must be employed (Section II,I). Experiments in our laboratory indicate that the ts mutations in each of about twelve lines of CHO and L cells examined behave as recessives, as shown by crossing the ts lines with a non-ts line carrying an auxotrophic or drug-resistance marker. Many of the ts mutations will complement each other, and, hence, presumably represent alterations in different genes. Two clones from one selection appeared *not* to complement, which was not surprising since both lines were derived from the same three-cycle selection and their ts phenotypes appeared very similar. This result illustrates the potential problem of obtaining sibling clones of the same mutant from a culture carried through a multistep selection.

Another pertinent characterization of ts isolates is to examine the karyotypes of the mutants. A determination of the distribution of chromosome numbers of six of our ts lines of CHO cells has shown that, with one exception, the mutants have all retained the modal chromosome number (21) of the wild-type parental line. The exceptional mutant, which appears to have lost one of its smaller chromosomes, may be analogous to the K1 subclone isolated by Kao and Puck (1968). These results indicate that the ts phenotypes are not associated with major upheavals in the karyotype.

V. Mutant Frequency and the Effects of Mutagenic Agents

Measurement of the frequency of mutants in a cell culture under various conditions is a critical aspect of somatic cell genetics. First, observations on the frequency and statistical distribution of spontaneous "mutants" are necessary to characterize the variability of the cells and to substantiate the *random* rather than the *adaptive* origin of the mutants. Second, evidence as to whether the frequency of a particular phenotype is significantly enhanced by known mutagens is at present an important criterion by which to assess the likelihood that it results from a structural alteration in DNA. Third, granted that the frequency of occurrence of a mutant phenotype is a reliable index of genetic variation in a culture, assay of these mutants may be utilized as a test system to evaluate the potential mutagenicity to animal cells of various environmental agents. At the outset of this discussion the reader is reminded of the distinction between the *frequency of mutants* (or *mutant frequency*) in a culture, which is the fraction of the total cell population exhibiting a particular mutant phenotype, and the *mutation rate*—defined as the average number of mutations to the particular character occurring per cell per generation.

A.   Frequency of Spontaneous Mutants in a Population

The frequencies at which various mutant phenotypes occur in somatic cell cultures range from $\gtrsim 10^{-3}$ to $< 10^{-7}$. Of course the frequency will vary considerably with the nature of the mutant phenotype being assayed. Even for a particular phenotype, however, variations of two or more orders of magnitude in the frequency of mutants can occur not only between different cell lines but also in the same line assayed at different times.

1.   Changes in Mutant Frequency

Under nonselective conditions the frequency of mutants in a large, recently cloned population will in theory increase linearly with the number of generations of growth, due to the accumulation of new mutations at a constant rate. (Typical mutation rates of somatic cells in culture are in the range of 10^{-4} to 10^{-8} mutations per cell per generation.) Naturally, if the number of cells transferred to a subculture is relatively small, the mutant subpopulation may be diluted out; but with continual serial subcultures the probability of a mutant subpopulation becoming established in the culture will increase. One might expect that in a culture of sufficient size a constant frequency of mutants would ultimately be attained, perhaps when an equilibrium is achieved between forward and reverse mutations. Actually, oscillations in the frequency of mutants may be observed. This is presumably due to uncontrolled variation in the advantage of the wild type relative to the mutant phenotype in the culture. A general example of this is the phenomenon of *periodic selection,* whereby "more fit" cell types of the nonmutant population occasionally overgrow the culture (see Harris, 1964).

These considerations are illustrated by the data in Fig. 5, which describe the frequencies of mutants to ouabain-resistance (Section III,D) and to proline prototrophy (Kao and Puck, 1967) that we observed at various times in two different cultures of CHO cells. The frequencies of the mutant phenotypes varied over a 20- to 100-fold range during the intervals over which the cultures were sampled. (We have observed differences in mutant frequency of up to 1000-fold between freshly cloned populations and populations that had been maintained in serial culture for several months.) The general trend of the data in Fig. 5 is clearly toward higher mutant frequencies. Sometimes, however, there was a substantial decrease in the frequency of a marker between two consecutive samplings, and the changes in frequencies of the two markers assayed with the same culture (Experiment II) were often not correlated. Clearly, the frequency of a mutant phenotype can be variable, and in any case the mutant frequency must be viewed in the context of the culture history.

These observations also serve to reemphasize the point made in Section II,G concerning precautions that should be taken to preserve culture uniformity. In order to ensure the most homogeneous possible culture it is important to work with a recently cloned population. Often, accumulated subpopulations of *particular* mutant or revertant phenotypes can be removed by subculturing to a very low number of cells.

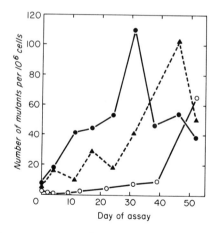

Fig. 5. Mutant frequencies assayed in two different cultures of CHO cells as a function of time. Suspension cultures containing $\gtrsim 10^7$ cells were maintained in exponential growth at 37°C (doubling time ~15 hours) and periodically assayed using the selective conditions described in the footnote to Table II. Experiment I: (○) ouabain-resistant mutants. Experiment II: (●) ouabain-resistant mutants; (▲) revertants to proline prototrophy.

2. Measurement of Mutation Rates

There are two general ways in which the mutation rate for a particular marker could be determined. By plotting the frequency of mutants in a freshly cloned population as a function of number of generations of growth one can sometimes obtain an approximately straight line whose slope, expressed in units of number of mutants per number of wild-type cells per generation, is the mutation rate. From the immediately preceding discussion it will be obvious that because of uncontrolled variation in the mutant frequency, this method can at best yield only a semiquantitative estimate of the mutation rate. Furthermore, it is based on the assumption that neither mutant nor wild-type cells have a selective growth advantage, so that increases in the frequency of mutants in a culture would be an accurate indication of the rate at which they are generated. However, many mutant lines have doubling times longer than that of the wild type. An adaptation of this approach, employing a back selection to eliminate preexisting mutants and then determining the mutant frequency following several generations of growth under nonselective conditions (Szybalski et al., 1962), may be satisfactory in some cases.

The best procedures for determining mutation rates are based upon the fluctuation test of Luria and Delbrück (1943) because the results are more reliable quantitatively, and this method also enables one to ascertain the random origin of the altered phenotype. A number of parallel cultures are initiated from low inocula, grown without dilution to high cell number, and assayed for the frequency of the mutant phenotype. The measured variance of the mutant frequencies assayed in such parallel replicate cultures is typically quite large compared with the variance of replicate assays of the same mass culture. This type of result establishes the random as opposed to adaptive occurrence of the alteration. By appropriate statistical treatment of the data from parallel cultures an estimate may be made of the mutation rate (Lea and Coulson, 1949; Lieberman and Ove, 1959b; Kondo, 1972). Depending upon the method of calculation, this measurement may also suffer from the assumption of uniform growth properties of the mutant and wild-type cells. This assumption can be avoided if it is possible to base the calculation upon the fraction of replicate cultures in which no mutants occur (Lea and Coulson, 1949; Szybalski, 1959).

Since the fluctuation test is based upon a determination of the statistical distribution of mutants in cultures grown for a fixed number of generations, the main experimental problem is to ensure that the replicate cultures undergo the same amount of growth. If they are grown in monolayer culture, intermediate trypsinization will probably be necessary to disperse the colonies and to minimize differences between the cultures that might arise from heterogeneous cell densities. With lines like CHO that grow in suspension from small inocula the replicates might be initiated as dilute suspension cultures.

B. Effects of Mutagenic Agents on Mutant Frequency

A number of somatic cell markers have now been shown to be mutagen-inducible, although initial efforts to detect an effect of mutagens on somatic cells in culture were unfortunately not successful. Szybalski and co-workers examined a variety of agents mutagenic in bacterial systems for their efficacy in enhancing the frequency of forward mutations to AG-resistance and reversions from azahypoxanthine resistance in D98 cells, and observed only slight or negative effects (Szybalski et al., 1964; Szybalski, 1964). This study did not include the potent chemical mutagens ethyl methanesulfonate (EMS), methyl methanesulfonate (MMS), and N-methyl-n'-nitro-N-nitrosoguanidine (MNNG). It remains unclear whether the negative results reflected an experimental assay system of inadequate sensitivity, or, as suggested by the authors, that the predominant mechanism underlying the generation of spontaneous somatic cell mutations is unaffected by some mutagens.

Chu and Malling (1968) demonstrated that under optimal conditions the frequency of AG-resistant cells in V79 hamster cultures could be increased ten- to fiftyfold by exposure to EMS, MMS, and MNNG. In the same system, it was shown that the frequency of forward mutations to AG-resistance could be enhanced up to tenfold by UV-irradiation and up to 100-fold by X-irradiation (Bridges et al., 1970; Bridges and Huckle, 1970; Chu, 1971b; Shapiro et al., 1972). The chemical- and radiation-induced mutant clones displayed a wide range of reversion frequencies; many resistant lines seemed to be preferentially revertible by either the alkylating agents EMS or MNNG or by the frame-shift mutagen ICR-170, implying that the mutations were due to either DNA base pair substitution or base pair insertion or deletion, respectively (Chu, 1971a,b). Albertini and DeMars (1973) have shown that the frequency of AG-resistant cells in human diploid fibroblast cultures can be increased by X-irradiation. Sato et al. (1972) found that MNNG and particularly EMS were potent inducers of thioguanine-resistant cells in one line of human lymphoblasts.

The frequencies of forward and back mutations involving nutritional auxotrophy in CHO cells are also sensitive to a variety of mutagens. Kao and Puck (1968, 1969) demonstrated the utility of EMS and MNNG for inducing mutations to auxotrophy and reversions to prototrophy for proline. They later reported evidence for the induction of auxotrophs by ICR-191, ultraviolet, X-rays, and several carcinogens, but not by hydroxylamine or caffeine (Kao and Puck, 1969, 1971). EMS and MNNG appeared to be the most effective mutagens, but the spectra of auxotrophic types induced by the different agents varied. The efficacies of the agents in producing chromosome aberrations and inducing mutations were not always correlated.

In this laboratory we have examined the effectiveness of two chemical mutagens on CHO cells for inducing both reversion of the conditional lethal marker tsH1 (Thompson and Stanners, 1972) and forward mutations to ouabain-resistance and colchicine-resistance (Till et $al.$, 1973). Representative results for the tsH1 mutant are shown in Experiment A in Table II; the increase in frequency of colonies at the nonpermissive

TABLE II

EFFECTS OF MUTAGENS UPON MUTANT FREQUENCIES IN CHO CELLS

Experiment[a]	Marker assayed[b]	Frequency of mutants[c]		
		No mutagen	EMS[d]	MNNG[e]
A	tsH1$^+$	3×10^{-7}	7×10^{-6}	6×10^{-6}
B	PRO$^+$	7×10^{-6}	2×10^{-4}	8×10^{-5}
	OUA^R	2×10^{-6}	1×10^{-4}	3×10^{-6}
	CCH^R	3×10^{-5}	2×10^{-4}	—
	AUXB1$^+$	6×10^{-7}	8×10^{-7}	3×10^{-7}

[a] The experiments were conducted according to our usual protocol, outlined in Section V,C,2. Parallel cultures of the CHO cell line tsH1 (Section IV,C,2) were used for experiment A. Experiment B was carried out with a cell line isolated and designated AUXB1 by M. McBurney of this laboratory; in addition to the PRO$^-$ marker it has a triple requirement for glycine, thymidine, and adenosine.

[b] Mutants carrying the indicated marker were assayed by colony formation under the following selective conditions. tsH1$^+$ (temperature resistance): α medium at 38.5°C; PRO$^+$ (proline prototrophy): Dα medium (Section IV,B,1) lacking proline supplemented with glycine, thymidine, and adenosine; OUA^R (ouabain resistance): α medium with 1 mM K$^+$ plus 3 mM ouabain; CCH^R (colchicine resistance): α medium plus 0.10 μg/ml colchicine; AUXB1$^+$ (prototrophy for glycine, thymidine, adenosine): Dα medium. α Medium was supplemented with 10% fetal calf serum, Dα medium with dialyzed serum. The tsH1$^+$ and CCH^R markers were assayed 8 days after mutagen treatment. The PRO$^+$, OUA^R, and AUXB1$^+$ markers were assayed 11 days after treatment, from the same populations.

[c] The number of mutant colonies per cell plated, corrected for plating efficiency under nonselective conditions.

[d] Experiment A: 300 μg/ml for 16 hours, 60% survival.
   Experiment B: 150 μg/ml for 18 hours, 5% survival.

[e] Experiment A: 0.1 μg/ml for 3 hours, 10% survival.
   Experiment B: 0.03 μg/ml for 2 hours, 50% survival.

temperature induced by EMS or MNNG is consistent with the belief that such conditional lethal mutants generally result from point mutations, probably base pair substitutions.

Experiment B in Table II demonstrates the efficacy of EMS and MNNG in enhancing the frequency of ouabain-resistant and colchicine-resistant mutants and in reverting the proline auxotrophy of "wild-type" CHO cells (Kao and Puck, 1967) and the AUXB1 marker isolated in this laboratory. In contrast to EMS, MNNG seems to have little effect upon the frequency of ouabain-resistant cells. In contrast to the proline marker, neither mutagen affects the AUXB1 phenotype. As others have reported for assays of AG-resistant phenotypes (Chu and Malling, 1968; Bridges and Huckle, 1970), we find that increasing the strength of the selection for ouabain-resistance results in lower absolute frequencies of both spontaneous and induced mutants but does not alter the magnitude of the mutagenic effect (Baker et al., 1973).

The recent observations surveyed above establish that chemical and physical agents of known mutagenic potential (Freese, 1971) do increase the frequencies of many mutant phenotypes in somatic cells. This evidence supports the attribution of the mutant phenotypes to specific alterations in the DNA. Some experimental aspects of such mutagenesis assays are discussed in Section V,C, below, where we consider utilization of somatic cells for assay of potential but as yet uncharacterized mutagens.

C. Detecting Mutagenic Agents

Sophisticated biological assays have been devised for detection of mutagenic effects in a variety of organisms (see Hollaender, 1971), but studies based on cultured somatic cells will no doubt be of special significance for evaluating the specific genetic impact of environmental agents upon man and other higher animals. In several recent studies somatic cell genetic markers have been successfully employed for evaluating the mutagenicity of various carcinogens (Kao and Puck, 1971; Huberman et al., 1971, 1972). The application of somatic cells for this purpose awaited the availability of suitable markers with demonstrated sensitivity to known mutagens. Although present knowledge concerning somatic cell mutations is generally superficial, there is sufficient information to warrant discussion of the important variables of mutagen assay systems.

1. CHOICE OF MARKERS

The qualifications of particular genetic markers for use in detecting mutagenic agents may be assessed in terms of the types of genetic alterations that they might indicate and the sensitivity and convenience of the experimental assays.

With regard to the first of these criteria, mutagenesis assays in somatic cells as in other systems can be based on measurements of either forward or reverse mutations. The advantage of assaying forward as opposed to reverse mutations is the potential for detecting a broad spectrum of genetic lesions that cause functional inactivation or alteration of the relevant gene product(s). Thus mutations to auxotrophy or to AG-resistance due to HGPRTase deficiency could reflect a variety of mutagen-induced lesions ranging from extensive deletions to single base substitutions. Phenotypes such as temperature sensitivity, and perhaps ouabain resistance, might generally indicate point mutations since they imply the presence of a protein that is altered but still functional. It is quite advantageous if a back selection against the induced phenotype is available, for then the nature of particular mutagen-induced forward mutations can be analyzed by examining the effectiveness of other mutagens, with known specificity, in promoting reversion (see Drake, 1970; Chu, 1971a,b).

In the case of reversions the alteration at the mutant locus necessary to restore or approximate the relevant wild-type character may be quite specific. Assays of reverse mutation rates for particular, well-characterized mutant alleles could be utilized to evaluate the types of alterations induced by various mutagens. Among the mutagen-induced AG^R V79 hamster clones isolated by Chu and co-workers are some that may be suitable for this purpose, since they appear to be selectively revertible by EMS, or ICR-170 (Chu, 1971a,b), as noted in Section V,B. However, we must anticipate that in many instances the interpretation of reversion data in somatic cells, as in other genetic systems, will be complicated by the occurrence of intragenic and intergenic suppressor mutations (Gorini and Beckwith, 1966).

The second general criterion to be considered in choosing particular markers for mutagen studies is the ease and sensitivity of the selective assays. Ideally, the induced phenotype should be detectable in a single step with a reliability that is not significantly affected by small variations in the selective conditions. Quantitation of forward mutations to auxotrophy or temperature sensitivity will generally not be feasible since detection of these phenotypes requires the testing of numerous clones following an intermediate selective step (Kao and Puck, 1969; Thompson et al., 1971). Assays for reversion of these characters and for many types of forward mutations to drug resistance require only a single-step selection; but they often encounter the problem of density dependence for detection of the altered phenotype, which limits the number of cells that can be conveniently assayed. To our present knowledge only assays for revertants of *ts*H1 (Section IV,C) and for ouabain-resistant mutants (Section III,D) yield results that are reliable over a broad range of cell densities (up to $\simeq 5 \times 10^6$ cells/100-mm culture dish).

None of the somatic cell markers available at the moment is perfectly suited for mutagen detection. The assay for AG-resistance has been the best available, and the most widely used, since it involves single-step selections for forward mutations for which back selection may be possible; its disadvantage is the density dependence of the plating assay. For detecting forward point mutations the assay for ouabain resistance might be superior, but no back selection is available for these mutants. Different mutant genotypes will almost surely be induced with different efficiencies relative to their spontaneous mutation rates (Albertini and DeMars, 1973), since different marker phenotypes are induced at different relative frequencies by various mutagens (Kao and Puck, 1969; Table II above). As yet, however, there is insufficient comparative information concerning the various markers for forward mutations to indicate which are inherently more sensitive to particular types of mutagens. In view of these considerations it is desirable to employ a number of different markers even in assays for specific types of alterations. We have found it convenient to utilize multiply-marked CHO lines, permitting a spectrum of assays for both reversions and forward mutations (cf. Table II above).

2. PROCEDURES FOR EXPOSURE AND ASSAY

Factors which must be considered in establishing protocols for mutagenesis assays are dosage of the potentially mutagenic agent, expression time for induced mutations, and conditions for optimal detection of mutant phenotypes.

A dose-response curve describing cell survival as a function of dose of the potential mutagen must be obtained in order to determine the kinetics and dose range for cytotoxicity (e.g., see Kao and Puck, 1969). Assays for mutation induction have often been carried out for a mutagen dose that results in 10–50% survival with a view to ensuring an effective exposure while minimizing the possibility of selective killing of mutant or wild-type cells. [An appropriate though often omitted control experiment is a comparison of the toxicity of the agent for the wild-type with that for the induced mutant populations (Albertini and DeMars, 1973).] In assays for induction of AG-resistant phenotypes by various mutagens, the mutant frequency per surviving cell has generally been found to increase with decreasing cell survival (Bridges and Huckle, 1970; Huberman *et al.*, 1971; Sato *et al.*, 1972). Moreover, with some agents, such as

X-rays (Bridges and Huckle, 1970; Albertini and DeMars, 1973), the frequency of mutants induced per unit dose increases with dose. Thus in order to obtain the most conclusive result it is advisable to test for a mutagenic effect over a range of doses. In addition, comparison of the dose-response kinetics for cell killing and mutation induction may indicate whether there could be a relation between lethal and mutagenic lesions (Haynes et al., 1968; Drake, 1970).

In order to maximize the sensitivity of tests for mutagenesis, it is clearly important that experiments be carried out on cultures containing the minimum possible frequency of preexisting spontaneous mutants so that low frequencies of induced mutants will be detectable. Preferably, cultures derived from a recent cloning or from a frozen stock that has been pretested should be used. Even in cases where a back selection seems to be applicable it may be an unnecessary risk to rely upon it for elimination of preexisting mutants, because some mutant types might be resistant (e.g., HAT^R-AG^R phenotypes; Section III,B,1).

After treatment with a potential mutagen, the cells must be maintained for a period under nonselective conditions to permit recovery from sublethal damage, fixation of any mutations, and expression of mutant genes. By assaying a culture at various times after mutagen exposure, the optimal time for detection of induced mutations can be established for a particular system. For example, Chu and Malling (1968; Chu, 1971a) determined that the best time to allow for expression of AG-resistant phenotypes in their system was approximately 42 hours. The optimal interval will probably vary for different mutant loci, assay conditions, and types of mutagenic agents; this information is not yet available.

Mutagenesis assays may be carried out either on cells attached to a substrate in fixed position throughout the procedure (in situ) or on pooled populations of cells mixed in monolayer or suspension culture following treatment. With the in situ procedure described by Chu and Malling (1968; see Chu, 1971a, for a detailed discussion), the cells are attached to a substrate before being exposed to the mutagenic agent. Following a 2-hour treatment with the mutagen, the monolayer is rinsed and nonselective medium is added for the expression interval; then the medium is again removed and replaced with selective medium for assay of mutant colonies. The principal merit of this approach is that individual colonies of induced mutant cells should each represent independent mutations. Thus differential growth rates of different mutant clones will not bias the frequency with which they are detected. There are two potential disadvantages to this general method, however. First, its success is dependent upon accomplishing the necessary media changes without initiating the formation of a significant number of satellite colonies, which may be a problem with cell lines that attach loosely. Second, the growth which occurs on the assay plate prior to application of the selective agent may interfere with evaluation of the optimal expression time and maximal increases in mutant frequencies. This problem could be significant in assays of markers such as AG-resistance where detection of the mutant phenotype is often affected by cell density (see Section III,B,2) (Chu, 1971a). Nevertheless, utilization of the above experimental design for detecting mutagenic effects upon forward and reverse transitions involving AG-resistance has been quite successful (Chu and Malling, 1968; Bridges and Huckle, 1970; Chu, 1971a; Huberman et al., 1971; Albertini and DeMars, 1973).

The general procedure for detection of mutagenic effects employed by Kao and Puck (1969, 1971) and by ourselves (Section V,B above) differs in that the treated cells are pooled prior to assay of mutant frequencies. Kao and Puck have executed the steps involving mutagen exposure and mutation expression on cells attached in monolayer culture and have then subcultured the cells in order to perform the BrdU-light selection for mutations to auxotrophy. For cell lines that grow in suspension, we have generally utilized suspension culture for all steps prior to plating for assay of mutants. For example, exponentially growing suspension cultures containing $1-2 \times 10^5$ cells/ml have been exposed to mutagen for 2–18 hours, washed by centrifugation, resuspended in nonselective media, cultured in suspension until they regain a normal growth rate (2–10 days), and then plated. An obvious advantage of pooling the mutagen-treated cells is the greatly increased practicality of experiments that involve assays for several markers, particularly when mutagen dose or expression time is also a variable. In addition, one has better control of the assay conditions. Clearly, in any situation where the test population is subcultured after mutagen treatment, care must be taken to preserve a sufficient number of viable cells that the frequency of mutants to be assayed is not distorted by random partitioning of a small number of mutant cells.

The procedures that have been reported to date for somatic cells have been designed for the detection of mutagenic effects following a single, acute exposure to the agent under test. Consideration should also be given to quantitating the mutagenic effect of continuous low-level exposure to an agent, since this situation may more closely parallel the case with many environmental mutagens. Such tests could be conducted, for example, by performing fluctuation analyses to determine mutation rates in both the presence and the absence of the potential mutagen.

VI.   The Nature of Somatic Cell Mutants Isolated in Culture

In concluding we wish to make some additional comments with reference to the criteria set forth in the introduction for ascribing a genetic basis to the somatic cell "mutations" observed in culture. We have discussed in the preceding sections methodologies for the isolation and approach to characterization of three major classes of altered phenotypes of somatic cells—*ts* conditional lethality, nutritional auxotrophy, and drug resistance. We have attempted to deal with certain problems of experimental design and interpretation which are potentially troublesome and may have helped to foster skepticism as to the genetic basis of presumptive somatic cell mutants. On the whole, we feel that recent progress warrants an optimistic view of the possibilities for developing workable genetic systems using cultured mammalian cells.

Among the various types of mutant isolates whose selections have been discussed above we have seen no evidence of instability in basic phenotype during long-term growth in the absence of selective pressures. In addition to examining temporal stability in a population, however, one may test whether a marker is uniformly expressed by subclones derived from the mutant population. Most presumptive somatic cell mutants reported in the literature have not been systematically characterized in this respect. We have tested the properties of subclones of several mutant lines of L cells temperature-sensitive for growth or for ouabain resistance. Of many subclones tested, all have bred true qualitatively in the sense of expressing the temperature-sensitive phenotype. On the other hand, we have consistently observed quantitative differences among subclones, which we have termed "subclone variation" (Thompson *et al.*, 1970). Because of the karyotypic heterogeneity characteristic of L cells we think it is plausible that this type of second-order instability could reflect rearrangements, or a variation in the number of copies, of the genetic loci affecting the quantitative expression of the mutant character. However, Kraemer *et al.* (1972) have recently presented compelling evidence in support of the unorthodox idea that the variations in karyotype associated with heteroploidy are *not* a reflection of abnormal variations in the amount of genetic material.

If subclone variability *is* a consequence of karyotypic changes, then it should be minimized in cells that maintain a relatively stable karyotype in culture. Recent observations based on Giemsa-banding analysis (Deaven and Petersen, 1973) indicate uniformity of karyotype in CHO for those cells having the modal number of chromosomes, suggesting a stabilized configuration with respect to rearrangements. Analyses we have made on subclones of *ts* mutant lines of CHO indicate that for some mutants no variation can be detected whereas with others there may be significant differences among subclones. Further studies are clearly necessary for an understanding of the degree of subclone variation in CHO cultures, its relationship to subtle changes in karyotype, and its specific implications for the study of mutant markers in this line.

If a marker phenotype under consideration can arise from point mutation, it should be inducible (and in general revertible) by agents that act preferentially on DNA and are known to be potent mutagens in other biological systems. As discussed above (Section V,B), positive evidence on this count can be cited for most types of somatic cell genetic markers, including the ones whose selection we have described.

The association of an altered phenotype with an altered gene product constitutes strong evidence that a mutated gene is being expressed. In the case of some of the drug-resistance and auxotrophic markers we have described, it is possible *a priori* that the altered enzyme activity could be due to a nonmutational change in the *expression* of a gene rather than due to a mutation in the gene. However, the many temperature-sensitive lines with varied but specific phenotypes that have now been obtained are difficult to explain solely on the basis of nonmutational changes. Furthermore, several recent developments have provided more direct indications that altered gene products appear in different drug-resistant lines of somatic cells. Chan *et al.* (1972) have presented strong evidence that a CHO line selected for resistance to α-amanitin possesses an altered form of DNA-dependent RNA polymerase II. Beaudet (cited in Gillin *et al.,* 1972) has detected different immunologically active but enzymatically inactive HGPRTase proteins in V79 Chinese hamster cell lines selected for resistance to 8-azaguanine. Also in Chinese hamster cells, Albrecht *et al.* (1972) have demonstrated that in some lines resistant to inhibitors of folic acid metabolism there occurs a new form of dihydrofolate reductase which shows alterations in its pH maxima and in its sensitivities to the various inhibitors. In the near future we should expect to see further substantiation of the presence of altered proteins in somatic cell mutants.

A persistent question in connection with the nature of many somatic cell mutants is why they appear at relatively high frequency, typically several orders of magnitude higher than those found in microbial populations. Actually, the mutation rates of cultured somatic mammalian cells are usually comparable with the estimates derived from observations on humans (Vogel, 1965) or mice (Russell, 1963; Schlager and Dickie, 1971) which often range from 10^{-4} to 10^{-6}. A more problematical point is the fact that most of the mutant phenotypes isolated in somatic cells behave as recessives (in the hybrid configuration). Several general mechanisms may be envisioned to account for this phenomenon of unmasked recessive mutations: (1) hemizygosity resulting from deletion of chromosomal regions (Kao and Puck, 1968; Deaven and Petersen, 1973); (2) inactivity of other alleles due to repression (Nanney, 1968); (3) segregation of heterozygous alleles following mutation (Martin and Sprague, 1969); and (4) multiple mutations. The latter mechanism appears unlikely as a general explanation. The hypothesis of partial hemizygosity is perhaps the simplest conceptually, but in CHO cells the cytogenetic evidence suggests there is very little loss of chromatin (Kao and Puck, 1969; Deaven and Petersen, 1973). CHO cells have only 4% less DNA per cell than normal Chinese hamster cells (Kraemer *et al.*, 1972); these observations are difficult to reconcile with the variety of types of auxotrophic mutants which can be isolated with these cells. On the other hand, an explanation based on allelic repression could account for the relatively high frequencies of revertant-like cells (10^{-3}–10^{-4}) which occur in some of the *ts* lines of both L cells and CHO cells; phenotypic reversion would then sometimes be a reflection of derepression. The chromosomal rearrangements which have occurred in CHO cells might cause repression through position effects, which are well known in *Drosophila* (Becker, 1966) and mice (Baker, 1968). It may be significant that to a first approximation CHO cells appear to have a haploid set of intact chromosomes (Deaven and Petersen, 1973).

Finally, it might be emphasized that in studying mutants of cultured mammalian cells it will be particularly important for the geneticist to work with a minimum number of preconceptions about the behavior of the system. The mammalian genome is structurally complex in comparison with that of microorganisms; there may also be novel mechanisms of gene expression which are unique to higher animals. Thus we do not really know what to expect in terms of mutation rates and the stability and expression of mutant genotypes. In seems certain, however, that the study of somatic cell mutants will be rewarding and will yield many surprises.

Acknowledgments

We are grateful to Louis Siminovitch for his invaluable editorial comments and suggestions during the preparation of this article, and to Gordon Whitmore and James Till for reading and criticizing the manuscript. Many of our colleagues contributed substantially to the experimental work and ideas discussed here. These include Ralph Mankovitz, Vic Ling, Michael McBurney, David Ness, John Harris, Don Brunette, and Clifford Stanners. We thank W. C. Dewey, who provided our initial culture of CHO cells. This work was supported by the National Cancer

Institute of Canada, the Medical Research Council of Canada, and the National Institutes of Health of the United States.

REFERENCES

Albers, R. W., Koval, G. J., and Siegel, G. J. (1968). *Mol. Pharmacol.* **4**, 324.

Albertini, R. J., and DeMars, R. (1970). *Science* **169**, 482.

Albertini, R. J., and DeMars, R. (1973). *Mutat. Res.* (in press).

Albrecht, A. M., Biedler, J. L., and Hutchinson, D. J. (1972). *Cancer Res.* **32**, 1539.

Ashkenazi, Y. E., and Gartler, S. M. (1971). *Exp. Cell Res.* **64**, 9.

Atkins, J. H., and Gartler, S. M. (1968). *Genetics* **60**, 781.

Aviv, D., and Thompson, E. B. (1972). *Science* **177**, 1201.

Bach, M. K. (1969). *Cancer Res.* **29**, 1036.

Baker, P. F., and Willis, J. S. (1970). *Nature (London)* **226**, 521.

Baker, R. M., and Till, J. E. (1971). *Abstr., 15th Annu. Meet. Biophys. Soc., New Orleans* p. 283a

Baker, R. M., Brunette, D. M., Mankovitz, R., Thompson, L. H., Whitmore, G. F., Siminovitch, L., Till, J. E. (1973). In preparation.

Baker, W. K. (1968). *Advan. Genet.* **14**, 133.

Barban, S. (1962). *J. Biol. Chem.* **237**, 291.

Barski, G. S., Sorieul, S., and Cornefert, F. (1960). *C. R. Acad. Sci.* **251**, 1825.

Becker, H. J. (1966). *Curr. Top. Develop. Biol.* **1**, 155.

Benke, P. J., and Herrick, N. (1972). *Amer. J. Med.* **52**, 547.

Bennett, L. L., Vail, M. H., Chumley, S., and Montgomery, J. A. (1966). *Biochem. Pharmacol.* **15**, 1719.

Borisy, G. G., and Taylor, E. W. (1967a). *J. Cell Biol.* **34**, 525.

Borisy, G. G., and Taylor, E. W. (1967b). *J. Cell Biol.* **34**, 535.

Bosmann, H. B. (1971). *Nature (London)* **233**, 566.

Breslow, R. E., and Goldsby, R. A. (1969). *Exp. Cell Res.* **55**, 339.

Bridges, B. A., and Huckle, J. (1970). *Mutat. Res.* **10**, 141.

Bridges, B. A., Huckle, J., and Ashwood-Smith, M. J. (1970). *Nature (London)* **226**, 184.

Brinkley, B. R., Stubblefield, E., and Hsu, T. C. (1967). *J. Ultrastruct. Res.* **19**, 1.

Brockman, R. W. (1963). *Advan. Cancer Res.* **7**, 129.

Brunette, D. M. (1972). Ph.D. Thesis, Universiy of Toronto.

Brunette, D. M., and Till, J. E. (1972). *Abstr., 16th Annu. Meet. Biophys. Soc., Toronto* p. 58a

Cass, C. E. (1972). *J. Cell. Physiol.* **79**, 139.

Cereijido, M., and Rotunno, C. A. (1970). "Introduction to Biological Membranes." Gordon & Breach, New York.

Chan, V. L., Whitmore, G. F., and Siminovitch, L. (1972). *Proc. Nat. Acad. Sci. U. S.* **69**, 3119.

Choi, K. W., and Bloom, A. D. (1970). *Science* **170**, 89.

Chu, E. H. Y. (1971a). In "Chemical Mutagens, Principles and Methods for Their Detection" (A. Hollaender, ed.), Vol. II, pp. 411–444. Plenum, New York.

Chu, E. H. Y. (1971b). *Mutat. Res.* **11**, 23.

Chu, E. H. Y., and Malling, H. V. (1968). *Proc. Nat. Acad. Sci. U. S.* **61**, 1306.

Chu, E. H. Y., Brimer, P., Jacobson, K. B., and Merriam, E. V. (1969). *Genetics* **62**, 359.

Clive, D., Flamm, W. G., Machesko, M. R., and Bernheim, N. J. (1972). *Mutat. Res.* **16**, 77.

Coffino, P., Baumal, R., Laskov, R., and Sharff, M. D. (1972). *J. Cell Physiol.* **79**, 429.

Cox, R. P., Krauss, M. R., Balis, M. E., and Dancis, J. (1972). *Exp. Cell Res.* **74**, 251.

Cronan, J. E., Jr., Ray, T. K., and Vagelos, P. R. (1970). *Proc. Nat. Acad. Sci. U. S.* **65**, 737.

Dancis, J., Cox, R. P., Berman, P. H., Jansen, V., and Balis, M. E. (1969). *Biochem. Genet.* **3**, 609.

Davidson, R. L. (1971). *In Vitro* **6**, 411.

Davidson, R. L. (1972). *Proc. Nat. Acad. Sci. U. S.* **69**, 951.

Deaven, L. L., and Petersen, D. F. (1973). *Chromosoma* (in press).

DeMars, R. (1971). *Fed. Proc., Fed. Amer. Soc. Exp. Biol.* **30**, 944.

DeMars, R., and Held, K. R. (1972). *Humangenetik* **16**, 87.

DeMars, R., and Hooper, J. L. (1960). *J. Exp. Med.* **111**, 559.

Djordjevic, B., and Szybalski, W. (1960). *J. Exp. Med.* **112**, 509.

Drake, J. W. (1970). "The Molecular Basis of Mutation." Holden-Day, San Francisco, California.

Dubbs, D. R., and Kit, S. (1964). *Exp. Cell Res.* **33**, 19.

Dunham, P. B., and Hoffman, J. F. (1970). *Proc. Nat. Acad. Sci. U. S.* **66**, 936.

Felix, J. S., and DeMars, R. (1971). *J. Lab. Clin. Med.* **77**, 596.

Foley, G. E., Lazarus, H., Farber, S., Uzman, B. G., Boone, B. A., and McCarthy, R. E. (1965). *Cancer* **18**, 522.

Ford, D. K., and Yerganian, G. (1958). *J. Nat. Cancer Inst.* **21**, 393.

Fougère, C., Ruiz, F., and Ephrussi, B. (1972). *Proc. Nat. Acad. Sci. U. S.* **69**, 330.

Fox, M. (1971). *Mutat. Res.* **13**, 403.

Freed, J. J., and Mezger-Freed, L. (1970). *Proc. Nat. Acad. Sci. U. S.* **65**, 337.

Freese, E. (1971). *In* "Chemical Mutagens, Principles and Methods for Their Detection" (A. Hollaender, ed.), Vol. I, pp. 1–56. Plenum, New York.

Gartler, S. M., and Pious, D. A. (1966). *Humangenetik* **2**, 83.

Gillin, F. D., Roufa, D. J., Beaudet, A. L., and Caskey, C. T. (1972). *Genetics* **72**, 239.

Glade, P. R., Kasel, J. A., Moses, H. L., Whang-Peng, J., Hoffman, P. F., Kammermeyer, J. K., and Chessin, L. N. (1968). *Nature* (*London*) **217**, 564.

Glynn, I. M. (1964). *Pharmacol. Rev.* **16**, 381.

Goldsby, R. A., and Zipser, E. (1969). *Exp. Cell Res.* **54**, 271.

Gorini, L., and Beckwith, J. R. (1966). *Annu. Rev. Microbiol.* **20**, 401.

Graham, F. L., and Whitmore, G. F. (1970). *Cancer Res.* **30**, 2627.

Green, H. (1969). *Wistar Inst. Symp. Monogr.* **9**, 51.

Ham, R. G. (1972). *Methods Cell Physiol.* **5**, 37.

Harris, H., and Watkins, J. F. (1965). *Nature* (*London*) **205**, 640.

Harris, M. (1964). "Cell Culture and Somatic Variation." Holt, New York.

Harris, M. (1967). *J. Nat. Cancer Inst.* **38**, 185.

Harris, M. (1971). *J. Cell. Physiol.* **78**, 177.

Hayflick, L. (1965). *Exp. Cell Res.* **37**, 614.

Haynes, R. H., Baker, R. M., and Jones, G. E. (1968). *In* "Energetics and Mechanisms in Radiation Biology" (G. O. Philips, ed.), pp. 425–465. Academic Press, New York.

Hollaender, A., ed. (1971). "Chemical Mutagens, Principles and Methods for Their Detection," Vols. I and II. Plenum, New York.

Hsu, T. C., and Somers, C. E. (1961). *Proc. Nat. Acad. Sci. U. S.* **47**, 396.

Hsu, T. C., and Somers, C. E. (1962). *Exp. Cell Res.* **26**, 404.

Huberman, E., Aspiras, L., Heidelberger, C., Grover, P. L., and Sims, P. (1971). *Proc. Nat. Acad. Sci. U. S.* **68**, 3195.

Huberman, E., Donovan, P. J., and DiPaolo, J. A. (1972). *J. Nat. Cancer Inst.* **48**, 837.

Kao, F. T., and Puck, T. T. (1967). *Genetics* **55**, 513.

Kao, F. T., and Puck, T. T. (1968). *Proc. Nat. Acad. Sci. U. S.* **60**, 1275.

Kao, F. T., and Puck, T. T. (1969). *J. Cell. Physiol.* **74**, 245.

Kao, F. T., and Puck, T. T. (1970). *Nature (London)* **228**, 329.

Kao, F. T., and Puck, T. T. (1971). *J. Cell. Physiol.* **78**, 139.

Kao, F. T., and Puck, T. T. (1972). *J. Cell. Physiol.* **80**, 41.

Kao, F. T., Johnson, R. T., and Puck, T. T. (1969a). *Science* **164**, 312.

Kao, F. T., Chasin, L., and Puck, T. T. (1969b). *Proc. Nat. Acad. Sci. U. S.* **64**, 1284.

Kaplan, S., and Anderson, D. (1968). *J. Bacteriol.* **95**, 991.

Kim, S. C. (1969). M.Sc. Thesis, Dept. Med. Biophys., University of Toronto.

Kit, S., Dubbs, D. R., Piekarski, L. J., and Hsu, T. C. (1963). *Exp. Cell Res.* **31**, 297.

Kit, S., Dubbs, D. R., and Frearson, P. M. (1966). *Int. J. Cancer* **1**, 19.

Kondo, S. (1972). *Mutat. Res.* **14**, 365.

Kraemer, P. M., Deaven, L. L., Crissman, H. A., and Van Dilla, M. A. (1972). *Advan. Cell Mol. Biol.* **2**, 47.

Krenitsky, T. A., Papaioannou, R., and Elion, G. B. (1969). *J. Biol. Chem.* **244**, 1263.

Krooth, R. S., and Sell, E. K. (1970). *J. Cell. Physiol.* **76**, 311.

Kusano, T., Long, C., and Green, H. (1971). *Proc. Nat. Acad. Sci. U. S.* **68**, 82.

Kyte, J. (1971). *J. Biol. Chem.* **246**, 4157.

Lea, D. E., and Coulson, C. A. (1949). *J. Genet.* **49**, 264.

Levisohn, S. R., and Thompson, E. B. (1972). *Nature (London) New Biol.* **235**, 102.

Lieberman, I., and Ove, P. (1959a). *Proc. Nat. Acad. Sci. U. S.* **45**, 867.

Lieberman, I., and Ove, P. (1959b). *Proc. Nat. Acad. Sci. U. S.* **45**, 872.

Ling, V., and Thompson, L. H. (1973). In preparation.

Littlefield, J. W. (1963). *Proc. Nat. Acad. Sci. U. S.* **50**, 568.

Littlefield, J. W. (1964a). *Nature (London)* **203**, 1142.

Littlefield, J. W. (1964b). *Cold Spring Harbor Symp. Quant. Biol.* **29**, 161.

Littlefield, J. W. (1964c). *Science* **145**, 709.

Littlefield, J. W. (1965). *Biochim. Biophys. Acta* **95**, 14.

Littlefield, J. W., and Basilico, C. (1966). *Nature (London)* **211**, 250.

Luria, S. E., and Delbrück, M. (1943). *Genetics* **28**, 491.

Martin, G. M., and Sprague, C. A. (1969). *Science* **166**, 761.

Mayhew, E., and Levinson, C. (1968). *J. Cell. Physiol.* **72**, 73.

Mazur, P., Leibo, S. P., and Chu, E. H. Y. (1972). *Exp. Cell Res.* **71**, 345.

Meiss, H. K., and Basilico, C. (1972). *Nature (London) New Biol.* **239**, 66.

Mezger-Freed, L. (1971). *J. Cell Biol.* **51**, 742.

Mezger-Freed, L. (1972). *Nature (London) New Biol.* **235**, 245.

Miech, R. P., York, R., and Parks, R. E., Jr. (1969). *Mol. Pharmacol.* **5**, 30.

Miller, O. J., Allderdice, P. W., Miller, D. A., Breg, W. R., and Migeon, B. R. (1971). *Science* **173**, 244.

Moore, G. E., and McLimans, W. F. (1968). *J. Theor. Biol.* **20**, 217.

Moore, G. E., Gerner, R. E., and Franklin, H. A. (1967). *J. Amer. Med. Ass.* **199**, 519.

Morris, N. R., and Fischer, G. A. (1960). *Biochim. Biophys. Acta* **42**, 183.

Morrow, J. (1970). *Genetics* **65**, 279.

Morrow, J. (1972). *Genetics* **71**, 429.

Murray, A. W. (1971). *Annu. Rev. Biochem.* **40**, 811.

Nagy, M. (1970). *Biochim. Biophys. Acta* **198**, 471.

Naha, P. M. (1969). *Nature (London)* **223**, 1380.

Nanney, D. L. (1968). *Annu. Rev. Genet.* **2**, 121.

Nyhan, W. L. (1968). *Fed. Proc., Fed. Amer. Soc. Exp. Biol.* **27**, 1034.

Orkin, S. H., and Littlefield, J. W. (1971). *Exp. Cell Res.* **69**, 174.

Overath, P., Schairer, H. U., and Stoffel, W. (1970). *Proc. Nat. Acad. Sci. U. S.* **67**, 606.

Overath, P., Hill, F. F., and Lamnek-Hirsch, I. (1971). *Nature (London) New Biol.* **234**, 264.

Peery, A., and LePage, G. A. (1969). *Cancer Res.* **29**, 617.

Petersen, D. F., Anderson, E. C., and Tobey, R. A. (1968). *Methods Cell Physiol.* **3**, 347.

Puck, T. T., and Kao, F. T. (1967). *Proc. Nat. Acad. Sci. U. S.* **58**, 1227.

Puck, T. T., Cieciura, S. J., and Robinson, A. (1958). *J. Exp. Med.* **108**, 945.

Puck, T. T., Sanders, P. C., and Petersen, D. (1964). *Biophys. J.* **4**, 441.

Puck, T. T., Wuthier, P., Jones, C., and Kao, F. T. (1971). *Proc. Nat. Acad. Sci. U. S.* **68**, 3102.

Rao, P. N., and Johnson, R. T. (1972). *Methods Cell Physiol.* **5**, 75.

Ruddle, F. H. (1972). *Advan. Human Genet.* **3**, 73.

Ruddle, F. H., Chapman, V. M., Ricciuti, F., Murnane, M., Klebe, R., and Meera Khan, P. (1971). *Nature (London) New Biol.* **232**, 69.

Russell, W. L. (1963). *In* "Repair from Genetic Radiation Damage" (F. H. Sobels, ed.), pp. 205–217. Macmillan, New York.

Sato, K., Slesinski, R. S., and Littlefield, J. W. (1972). *Proc. Nat. Acad. Sci. U. S.* **69**, 1244.

Scheffler, I. E., and Buttin, G. (1973). *J. Cell Physiol.* (in press).

Schlager, G., and Dickie, M. M. (1971). *Mutat. Res.* **11**, 89.

Seegmiller, J. E., Rosenbloom, F. M., and Kelley, W. N. (1967). *Science* **155**, 1682.

Shapiro, N. I., Khalizev, A. E., Luss, E. V., Manuilova, E. S., Petrova, O. N., and Varshaver, N. B. (1972). *Mutat. Res.* **16**, 89.

Smith, B. J., and Wigglesworth, N. M. (1972). *J. Cell. Physiol.* **80**, 253.

Smith, D. B., and Chu, E. H. Y. (1972). *Cancer Res.* **32**, 1651.

Sonneborn, T. M. (1970). *Symp. Int. Soc. Cell Biol.* **9**, 1.

Stanners, C. P., Eliceiri, G. L., and Green, H. (1971). *Nature (London) New Biol.* **230**, 52.

Steel, C. M. (1971). *Nature (London)* **233**, 555.

Stubblefield, E. (1966). *J. Nat. Cancer Inst.* **37**, 799.

Subak-Sharpe, H., Bürk, R. R., and Pitts, J. D. (1969). *J. Cell Sci.* **4**, 353.

Suzuki, D. T. (1970). *Science* **170**, 695.

Suzuki, F., Kashimoto, M., and Horikawa, M. (1971). *Exp. Cell Res.* **68**, 476.

Szybalski, W. (1959). *Exp. Cell Res.* **18**, 588.

Szybalski, W. (1964). *Cold Spring Harbor Symp. Quant. Biol.* **29**, 151.

Szybalski, W., and Smith, M. J. (1959). *Proc. Soc. Exp. Biol. Med.* **101**, 662.

Szybalski, W., Szybalska, E. H., and Ragni, G. (1962). *Nat. Cancer Inst., Monogr.* **7**, 75.

Szybalski, W., Ragni, G., and Cohn, N. K. (1964). *Symp. Int. Soc. Cell Biol.* **3**, 209.

Taylor, E. W. (1965). *J. Cell Biol.* **25,** 145.

Taylor, M. W., Souhrada, M., and McCall, J. (1971). *Science* **172,** 162.

Terasima, T., and Tolmach, L. J. (1961). *Nature (London)* **190,** 1210.

Thompson, L. H., and Stanners, C. P. (1972). *Abstr., 12th Annu. Meet. Amer. Soc. Cell Biol., St. Louis* p. 260a.

Thompson, L. H., Mankovitz, R., Baker, R. M., Till, J. E., Siminovitch, L., and Whitmore, G. F. (1970). *Proc. Nat. Acad. Sci. U. S.* **66,** 377.

Thompson, L. H., Mankovitz, R., Baker, R. M., Wright, J. A., Till, J. E., Siminovitch, L., and Whitmore, G. F. (1971). *J. Cell. Physiol.* **78,** 431.

Till, J. E., Baker, R. M., Brunette, D. M., Ling, V., Thompson, L. H., and Wright, J. A. (1973). *Fed. Proc., Fed. Amer. Soc. Exp. Biol.* **32,** 29.

Tocchini-Valentini, G. P., and Felicetti, L., and Rinaldi, G. M. (1969). *Cold Spring Harbor Symp. Quant. Biol.* **34,** 463.

van Zeeland, A. A., van Diggelen, M. C. E., and Simons, J. W. I. M. (1972). *Mutat. Res.* **14,** 355.

Vaughn, M. H., Jr., and Steele, M. W. (1971). *Exp. Cell Res.* **69,** 92.

Vogel, F. (1965). *Genet. Today, Proc. Int. Congr., 11th, 1963* Vol. 3, p. 833.

Whitmore, G. F., and Gulyas, S. (1966). *Science* **151,** 691.

Whitmore, G. F., and Till, J. E. (1964). *Annu. Rev. Nucl. Sci.* **14,** 347.

Wolpert, M. K., Damle, S. P., Sznycer, E., Agrawal, K. C., and Sartorelli, A. C. (1971). *Cancer Res.* **31,** 1620.

Wright, J. A. (1973). *J. Cell Biol.* (in press).

Yamamoto, M., Endo, H., and Kuwano, M. (1972). *J. Mol. Biol.* **69,** 387.

Yu, C. K., and Sinclair, W. K. (1964). *Can. J. Genet. Cytol.* **6,** 109.

Chapter 8

Isolation of Metaphase Chromosomes, Mitotic Apparatus, and Nuclei

WAYNE WRAY

*Department of Cell Biology, Baylor College of Medicine,
Houston, Texas*

I. Introduction

The methods employed for the mass isolation of any cell organelle are limited by the physical and chemical nature of the structure, the adherence of contaminating materials from the cell, and the final desired state of the isolated component. Isolation of subcellular components for experimental study has generally been aimed at a specific cell organelle with little or no consideration of the slight modifications which might allow nearly identical isolation conditions for a number of related structures. The advantage of subjecting biochemical isolations of organelles to similar conditions is that the observations may be compared without concern that differences in experimental parameters may prevent parallel or related observations. This review describes a general isolation procedure which is rapid and applicable to the isolation of chromosomes, mitotic apparatus, or nuclei.

A. Chromosomes

Isolation procedures for mammalian metaphase chromosomes may be catalogued by the pH of the isolation medium. Most of the methods (Prescott and Bender, 1961; Somers et al., 1963; Chorazy et al., 1963; Cantor and Hearst, 1966; Huberman and Attardi, 1967; Franceschini and Giacomoni, 1967; Mendelsohn et al., 1968; Burkholder and Mukherjee, 1970) utilize low pH, generally between 3.0 and 3.7. According to Huberman and Attardi (1966) lowering the pH has the effect of increasing the contraction of the chromosomes, which could be partially responsible for increased resistance of the isolated chromosomes to mechanical damage at low pH, and could be the result of denaturation and precipitation of some chromosomal proteins. The use of low pH introduces the possibility of undesirable side effects such as the extraction of histones. Huberman and Attardi (1966) found that most histones are not extracted by the conditions of the various chromosome isolations, but some lysine-rich histones in HeLa chromosomes are extracted. The use of a pH below 6.0 causes the formation of aggregates of materials of high RNA content which are difficult to remove from the final preparation (Hamilton and Peterman, 1959). These aggregates are probably composed of ribosomes, which clump at lower pH values and sediment easily at low centrifugal forces.

Chromosome isolations at neutral pH according to Hearst and Botchan (1970) should be viewed as a major advantage over acidic extraction conditions. Stabilization with $10^{-3} M$ $ZnCl_2$ has been used for isolation of chromosomes at a neutral pH (Maio and Schildkraut, 1967). In neutral solutions $ZnCl_2$ forms amphoteric ions which seem to stabilize the chromosomes, but the ions also extract nuclear ribosomes and nonchromosomal proteins and dissolve the nuclear membrane (Frenster, 1963). The method of Wray and Stubblefield (1970) presented here utilizes hexylene glycol as a stabilizing agent at the nearly neutral pH of 6.5. The rationale for using hexylene glycol to stabilize metaphase chromosomes is attributed to Kane (1965) who stated, ". . . the addition of glycol to water reduces the dielectric constant and increases the electrostatic free energy of the protein molecules, thus reducing their solubility. The addition of glycol to water also influences the hydrophobic interactions between the non-polar groups of the protein molecules, which play an important role in controlling the conformation of protein molecules in solution."

B. Mitotic Apparatus

The mitotic apparatus of sea urchin eggs was isolated by Kane (1965) using a 1 M solution of hexylene glycol buffered in a pH range of 6.2 to 6.4. Cytolysis of the cells occurs hypotonically, with the intact mitotic apparatus released by very mild agitation. The fact that the cells tolerate 0.1 M hexylene glycol during division and development suggested that there was no direct chemical effect on the mitotic apparatus.

The mammalian mitotic apparatus has been isolated by Sisken *et al.* (1967) using a technique based on the Kane procedure. The modifications consisted of omitting the phosphate buffer, and adding $10^{-4} M$ $CaCl_2$, which tended to enhance the stability of the mitotic apparatus. The solution was left unbuffered because the addition of phosphate or "tris" prevented lysis of the cells.

Both Kane and Sisken noted that the chromosomes were stable in the isolated mitotic apparatus, and Sisken observed that nuclei could also be isolated by the same procedure used for isolation of the mitotic apparatus. Wray and Stubblefield (1970) confirmed and extended these observations. They noted that under conditions where spindle microtubules are not stable, free chromosomes can be isolated from mitotic cells. Nuclei may also be obtained from the interphase cells.

C. Nuclei

The common nuclear isolation methods use either aqueous or non-aqueous homogenization media, the choice of which depends upon the nature of the end products desired. The nonaqueous organic solvent techniques unavoidably damage the nuclear membrane, whereas the aqueous methods allow a leakage of soluble nuclear components. A theoretically perfect nuclear isolation technique should yield a product which contains no cytoplasmic contamination and has lost no nuclear components during the isolation. As yet, no isolation method described meets all of the theoretical considerations. Excellent reviews by Wang (1967), Busch (1967), and Muramatsu (1970) have been published which discuss and evaluate in detail several isolation procedures and their relative advantages. Muramatsu (1970) notes that nuclear isolation procedures may be specific for the tissue or cell type. The method presented here has been successful for all tissues and cell types in which it has been tried.

D. System Universality

The methods given in this review are only for the isolation of chromosomes, mitotic apparatus, and nuclei from Chinese hamster fibroblasts, but the buffer system described below may be the nearest approximation to a universal isolation medium yet developed. Good preparations of many cell organelles may be obtained from all organs, tissues, and tissue culture lines that I have tried. Some require modification (i.e., HeLa chromosomes are easier to isolate at pH 6.7), but most are not that sensitive. Cellular organelles such as ribosomes, polysomes, lysosomes, centrioles, mitochondria, and membranes have been tested, and all expected *in vitro* activities which have been checked were found, although extensive biochemistry has not yet been attempted. I would hope that laboratories which are routinely isolating a particular cellular organelle from any source would be willing to try these methods and/or buffer system, possibly to modify them further, and compare the preparations with those routinely used.

II.   General Methods

A.   Tissue Culture and Cell Line

The experiments presented in this review were performed using the cloned strain Don-C (Stubblefield, 1966) which was derived from line Don (Hsu and Zenses, 1964), a normal diploid lung culture from a male Chinese hamster. The strain Don-C has a portion of the autosome 1b translocated to the Y chromosome, and a segment of chromosome 8 translocated to the short arm of chromosome 9. The Don-C cells were cultured in monolayers in 600-ml bottles rotating on a wheel at one revolution every 15 minutes. Each bottle contained 50 ml of McCoy's medium 5a supplemented with 0.8 gm/liter lactalbumin hydrolyzate and 20% fetal calf serum. The pH of the medium was buffered by a bicarbonate system requiring 10% CO_2 in the atmosphere. The Don-C strain has a cell cycle of approximately 12 hours. Cultures were subdivided every 24 hours.

B. Cell Synchronization

In order to increase the number of cells in mitosis at the time of harvest, mitotic cells were accumulated with Colcemid (0.06 μg/ml) and selectively trypsinized away from the remaining interphase cells (Stubblefield and Klevecz, 1965). The mitotic cell populations obtained by this method were routinely 94% to 97% metaphase cells.

C. Isolation Buffers

The chromosome isolation buffer (Wray and Stubblefield, 1970) contains 1.0 M hexylene glycol (2-methyl-2,4-pentanediol), Eastman Organic; $5 \times 10^{-4} M$ CaCl$_2$; and $10^{-4} M$ piperazine-N,N'-bis(2-ethane sulfonic acid) monosodium monohydrate ("PIPES"), Calbiochem; at a pH of 6.5. The buffer is most conveniently made by a 1 to 10 dilution of $5 \times 10^{-3} M$ CaCl$_2$, $10^{-3} M$ PIPES (10× buffer), which has been adjusted to a pH of 6.5 with 1.0 N sodium hydroxide. Hexylene glycol is then added after dilution. It is important that the pH be adjusted before the addition of hexylene glycol! Apparently some pH electrodes respond sluggishly if the solution contains an appreciable amount of this organic liquid.

Chromosomes, mitotic apparatus, and nuclei are all stable in the chromosome isolation buffer, but it was found that for Chinese hamster cells slight modifications made in the buffer system for mitotic apparatus and for nuclei augmented cell breakage and decreased the amount of contamination of the product. For the mitotic apparatus buffer the PIPES concentration was changed to $5 \times 10^{-5} M$. In the nuclei buffer the hexylene glycol was reduced to 0.5 M, the PIPES was also $5 \times 10^{-5} M$, and the CaCl$_2$ was increased to $10^{-3} M$. In these isolation buffers the pH is relatively neutral, heavy amphoteric metals such as zinc are absent, and the chromosomes, mitotic apparatus, and nuclei are stabilized but not fixed.

D. Other Methods

DNA (Burton, 1956), RNA (Hurlbert et al., 1954), and protein (Lowry et al., 1951) were measured colorimetrically. Whole chromosomes could be studied using electron microscopy after air drying or critical-point drying unstained preparations on collodion- or Formvar-coated grids. Sectioned material was prepared by methods described by Brinkley (1965), except that the initial fixation was always made in 3% glutaraldehyde in the chromosome isolation buffer.

III. Isolation Methods and Properties

A flow chart which outlines all steps in these isolation procedures and indicates pertinent explanations for some of these steps is shown in Fig. 1.

A. Chromosome Isolation

The starting material for chromosome isolation is a relatively pure mitotic cell population. Cooling to 4°C in fresh media inactivates trypsin, dilutes any remaining Colcemid, and causes many microtubules remaining after Colcemid treatment to disassemble spontaneously (Inoue, 1964), thereby helping to minimize aggregation of chromosomes and contamination. The cells are then centrifuged, the medium removed with a pasteur pipette, and the cells washed rapidly in 4°C chromosome isolation buffer. After centrifugation, the pellet is gently resuspended in about 50 volumes of the cold buffer and incubated in a 37°C water bath for 10 to 15 minutes. Following incubation, the cells may be broken by homogenization. However, for small volumes of cells, gentle syringing through a 22-gauge needle breaks the cell membrane and frees the chromosomes into the buffer. The number of times one passes the solution through the needle and the amount of force necessary varies with cell line, cell incubation time, and thoroughness of the washing. This and all steps of these procedures should be monitored with phase-contrast microscopy. It is better to syringe too gently than too vigorously, since syringing too vigorously damages chromosome morphology and is irrevocable. Do not allow the temperature to drop until the desired breakage is accomplished. After the cells are broken and the chromosomes liberated, further operations may be done in the cold.

The most convenient equipment for these isolation procedures with a cell pellet of 0.1-ml volume are 15-ml Pyrex or plastic conical centrifugation tubes, 5-ml plastic B-D syringes with 1.5-inch 22-gauge needles and a clinical centrifuge. For large volumes of cells, comparable results may be obtained with a Dounce glass homogenizer from Kontes. To facilitate removal of nuclei and unbroken cells, Nucleopore membrane filters (8 μ and 5 μ pore sizes) from Wallabs, Inc., San Rafael, California may be used. The entire procedure is conveniently done within 1 hour. The appearance of the cells and chromosomes before, during, and after the isolation procedure is shown in the phase-contrast micrographs of Fig. 2. Differential centrifugation yields a chromosome preparation like that shown in Fig. 2d. Figure 3 demonstrates the ultrastructure of typical

Metaphase Cells

1. Centrifuge 1000 rpm; 2 min.
2. Suspend in media to inactivate trypsin.
3. Cool 20 min or longer at 4°C to dissolve mitotic apparatus.
4. Centrifuge 1000 rpm; 2 min.
5. Wash in 4°C isolation buffer, pH 6.5
   1.0 M hexylene glycol
   0.5 mM CaCl$_2$
   0.1 mM PIPES
6. Centrifuge 2000 rpm; 3 min.
7. Suspend gently in cold chromosome buffer and incubate in a water bath at 37°C, 10 min.
8. Syringe gently through 22-gauge needle to lyse cells.
9. Centrifuge 3000 rpm; 5–10 min.
ISOLATED CHROMOSOMES

Metaphase Cells

1. Centrifuge 1000 rpm; 2 min.
2. Suspend in media to inactivate trypsin.
3. Incubate 20 min at 37°C to allow formation of mitotic apparatus.
4. Centrifuge 1000 rpm; 2 min.
5. Wash in 25°C isolation buffer, pH 6.5
   1.0 M hexylene glycol
   0.5 mM CaCl$_2$
   0.05 mM PIPES
6. Centrifuge 2000 rpm; 3 min.
7. Suspend gently in 25°C mitotic apparatus buffer
8. Swirl on Vortex mixer at high speed or syringe gently to lyse cells.
9. Centrifuge 3000 rpm; 5 min.
ISOLATED MITOTIC APPARATUS

Interphase Cells

1. Centrifuge 1000 rpm; 2 min.
2. Suspend in media to inactivate trypsin.
3. Centrifuge 1000 rpm; 2 min.
4. Wash in 25°C isolation buffer, pH 6.5
   0.5 M hexylene glycol
   1.0 mM CaCl$_2$
   0.05 mM PIPES
5. Centrifuge 2000 rpm; 3 min.
6. Suspend gently in nuclei buffer.
7. Syringe through 22-gauge needle to lyse cells.
8. Centrifuge 2000 rpm; 3 min.
ISOLATED NUCLEI

FIG. 1. Flow chart delineating the procedures developed during the course of this investigation. Starting with an exponentially growing culture, either chromosomes or mitotic apparatus may be isolated from the metaphase population after blocking the cell with Colcemid (0.06 μg/ml; 3 hours) followed by differential trypsinization to increase the yield of mitotic cells. Nuclei may be isolated either directly from the exponential culture or from the interphase cells remaining after differential trypsinization to remove the metaphase cells. If an exponential culture is chosen for nuclei isolation, the cells may be removed without the use of trypsin. The culture flask is purged with air to remove the carbon dioxide, which would influence the pH of the isolation buffer. This is followed by removing the media, washing the culture with nuclei isolation buffer twice, and allowing the cells to stand at room temperature in a minimal volume (5 ml) of buffer until they detach from the glass.

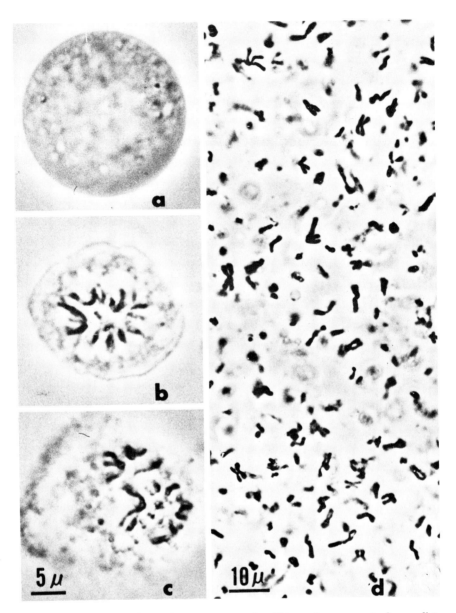

FIG. 2. In (a) is shown the appearance of a Chinese hamster metaphase cell in chromosomal isolation buffer before equilibration. The chromosomes are not visible until the cell membrane ruptures. After a 5-minute incubation at 37°C the chromosomes can be seen as in (b). Upon further incubation the cell disintegrates (c) releasing the chromosomes which can then be freed of contaminating cytoplasm by gentle shearing forces. (d) A chromosome suspension prepared according to the methodology described here and in the text. Phase contrast. Reference bar: a, b, and c, 5 μ; d, 10 μ. (Stubblefield and Wray, 1971.)

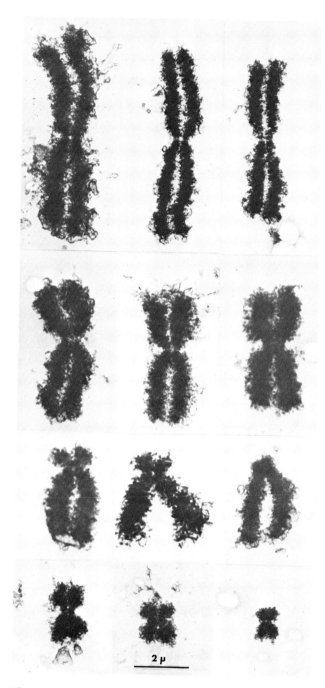

Fig. 3. Electron micrographs of typical Chinese hamster chromosomes. Reference bar, 2.0 μ. Anderson critical-point dried, unstained. (Stubblefield and Wray, 1971.) ×7500.

chromosomes which are unstained and critical-point dried on a Formvar film.

1. CHROMOSOME FRACTIONATION

The isolation of metaphase chomosomes leads to the next logical step, the fractionation of these chromosomes into groups according to their size. Sedimentation velocity sucrose gradients of Chinese hamster chromosome suspensions give patterns like that seen in Fig. 4. The contaminating whole cells and nuclei sediment to the 85% sucrose-buffer cushion, and mitotic apparatus and aggregates of chromosomes also sediment far down the gradient. The broad absorbance peak which occurs from the middle of the gradient to near the top is a rough distribution of chromosomes according to size. For Chinese hamster cells the large metacentric chromosomes (A group) which include the 1's and 2's are sedimented the greatest distance into the gradient. The medium sized metacentric chromosomes (B group) are the X, Y, 4's, and 5's. The majority of these are located above the A group and below the acrocentric chromosomes (6's,

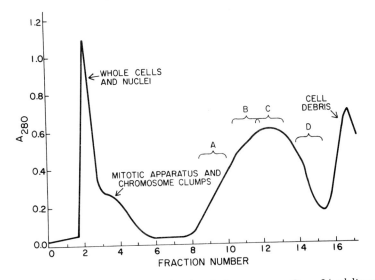

FIG. 4. Sucrose velocity gradient of isolated chromosomes. On a 24-ml linear 10 to 40% w/v buffered sucrose gradient with a bottom cushion of 85% sucrose-buffer, was layered 3 ml of an isolated chromosome suspension. The 85% sucrose-buffer cushion prevented a pellet of whole cells and nuclei which would occlude the puncture needle. Coating with Siliclad (Clay-Adams) minimized the adherence of chromosomes to the nitrocellulose centrifuge tubes and glassware. The gradient was centrifuged 90 minutes at 1000 rpm, in an HB-4 swinging bucket rotor in a Sorvall RC-2 B centrifuge, after which the tube was punctured and absorbance monitored by a Gilford Model 240 recording spectrophotometer.

7's, and 8's) which comprise the C group. The D group chromosomes, which are the small metacentric chromosomes (9's, 10's, and 11's), are sedimented the least distance into the gradient and run only slightly ahead of a large peak of ultraviolet-absorbing material comprised of light cellular debris. If the gradient is monitored closely by phase-contrast microscopy, four fractions corresponding to the four groups may be obtained which include the majority of the chromosomes in the appropriate group, but contaminated with those from adjoining groups. Analysis of the centrifugation pattern shows that the chromosomes which are separated by more than one group are not contaminants of each other (i.e., group B and group D). Further purification is accomplished by repeating the velocity sedimentation centrifugation on each of the obtained fractions.

2. CHROMOSOME COMPOSITION

Hearst and Botchan (1970) reviewed the gross chemical composition of chromosomes isolated by a variety of reported methods. RNA is reported on an almost equal weight basis with the DNA of the metaphase chromosome. It has been shown that this RNA present in isolated chromosome preparations is primarily the 18 S and 28 S rRNA (Huberman and Attardi, 1966; Maio and Schildkraut, 1967), although it has not been demonstrated that these RNAs are on the chromosome as ribosomes. Electron microscopy of these chromosome isolation "products" from the other methods has not been reported so that visual comparisons of methods may be evaluated for contamination only at the grossly inadequate light microscope level.

Mendelsohn et al. (1968) showed that contaminant RNA and protein remained near the top of a sucrose velocity gradient with the small chromosomes. The fractions enriched for larger chromosomes sedimented farther down the gradient and contained much less RNA and protein. The reduced amount of RNA in this A + B (large chromosome) fraction supports their previous finding (Salzman et al., 1966) that RNA is not an intrinsic component of metaphase chromosomes.

Table I shows the relative amounts of DNA, RNA, and protein present in total heterogeneous chromosome fractions isolated by this method compared to partially purified fractions prepared by Mendelsohn et al. (1968). It may be seen that the results obtained were very similar to those of the A + B fraction of Mendelsohn. In one instance RNA could not be detected in the preparation while both the DNA and protein compared favorably. Apparently there is little, if any, adventitiously adsorbed RNA or protein in these chromosomes which can be shown biochemically. Morphologically (Fig. 3) they are excellent.

TABLE I

RELATIVE MACROMOLECULAR CONTENTS OF ISOLATED CHROMOSOMES, MITOTIC APPARATUS, AND NUCLEI

Preparation	DNA[a]	RNA	Protein	Investigator
Chromosome fraction				
ABCD	100	0	209	Wray and Stubblefield (1970)
ABCD	100	17	220	Wray and Stubblefield (1970)
AB	100	18	240	Mendelsohn et al. (1968)
C	100	79	540	Mendelsohn et al. (1968)
D	100	140	850	Mendelsohn et al. (1968)
Mitotic apparatus	100	35	1502	Wray and Stubblefield (1970)
Nuclei				Mendelsohn et al. (1968)
Chinese hamster	100	31	272	Wray and Stubblefield (1970)
Rat liver	100	27	400	Busch (1967)

[a] The amount of DNA in each preparation was taken to be 100, and the relative amounts of RNA and protein were calculated. DNA values are not related from one preparation to another. Analyses were measured colorimetrically as referenced in the text. (Wray and Stubblefield, 1970.)

B. Mitotic Apparatus Isolation

The isolation of the mitotic apparatus differs slightly from the isolation of chromosomes, in that the Colcemid-treated mitotic cells are not refrigerated to dissolve the microtubules. Instead, after trypsinization they are incubated in fresh medium at 37°C for 20 minutes to allow complete formation of the mitotic spindle. Subsequently, the cells are centrifuged and washed at room temperature in buffer modified for mitotic apparatus isolation. The cells are again centrifuged, resuspended gently in 50 volumes of buffer, and allowed to equilibrate in the buffer for 5 to 10 minutes. The cells are then lysed by vigorous swirling on a vortex mixer or by gentle homogenization, or they may be sheared very gently through a 22-gauge needle. If the shearing forces are too vigorous, chromosomes may be released from the spindle. Monitoring by phase-contrast microscopy is essential.

The intact mitotic apparatus preparation as shown in Fig. 5 may be concentrated and purified by differential centrifugation or sedimentation velocity gradients similar to those for chromosomes. Figure 6 shows an electron micrograph of a thin section through a mitotic cell before isolation, and Fig. 7, for the purpose of comparison, shows a section through an isolated mitotic apparatus. It is quite apparent that most of the cell contents have been extracted, leaving chromosomes, microtubules, centrioles, and in many cases adhering membranes. The membranes are more

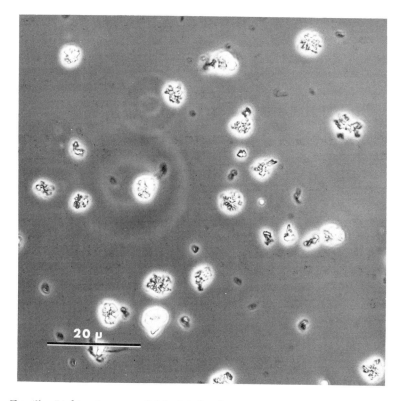

FIG. 5. Light microscope field of isolated mitotic apparatus. The orientations of the spindle are difficult to determine because the chromosomes are so large in comparison to the spindle itself. Reference bar, 20.0 μ. Phase contrast, no fixation or stain. ×800.

difficult to remove if the PIPES concentration is higher than $5 \times 10^{-5}\,M$. The microtubules in this cell line are unstable in $CaCl_2$ concentrations below $5 \times 10^{-4}\,M$. Figure 8 shows an electron micrograph of a whole-mount mitotic apparatus which has been critical-point dried. Table I shows the relative amount of DNA, RNA, and protein present in a mitotic apparatus preparation.

C. Nuclei Isolation

Rupture of contaminating interphase cells in the isolation of chromosomes or mitotic apparatus generally releases the nucleus as a contaminant of these preparations. As is the case with the mitotic apparatus the

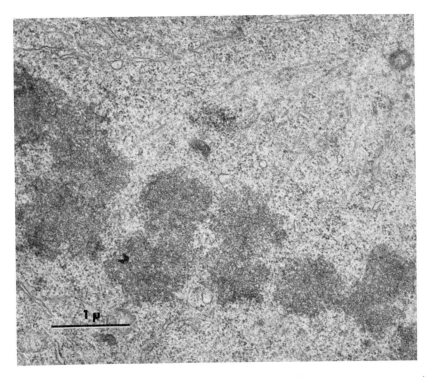

FIG. 6. Thin section through a whole mitotic cell. The orientation is approximately the same as in Fig. 7. Reference bar, 1.0 μ. Fixed in glutaraldehyde, postfixed in OsO$_4$, embedded in Epon, stained with uranium and lead salts. ×20,000.

interphase cells lyse more readily if the PIPES concentration is $5 \times 10^{-5} M$ or lower. However, raising the CaCl$_2$ concentration to $10^{-3} M$ seems to stabilize the nuclei against breakage during shearing. The hexylene glycol is reduced to 0.5 M in most preparations; this also seems to reduce cytoplasmic contamination.

Nuclei can be isolated directly from cell monolayers. It is necessary to purge the CO$_2$ from the culture atmosphere in order to prevent a drop in pH, which prevents cell lysis. Fibroblasts are washed with the nuclear isolation buffer. A second wash is left on the cells for about 10 minutes at room temperature, after which the cells spontaneously detach from the glass. The suspension is then sheared by syringing or homogenizing as in the case for chromosomes. Preparations should be routinely checked by phase-contrast microscopy. The analogous procedure also works for cell suspensions following trypsinization so that it is possible to isolate chromosomes or mitotic apparatus from the metaphase cells, and nuclei

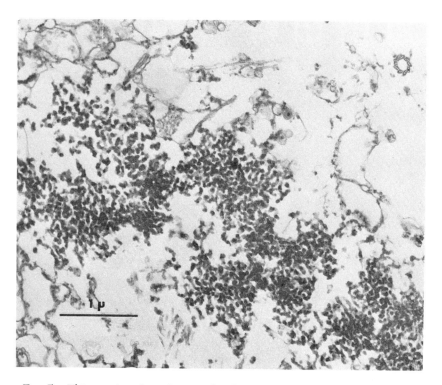

FIG. 7. Thin section through an isolated mitotic apparatus. Evident are chromosomes, microtubules, a centriole, and some membranes. Fixed with glutaraldehyde after 10 minutes in mitotic apparatus isolation buffer and prepared as in Fig. 6. Reference bar, 1.0 μ. ×20,000.

from the interphase cells of the same culture. To test the procedure on freshly excised tissue routinely used in many biochemical laboratories, rat liver was chosen, and the liver nuclei were found to be cleanly isolated following homogenization in the nuclei buffer.

Figure 9 shows Don-C nuclei isolated using this procedure. In Fig. 10 an electron micrograph of a thin section through such a nucleus before extraction is shown. Figure 11 shows a thin section through a nucleus after it has been extracted with buffer, and Fig. 12 shows a thin section of a nucleus after the cytoplasm has been stripped away. In isolated nuclei, most of the chromatin appears in the form of 250-Å fibers very similar to those comprising isolated chromosomes (Fig. 3).

Isolated Don-C nuclei appear to have a composition that compares favorably with nuclei from other sources and isolated with other techniques (Table I) (Busch, 1967). Although the protein values are lower

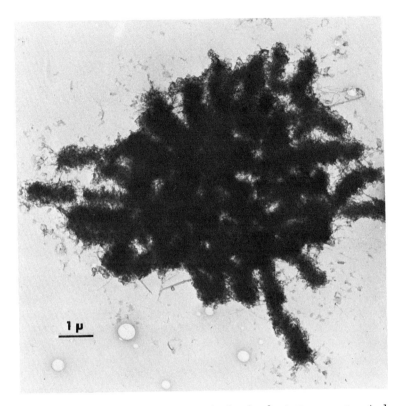

FIG. 8. Whole-mount electron micrograph of isolated mitotic apparatus. Anderson critical-point dried, no stain. Reference bar, 1.0 μ. ×9000.

than those reported, enzymatic studies reveal that the isolated nuclei contain DNA polymerase, RNA polymerase, and adenylate kinase. Table II shows these activities along with numerous cytoplasmic enzymes which are found associated only with the cytoplasmic fraction. It is obvious, also, that exposure to the buffer solution used does not destroy the activity of most enzymes. Determination of enzyme kinetic rates after isolation has not been attempted.

Studies of DNase activity in the cytoplasmic lysate indicate that this enzyme is not released when the cell is broken open (Table III). The DNA in chromosomes and nuclei prelabeled with thymidine-2-^{14}C appears to remain stable when incubated at 37°C in the chromosome isolation buffer. Added Mg^{++} ($1.6 \times 10^{-3} M$) enhances stability, whereas in the presence of added DNase (670 μg/ml) it greatly accelerates degradation. The nuclei are more resistant to added DNase, and the kinetics suggest that about two-thirds of the nuclei may be totally resistant, perhaps be-

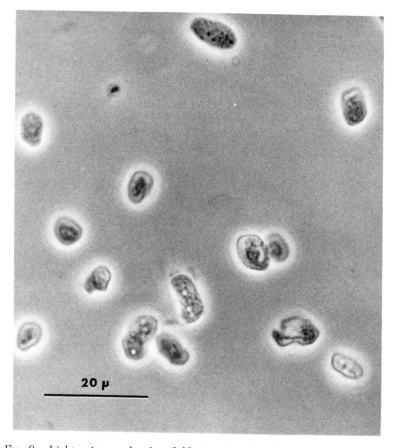

FIG. 9. Light micrograph of a field of isolated nuclei. Reference bar, 20.0 μ. Phase contrast, unstained. ×1400.

cause of an intact nuclear membrane. At any rate, it seems clear that the isolated preparations do not contain appreciable DNase activity.

Rigorous studies have not been attempted to determine the minimum concentrations of the buffer components. It has been determined, however, that the presence of hexylene glycol and calcium are essential for the stability of both chromosomes and the mitotic apparatus while nuclei demonstrate a requirement for calcium.

The relative nontoxicity of the buffer has been demonstrated by growing Chinese hamster fibroblasts in media containing 10% chromosome isolation buffer (Wray and Stubblefield, 1970). Since this dilution would contain 0.1 M hexylene glycol, results analogous to those of Kane (1965) may be inferred.

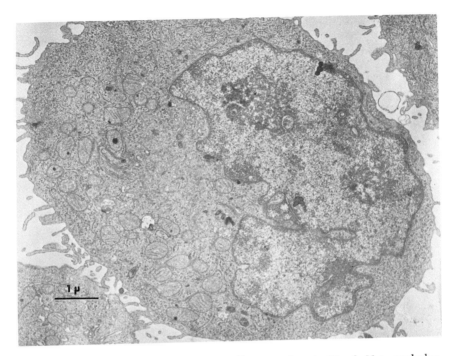

Fig. 10. Interphase Chinese hamster cell prepared as in Fig. 6. Note nucleolus architecture and finely dispersed chromatin fibers. Reference bar, 1.0 μ. ×12,000.

IV. Discussion

The isolation system of Wray and Stubblefield (1970) is derived from that used by Sisken *et al.* (1967) for the isolation of the mitotic apparatus. The selection of PIPES for the buffer, however, was due to the statements by Good *et al.* (1966) where it was emphasized that PIPES was a nonmetal binding buffer with a pK_a of 6.8 at 20°C. Subsequently Ris (1968) has demonstrated that the morphology of chromatin in electron microscope preparations depends upon the buffer chosen. His pictures suggest buffer artifacts when chelating buffers such as phosphate, Veronal acetate, cacodylate, and *s*-collidine are used; but when nonmetal binding buffers are used the normal configurations of nucleohistone fibers are exhibited.

It is logical to assume that the stability of isolated components should depend to some extent upon the approximation of the intracellular environment *in vitro*. This particular system does not mimic the intracellular

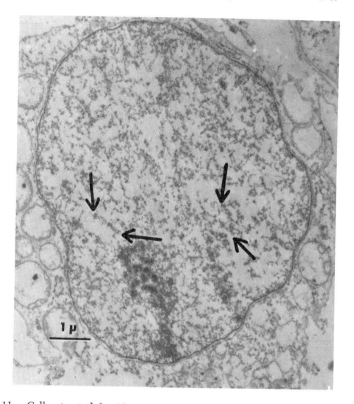

FIG. 11. Cell extracted for 10 minutes in nuclei isolation buffer, then prepared as in Fig. 10. The mitochondria are swollen and the cell membrane is ruptured, but the nucleus is intact. The nucleolus still resembles those seen in Fig. 10. The chromatin fibers appear to be in a state intermediate between the finely dispersed form seen in Fig. 10 and the thicker fibers visible in Fig. 12. The transition is probably accomplished by coiling (arrows). Reference bar, 1.0 μ. ×11,000.

conditions. However, using the rigorous guides of localization of enzyme activity, enzyme functionality, and morphological stability of components, it appears to act primarily as a support medium which preserves biological activity and prevents subcellular degradation of most of the cellular organelles.

Let it again be emphasized that the chromosomes, mitotic apparatus, and nuclei are all stable in the chromosome isolation buffer, and the only reason for the buffer modification in the isolation procedures for mitotic apparatus and nuclei in this cell line is the ease of isolation and decontamination of the component chosen as the end point.

Isolated chromosomes are stable morphologically for periods of several months at 4°C. They may also be quick frozen, stored, and thawed with

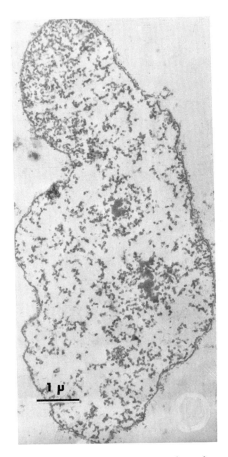

FIG. 12. Electron micrograph of a thin section through an isolated nucleus. A few heterochromatic regions are evident and there appears to be a minimum of contaminating ribosomes on the outer membrane. The majority of the chromatin is present as 250-Å fibers. Reference bar, 1.0 μ. ×12,000.

no apparent harm. The appearance of chromosomes when isolated by gentle shearing forces compares favorably with that demonstrated by many other laboratories. If the shearing force is too vigorous, however, stretched chromosomes may result (Stubblefield and Wray, 1971). The reasons for preincubating the whole cells at 37°C are that the cells break easier and that this incubation time allows the chromosomes to equilibrate with the buffer and not be shocked or stretched by the difference in environment upon breakage. If the cells are broken in a buffer lacking hexylene glycol, the chromosomes are unstable and gradually disintegrate

TABLE II
ENZYMES ASSOCIATED WITH CELL FRACTIONS[a]

Enzyme	Nuclei	Cytoplasm	Chromosome
DNA Polymerase	+	+	*
RNA Polymerase	+		*
Adenylate kinase	+	+	
Catalase	−	+	
Alkaline phosphatase	−	+	
Acid phosphatase	−	+	
Lactic dehydrogenase	−	+	
Alcohol dehydrogenase	−	+	
Phosphogluco mutase	−	+	
Creatine kinase	−	+	
6-Phosphogluconate dehydrogenase	−	+	
Glucose-6-P dehydrogenase	−	+	
Hexokinase	−	+	
Isocitric dehydrogenase	−	+	
α-Glycerol phosphate dehydrogenase	−	+	

[a] After isolation of nuclei, starch gel electrophoresis was performed on samples of nuclei and cytoplasm. Enzyme activity was then determined by appropriate staining reactions (Shaw and Koen, 1968). The presence (+) or absence (−) of enzyme is noted in the table. Both DNA polymerase and RNA polymerase activities (*) were found in isolated chromosomes without added template material in in vitro systems. (Wray and Stubblefield, 1970.)

TABLE III
DNASE DIGESTION OF CHROMOSOMES AND NUCLEI

Substrate[a]	Additions	Percent DNA hydrolyzed		
		In 0 minutes	In 15 minutes	In 30 minutes
Chromosomes	—	0.2	6.1	6.2
Chromosomes	+Mg++	—	0.4	1.1
Chromosomes	+DNase	—	43.1	65.5
Chromosomes	+Mg++ + DNase	—	85.3	94.6
Nuclei	—	0.3	1.1	1.5
Nuclei	+Mg++	—	0.5	0.6
Nuclei	+DNase	—	27.5	26.0
Nuclei	+Mg++ + DNase	—	37.8	36.5

[a] The reaction system contained in 0.3 ml either ^{14}C-chromosomes or ^{14}C-nuclei in buffer. Where indicated, $MgCl_2$ was added to a concentration of 1.6 mM and DNase was added to a concentration of 670 μg/ml. The hydrolysis of DNA was measured as increase of TCA nonsedimentable isotope into the supernatant. After centrifuging for 10 minutes at 10,000 rpm at 4°C, 0.1-ml samples were counted in an NCS-toluene scintillating fluid in a Beckman Liquid Scintillation Counter. (Wray and Stubblefield, 1970.)

over several hours' time. At Ca^{++} concentrations below $5 \times 10^{-4}\,M$, both chromosomes and nuclei rapidly dissolve.

The composition of the heterogeneous isolated chromosomes compares favorably with the most highly purified fraction of Mendelsohn et al. (1968) and appears to fulfill the prediction by Salzman and Mendelsohn (1968) that chromosomes will be ultimately isolated which contain no contaminant RNA.

When isolating the mitotic apparatus from Colcemid-blocked cells one must allow the premetaphase cells, which consist of condensed chromosomes in a spherical arrangement around the unseparated parent and daughter centriole (Brinkley et al., 1967), to reverse from the mitotic inhibitor. Incubation at 37°C for 20 minutes allows a normal mitotic spindle to form before the isolation is attempted. By carefully controlling the timing of the incubation one may select for the isolation of metaphase or anaphase configurations of the mitotic apparatus.

The isolation of the mitotic apparatus has as its biggest problem the limitation imposed by the physical nature of its composition. This large, unwieldy, and rather fragile structure is more subject to physical disruption by shear forces than either chromosomes or nuclei. Consequently the isolation must be more gentle; lowering the amount of PIPES tends to facilitate removal of the membranes which tend to contaminate the isolated mitotic apparatus.

In this system the microtubules are well defined and appear to be free of contaminating cellular material. The centrioles are demonstrated to be stable and to be physically attached to the ends of the spindle.

The nuclei when isolated were found to be free from cytoplasmic tags. There was one predominant chromatin fiber type of 250 Å, heterochromatic and euchromatic areas were observed, and the RNA content was found to be similar to the amount found in other cells. Differences in the amount of protein found in these nuclei could be attributed to contamination in other preparations, cell type deviations, or loss of intranuclear protein from these nuclei. The latter is questioned by enzymatic data which indicate that cytoplasmic enzymes are excluded from the nuclei as expected and that both DNA and RNA polymerase are found in the nuclear fraction. It would be expected that damaged nuclei would exchange proteins with the cytoplasm.

The advantages these procedures have over classical homogenization methods are: (1) the ready adaptation to as few as a million cells; (2) applicability to all tissue culture cell lines, organisms, and tissues; (3) the speed, simplicity, and minimum of equipment needed for the isolation techniques; and (4) the nearly identical isolation conditions for functionally related organelles.

ACKNOWLEDGMENTS

This work was supported by grants from the National Science Foundation, The National Cancer Institute, The American Cancer Society, and The National Institutes of Health. The author was a Damon Runyon Postdoctoral Fellow.

Sincere appreciation is extended to Dr. Elton Stubblefield for photographic assistance and helpful advice.

REFERENCES

Brinkley, B. R. (1965). *J. Cell Biol.* **27**, 411.

Brinkley, B. R., Stubblefield, E., and Hsu, T. C. (1967). *J. Ultrastruct. Res.* **19**, 1.

Burkholder, G. D., and Mukherjee, B. R. (1970). *Exp. Cell Res.* **61**, 413.

Burton, K. (1956). *J. Biochem. (Tokyo)* **62**, 315.

Busch, H. (1967). *In* "Methods in Enzymology" (L. Grossman and K. Moldave, eds.), Vol. 12, Part A, pp. 421–448. Academic Press, New York.

Cantor, K. P., and Hearst, J. E. (1966). *Proc. Nat. Acad. Sci. U. S.* **55**, 642.

Chorazy, M., Bendich, A., Borenfreund, E., and Hutchison, D. J. (1963). *J. Cell Biol.* **19**, 59.

Franceschini, P., and Giacomoni, D. (1967). *Atti Ass. Genet. Ital.* **12**, 248.

Frenster, J. H. (1963). *Exp. Cell Res., Suppl.* **9**, 235.

Good, N. E., Winget, G. D., Winter, W., Connolly, T. N., Izawa, S., and Singh, R. M. M. (1966). *Biochemistry* **5**, 467.

Hamilton, M. G., and Peterman, M. L. (1959). *J. Biol. Chem.* **234**, 1441.

Hearst, J. E., and Botchan, M. (1970). *Annu. Rev. Biochem.* **39**, 151–182.

Hsu, T. C., and Zenses, M. (1964). *J. Nat. Cancer Inst.* **32**, 857.

Huberman, J. A., and Attardi, G. J. (1966). *J. Cell Biol.* **31**, 95.

Huberman, J. A., and Attardi, G. J. (1967). *J. Mol. Biol.* **29**, 487.

Hurlbert, R. B., Schmitz, H., Brumm, A. F., and Potter, V. R. (1954). *J. Biol. Chem.* **209**, 23.

Inoue, S. (1964). *In* "Primitive Motile Systems in Cell Biology" (R. D. Allen and N. Kamiya, eds.), pp. 549–598. Academic Press, New York.

Kane, R. E. (1965). *J. Cell Biol.* **25**, 136.

Lowry, O. H., Rosebrough, N. J., Farr, A. L., and Randall, R. (1951). *J. Biol. Chem.* **193**, 265.

Maio, J. J., and Schildkraut, C. L. (1967). *J. Mol. Biol.* **24**, 29.

Mendelsohn, J., Moore, D. E., and Salzman, N. P. (1968). *J. Mol. Biol.* **32**, 101.

Muramatsu, M. (1970). *In* "Methods in Cell Physiology" (D. M. Prescott, ed.), Vol. 4, pp. 195–230. Academic Press, New York.

Prescott, D. M., and Bender, M. A. (1961). *Exp. Cell Res.* **25**, 222.

Ris, H. (1968). *J. Cell Biol.* **39**, 158a.

Salzman, N. P., and Mendelsohn, J. (1968). *In* "Methods in Cell Physiology" (D. M. Prescott, ed.), Vol. 3, pp. 277–292. Academic Press, New York.

Salzman, N. P., Moore, D. E., and Mendelsohn, J. (1966). *Proc. Nat. Acad. Sci. U. S.* **56**, 1449.

Shaw, C. E., and Koen, A. L. (1968). *In* "Chromatography and Electrophoresis (I. Smith, ed.), 2nd ed., pp. 325–364. Wiley (Interscience), New York.

Sisken, J. E., Wilkes, E., Donnelly, G. M., and Kakefuda, T. (1967). *J. Cell Biol.* **32**, 212.

Somers, C. E., Cole, A., and Hsu, T. C. (1963). *Exp. Cell Res., Suppl.* **9**, 220.

Stubblefield, E. (1966). *J. Nat. Cancer Inst.* **37**, 799.

Stubblefield, E., and Klevecz, R. (1965). *Exp. Cell Res.* **40,** 660.

Stubblefield, E., and Wray, W. (1971). *Chromosoma* **32,** 262.

Wang, T. Y. (1967). *In* "Methods in Enzymology" (L. Grossman and K. Moldave, eds.), Vol. 12, Part A, pp. 417–421. Academic Press, New York.

Wray, W., and Stubblefield, E. (1970). *Exp. Cell Res.* **59,** 469.

Chapter 9

Isolation of Metaphase Chromosomes with High Molecular Weight DNA at pH 10.5

WAYNE WRAY

*Department of Cell Biology, Baylor College of Medicine,
Houston, Texas*

I. Introduction

Biochemical and biophysical questions may be asked in order to advance the understanding of the architecture of the metaphase chromosome when one is able to obtain large populations of morphologically excellent metaphase chromosomes (Wray and Stubblefield, 1970). Knowledge of the structure of metaphase chromosomes has been inferred from both light and electron microscopy, but little data other than visual observation have been published. The unineme concept (one DNA duplex per chromatid) versus the multineme concept (more than one DNA

307

molecule per chromatid) has been widely debated, but the positions are not always firmly founded on data. The reader is urged to closely examine the following reviews to gain an insight into the issues and interpretations of data (Wolff, 1969; Prescott, 1970; Holliday, 1970; Ris and Kubai, 1970; Cohn, 1971; Crick, 1971; Watson, 1971; Thomas, 1971; Stubblefield, 1973; Huberman, 1973).

Since DNA is the most interesting molecule in the chromosome from a genetic standpoint, it was felt that the size distributions (molecular weight) present in metaphase chromosomes might lead to an insight as to their arrangement and/or packaging. Molecular weight may be estimated by the use of alkaline sucrose velocity gradients similar to the way radiation damage and repair processes have been analyzed (Humphrey et al., 1968). With the excellent methods of obtaining large populations of metaphase chromosomes and the ease of fractionation of these into size groups, the experimental design was to analyze and compare the molecular weight of the DNA in large and small chromosomes. Hopefully if the length of DNA is different, the emerging distributions will fall into a predictable or at least interpretable pattern. Several of the initial experiments gave the results that all of the DNA in isolated chromosome preparations regardless of chromosome size existed in 1×10^6 molecular weight (single-strand) pieces. This result prompted a study to ascertain whether any published chromosome isolation method would yield DNA of the high molecular weight (1×10^8 for single stranded) seen when metaphase cells are lysed directly on the gradient (Wray et al., 1972). The conditions which electron microscopists routinely use for chromosome preparation for structural studies also were examined to determine if this DNA was of a high molecular weight.

The lone method of chromosome isolation not discussed in Chapter 8 of this volume was developed by Corry and Cole (1968). This method is unique in that it utilizes the relatively high pH of 9.6. It also is somewhat obscure since it forms merely a small portion of a major paper on "Radiation-Induced Double-Strand Scission of the DNA of Mammalian Metaphase Chromosomes."

II. Cell Labeling

To label the DNA of cells, thymidine-^{3}H was added to the medium of logarithmically growing Chinese hamster fibroblast monolayer cultures, and the cultures were incubated for about 8 hours. For most experiments 0.1 μCi/ml of thymidine-^{3}H (0.05 μCi/μg) was used.

III. Alkaline Sucrose Gradient Techniques

The alkaline sucrose gradient technique originally described by Mc-Grath and Williams (1966) was modified by Humphrey *et al.* (1968) in order to measure the sedimentation characteristics of DNA from isolated metaphase chromosome populations, whole mitotic cells, and interphase cells. A volume of lysing solution (0.5 ml) was placed in the bottom of a polyallomer (Beckman) centrifuge tube, and then an equal volume of either whole cells or isolated chromosomes was pipetted into the lysing solution. The lysing solution consisted of 2% tri-iso-propylnaphthalene sulfonic acid and 1% *p*-aminosalicylic acid in a solvent of 6% secondary butyl alcohol in water which was then adjusted to pH 12.5 with 1.0 N NaOH. Exponential alkaline sucrose gradients (5 to 23%) were then poured under the lysing solution and centrifugation was carried out in a Spinco Model L ultracentrifuge at 22,500 rev/minute at 20°C for 4.5 hours, using an SW25.1 rotor.

The characteristics and calibration of the sucrose gradients for measurement of sedimentation coefficient and weight average molecular weight have been previously reported (Humphrey *et al.*, 1968).

IV. Molecular Weight Analysis of Reported Chromosome Isolation Procedures

The DNA was examined in preparations of chromosomes isolated by all of the published types (i.e., acid pH, neutral pH, etc.) of isolation (Somers *et al.*, 1963; Cantor and Hearst, 1966; Maio and Schildkraut, 1967; Corry and Cole, 1968; Wray and Stubblefield, 1970), with care taken to follow procedures exactly. Chromosomes were obtained from mitotic cells lysed on an air/water interface similar to the procedure for observation of whole mount chromosomes for electron microscopy (DuPraw, 1966; Sasaki and Norman, 1966; Abuelo and Moore, 1969).

V. Chromosome Isolation at pH 10.5

A. Procedure and Morphology

A series of experiments comparing different methods of metaphase chromosome isolation is summarized in Table I. As can be readily seen, the metaphase chromosomes isolated by all of the previously published methods except that of Corry and Cole (1968) contain degraded DNA. The methods of Sasaki and Norman (1966) and Maio and Schildkraut

TABLE I

COMPARISON OF THE METHODS FOR ISOLATING CHROMOSOMES

Method used	Weight average molecular weight (daltons $\times 10^6$)[a]	Peak molecular weight (daltons $\times 10^6$)
Interphase control	113	163
Mitotic control	137	122
50% Acetic acid (Somers et al., 1963)	36	16
Cantor and Hearst (1966)	16	2
Maio and Schildkraut (1967)	64	1
Wray and Stubblefield (1970)	11	10
Sasaki and Norman (1966)	32	17
Abuelo and Moore (1969)	12	3
Water-lysed mitotic cells (DuPraw, 1966)	20	11
Mitotic cells (hypotonic treated)	24	13
Corry and Cole (1968)	82	128

[a] The values given for this alkaline sucrose sedimentation procedure are an indication of degradation as compared to the control and are not to be taken as an absolute molecular weight.

(1967) yield chromosomes with some DNA of high molecular weight (weight average molecular weight is high), but most of the DNA is degraded to smaller molecules (peak molecular weight is low). Also, chromosome preparations prepared for electron microscopy by lysing mitotic cells on an air-water interface apparently contain degraded DNA. In fact, degradation of chromosomal DNA occurs any time the cell is subjected to hypotonic treatment. This degradation takes place any time the pH is either acid or near neutral, but is halted at high alkaline conditions. Although the chromosomes isolated by Corry and Cole (1968) contain a high molecular weight DNA, morphologically they are not pleasing and are difficult to isolate. Using high salt sucrose gradients at neutral pH their studies show that the molecular weight for the double-stranded DNA is approximately 2×10^8 daltons.

The buffer which we have developed for isolating chromosomes at pH 10.5 contains 1.0 M hexylene glycol, 2×10^{-3} M $CaCl_2$, and 1×10^{-3} M cyclohexylamino propane sulfonic acid (CAPS) from Schwartz/Mann. The buffer should be made fresh each time, since the pH tends to drift after a day, and the pH should be adjusted to 10.5 before addition of the hexylene glycol. The isolation procedure for mammalian chromosomes is given in the flow chart below and the appearance of chromosomes in the electron microscope is shown in Fig. 1. It is readily apparent that the procedure outlined in the flow chart is quite similar to the chromosome isolation procedure discussed in Chapter 8 in this volume. All of the comments and precautions included in that chapter apply also to the isolation of chromosomes at pH 10.5.

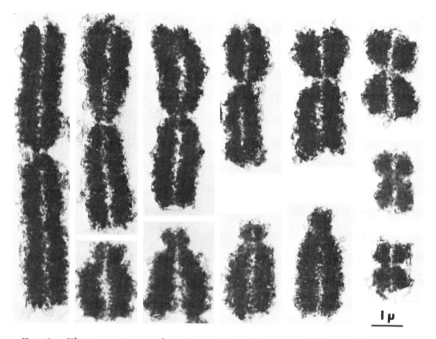

FIG. 1. Electron micrographs of a variety of unstained Chinese hamster chromosomes isolated at pH 10.5, then dehydrated and critical-point dried. Chromosomes isolated at neutral pH (Stubblefield and Wray, 1971) have a similar appearance (Wray *et al.*, 1972). Reference bar is 1 μ. ×8000.

Procedure for Chromosome Isolation at pH 10.5

Add 0.06 μg/ml Colcemid to an exponentially growing fibroblast culture for 3 hours and then differentially trypsinize to obtain metaphase cells.

1. Centrifuge 1000 rpm; 2 minutes.
2. Suspend in media to inactivate trypsin.
3. Cool 20 minutes or longer at 4°C to dissolve mitotic apparatus.
4. Centrifuge 1000 rpm; 2 minutes.
5. Wash in 4°C isolation buffer, pH 10.5
         1.0 M hexylene glycol (2-methyl-2,4-pentanediol)
         $2 \times 10^{-3}\ M$ CaCl$_2$
         $1 \times 10^{-3}\ M$ CAPS
6. Centrifuge 2000 rpm; 3 minutes.
7. Suspend gently in cold buffer and incubate in a water bath at 37°C for 10 minutes.
8. Syringe gently through 22-gauge needle to lyse cells.
9. Centrifuge 3000 rpm; 5–10 minutes to sediment chromosomes.

Chromosomes isolated at pH 10.5 were fractionated on a 20-ml linear sucrose-isolation buffer gradient (8–40%) by centrifuging 30 minutes at

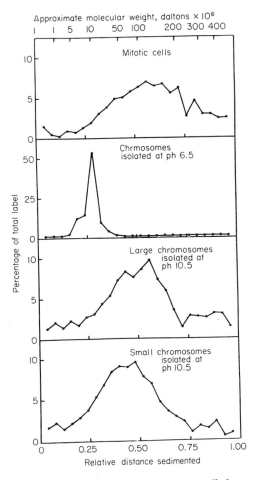

Fig. 2. Sedimentation profiles of denatured DNA on alkaline sucrose gradients. Mitotic or interphase cells lysed directly on the gradient give DNA with a high molecular weight distribution. The DNA of chromosomes isolated at neutral pH is clearly degraded. However the DNA from chromosomes isolated at pH 10.5 is of high molecular weight and is similar in size in both large and small chromosomes (Wray *et al.*, 1972).

2000 rpm (about 500*g*) in a Sorvall HB-4 rotor. Approximately 10 fractions of 2.0 ml each were removed from the top of the gradient. Fraction 4 (small chromosomes) was found to contain mostly chromosomes 9–11, while fraction 9 (large chromosomes) contained mostly chromosomes 1 and 2 contaminated by some chromosomes of the B group (X, Y, 4, and 5).

TABLE II
HIGH pH CHROMOSOME ISOLATION

DNA Source	Weight average molecular weight (daltons $\times 10^6$)[a]	Peak molecular weight (daltons $\times 10^6$)
Interphase control	113	163
Mitotic control	137	122
PIPES Buffer pH 10.0 (Wray et al., 1972)	60	38
CAPS Buffer pH 10.5 (large chromosomes) (Wray et al., 1972)	104	115
CAPS Buffer pH 10.5 (small chromosomes) (Wray et al., 1972)	84	77

[a] The values given for this alkaline sucrose sedimentation procedure are an indication of degradation as compared to the control and are not to be taken as an absolute molecular weight.

B. Molecular Weight Analysis

The results of experiments performed on Chinese hamster chromosomal DNA are shown in Fig. 2. Compared to the mitotic cell control lysed directly upon the gradient, which exhibits a DNA of high molecular weight, the chromosomes isolated at pH 6.5 contain a low molecular weight DNA. On the other hand, in chromosomes isolated at pH 10.5 the high molecular weight DNA is maintained, even after purification on sucrose gradients. Table II gives an indication of the weight average molecular weight and the peak molecular weight of both large chromosome and small chromosome fractions. The DNA of small chromosomes was not appreciably different in size from that of large chromosomes or the controls. An experiment using the same isolation procedure, but containing $1 \times 10^{-4} M$ piperazine-N,N'-bis (2-ethane sulfonic acid) monosodium monohydrate (PIPES) buffer at a pH of 10.0 gave molecular weight values between these results and those shown in Table I.

VI. Discussion

The physical constraints placed on DNA by its structure leave it extremely difficult to isolate in high molecular weight form (above 30 S). Excluding any enzymatic breakage, shear forces which result from practically any attempts to isolate and purify DNA leave it in a relatively low molecular weight form. The advantages therefore of being able to remove the DNA from cells as intact and discrete chromosomal packages are twofold. DNA from chromosomes isolated by the described procedure is obtained which may be larger than 70 S and is resistant to breakage even under rather severe handling conditions. This allows subsequent fractionation of the genome according to chromosome size with the DNA remaining in a "native" (i.e., unbroken) state. For many types of experiments this could be imperative. The advantages of the described system over that of Corry and Cole (1968) are ease of isolation and better morphology of the chromosomes. If, however, one is interested in chromosomal proteins, a pH of 10.5 is perhaps too drastic for some experiments and the PIPES buffer (pH 6.5) method of Wray and Stubblefield (1970) should be chosen.

Since there is breakage of the DNA accompanying all of the isolation procedures at both low and neutral pH, the evidence is circumstantial but there is a high probability that the alkali-inhibited degradative enzyme(s) responsible for this breakage is contained within the structure of the chromosome. This would likely be very similar to some of the repair enzymes which function during metaphase (Humphrey et al., 1968). There is also a high probability that the enzyme is an endonuclease, since experiments (Wray and Stubblefield, 1970) indicated that DNA from chromosomes isolated at pH 6.5 was not released into a supernatant fraction.

The ability to isolate mammalian metaphase chromosomes at a highly alkaline pH is in itself an exciting development, but the biological ramifications must now be carefully considered along with any data concerning the structure of the chromosome. The conclusions drawn from electron microscope evidence alone are suspect. Chromosomes may look "normal," contain intact fibers many microns long, and still have highly degraded DNA (Wray et al., 1972). The protein associated with the chromosome may be holding the fibers together, instead of the DNA as has been indicated (DuPraw, 1966). Observation of chromosomes after treatment with trypsin or Pronase apparently reveals long DNA fibers (Abuelo and Moore, 1969), but if the enzyme treatment is inadequate, some protein may still be responsible for the integrity of the fibers. The exposure of cells to hypotonic conditions, even without cell lysis (Table I), causes the breakage of DNA. Therefore many other cytological procedures such as in situ hybridization and similar light microscope studies are also affected.

The molecular weight of the single-stranded DNA of approximately 1×10^8 daltons (Wray et al., 1972) is about half that observed by Corry and Cole (1968) for double-stranded DNA. These lengths are still much shorter than the total length of DNA is even the smallest chromosome and would support the idea of a subunit structure for these chromosomes (Stubblefield and Wray, 1971).

ACKNOWLEDGMENTS

This work was supported by grants from the National Science Foundation, The National Cancer Institute, and the National Institutes of Health. The author was a Damon Runyon Postdoctoral Fellow.

Sincere appreciation is extended to Dr. Elton Stubblefield for photographic assistance and helpful advice, and to Dr. Ron Humphrey for advice on molecular weight determinations.

REFERENCES

Abuelo, J. G., and Moore, D. E. (1969). *J. Cell Biol.* **41**, 73.
Cantor, K. P., and Hearst, J. E. (1966). *Proc. Nat. Acad. Sci. U. S.* **55**, 642.
Cohn, N. K. (1971). *Acta Biotheor.* **21**, 41.
Corry, P. M., and Cole, A. (1968). *Radiat. Res.* **36**, 528.
Crick, F. (1971). *Nature (London)* **234**, 25.
DuPraw, E. J. (1966). *Nature (London)* **209**, 577.
Holliday, R. (1970). *Symp. Soc. Gen. Microbiol.* **20**, 359.
Huberman, J. A. (1973). *Annu. Rev. Biochem.* **42** (in press).
Humphrey, R. M., Steward, D. L., and Sedita, B. A. (1968). *Mutat. Res.* **6**, 459.
McGrath, R. A., and Williams, R. W. (1966). *Nature (London)* **212**, 534.
Maio, J. J., and Schildkraut, C. L. (1967). *J. Mol. Biol.* **24**, 29.
Prescott, D. M. (1970). *Advan. Cell Biol.* **1**, 57–117.
Ris, H., and Kubai, D. F. (1970). *Annu. Rev. Genet.* **4**, 263–294.
Sasaki, M. S., and Norman, A. (1966). *Exp. Cell Res.* **44**, 642.
Somers, C. E., Cole, A., and Hsu, T. C. (1963). *Exp. Cell Res., Suppl.* **9**, 220.
Stubblefield, E. (1973). *Int. Rev. Cytol.* **35** (in press).
Stubblefield, E., and Wray, W. (1971). *Chromosoma* **32**, 262.
Thomas, C. A. (1971). *Annu. Rev. Genet.* **5**, 237–256.
Watson, J. D. (1971). *Advan. Cell Biol.* **2**, 1–46.
Wolff, S. (1969). *Int. Rev. Cytol.* **25**, 279–296.
Wray, W., and Stubblefield, E. (1970). *Exp. Cell Res.* **59**, 469.
Wray, W., Stubblefield, E., and Humphrey, R. (1972). *Nature (London) New Biol.* **238**, 237.

Chapter 10

Basic Principles of a Method of Nucleoli Isolation

J. ZALTA AND J-P. ZALTA

Laboratoire de Chimie Biologique, Université Paul Sabatier, Toulouse, France

I. Introduction

The isolation of nucleoli requires, on the one hand, a differential treatment by means of physical or enzymatic agents which do not disrupt the nucleoli but disperse or destroy extranuclear chromatin, and, on the other hand, the separation of the nucleoli from the nuclear residue.

Many methods (Birnstiel *et al.*, 1961; Desjardins *et al.*, 1963; Maggio *et al.*, 1963; Monty *et al.*, 1956; Muramatsu *et al.*, 1963; Penman *et al.*, 1966; Poort, 1961) or modifications of these have been proposed. Some of them use sonic or ultrasonic oscillation effects (Maggio *et al.*, 1963; Muramatsu *et al.*, 1963), others the French press after calcium treatment (Desjardins *et al.*, 1963), or DNase after differential gelification of the chromatin by high ionic strength (Penman *et al.*, 1966). *The Nucleolus* contains an excellent technical review of the subject (Busch and Smetana, 1970).

It has frequently been found that good results are obtained from these techniques only by applying them to the biological material for which they were specially developed. Empirically, modification of these techniques tends to be difficult and unsatisfactory with other material.

It is well known that different factors affect structural components of isolated nuclei. Particularly, bivalent cations can act reversibly on nuclear morphology (see, for example, Anderson and Wilbur, 1951, 1952; Zajdela and Morin, 1952; Philpot and Stanier, 1956).

Hence, the differential behavior of the nucleoli and the extranuclear chromatin with respect to the concentration variation of divalent ions, Mg^{++} in particular, has been systematically examined, as well as the conditions of extranucleolar chromatin dispersion in the presence of these ions. Resulting from this work a new method, which combines high yield, structural integrity and biochemical preservation (Zalta et al., 1971), has been developed.

A discussion of each stage of the preparation of the nucleoli follows with a description of the technical details and a discussion of the underlying principles.

II. Determination of a Concentration Threshold of Mg Ions (Mg^{++}) to Obtain Differential Behavior of Nucleoli and Extranucleolar Chromatin

A. Isolation and Purification of Nuclei

1. MATERIAL AND METHODS

The cells used in the following experiments are ascites cells from a Zajdela's rat hepatoma strain (1963).

Wistar female rats are inoculated intraperitoneally with approximately 2×10^7 ascites cells. Four days later, the ascites are collected, washed several times at $4°C$ in saline phosphate buffer (PBS), and harvested by centrifugation. Nuclei are prepared according to a modification (Amalric et al., 1969) of a technique based on the use of nonionic detergents (Zalta et al., 1962). About 8×10^7 of washed cells are resuspended in 5 ml of NM medium: sucrose $0.25 M$; tris-HCl 10 mM, pH 7.4; $MgCl_2$ 2.5 mM; $CaCl_2$ 0.1 mM; polyvinylsulfate 20 μg/ml. To this suspension is added 0.5% (v/v) Cemulsol NPT 10; 0.5 mg of collagenase (bacterial B grade; Calbiochem, Los Angeles, California) dissolved in 0.5 ml NM and filtered through Millipore filter HA 25, and Celanol 251, 0.1% (v/v). Cemulsol NPT 10 and Celanol are obtained from Melle-Bezons, Neuilly-sur-Seine, France. Celanol 251 has to be omitted when working on RNA polymerase.

This suspension is homogenized for 2 minutes with an ultra-Turrax (Janke and Kunkel, Staufen, Germany) at a reduced speed of between 5,000 and 10,000 rpm. Each preparation is monitored under the phase-contrast microscope and the speed of the homogenizer is adapted accordingly. All the cells have to be opened and tags of cytoplasm removed around free nuclei. The nuclear suspension is then centrifuged at 600g. The pellet is resuspended in 5 ml of NM (without detergents and collagenase), homogenized for 2 minutes at the same speed used above, and then centrifuged at 600g. All these operations are performed at 2°C.

2. DISCUSSION

The method described here is particularly appropriate for the treatment of free cells (ascites, cultured cells, etc.). It can be adapted to practically any type of tissue; the composition of the medium of homogenization has to be such that the nucleoli are well preserved. The combined effect of Mn^{++} and Mg^{++} often favors this (see Section B,2). Use of detergents to remove any cytoplasmic contamination from the nuclear preparation is particularly valuable for the study of ribosome biogenesis, but, provided that the nucleoli are preserved, almost any nuclei preparation method can be employed.

B. Determination of an Mg^{++} Concentration Threshold for Nucleoli Preservation

1. METHOD

The previously prepared nuclei are resuspended in 5 ml of NM. NM medium *without* Mg^{++} is added to aliquots of this nuclear suspension in increasing amounts in order to lower the final Mg^{++} concentration. By means of a phase-contrast microscope it can be observed that:

(1) The lower the Mg^{++} concentration, the less dense are the nucleoplasm and chromatin.

(2) Below a certain concentration (Fig. 1), and depending on the

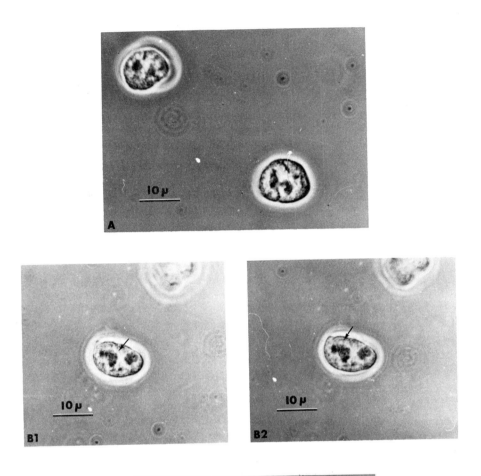

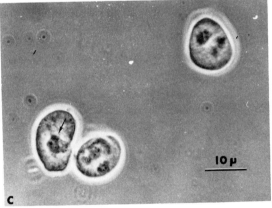

Fig. 1. (a) Isolated ascites nuclei, suspended in NM medium, MgCl₂ 2.5 mM. MgCl₂ concentration is lowered (see text) to 1.5 mM for (b1) 5 minutes, (b2) 10 minutes after dilution; and to 1.25 mM for (c) 2 minutes after dilution.

time of treatment at this concentration, the nucleoli cease to appear dense and refractive, and swell. Finally a loose outline containing granulations appears.

It can be noted that nucleolar behavior after isolation is similar with respect to a lowering of the Mg^{++} concentration.

This process can readily be reversed by increasing Mg^{++} concentration.

The threshold of Mg^{++} concentration for nucleoli preservation (TMg) is defined as the concentration below which the nucleoli lose their normal appearance. The TMg is simply and rapidly determined as above, but it is important that the measurements be made with the utmost precision. Depending on the experimental material, a variation in the concentration of as little as 0.1 to 0.2 mM is sufficient to trigger off nucleolar changes, or a reversal of it. For each type of cell we have studied, a TMg could be established, and generally before reaching this limit the extranucleolar chromatin loses its condensed aspect and appears loose. Thymus chromatin is an exception.

2. Discussion

Below the threshold, a relatively violent homogenization, sonication, or action of different dispersing agents of the chromatin causes the breakage of the nucleoli. Hence, a Mg^{++} concentration as great as or greater than TMg must be used in the preparation of nucleoli. A threshold can be determined for the divalent ions Ca^{++}, Mn^{++}, as well as for combinations of them.

Furthermore, for each given ion, Mg^{++} for example, the threshold seems to vary according to the method of nuclear isolation, and particularly to concentration of Mg^{++} in the medium of preparation. This is explained by the fact that nuclei contain Mg^{++} in varying amounts depending on the method utilized, and the nuclear dilution in a given medium determines the final concentration of this ion. A striking example is given by nuclei that have been prepared or washed in a medium containing a high concentration of Mg^{++} (and/or Ca^{++}). When they are subsequently suspended in a sucrose solution without one or both of these ions, the nucleolus swelling will take place depending on whether or not the resulting Mg^{++} concentration is below or above TMg of these nuclei. In order to have the conditions properly defined for TMg determination, the nuclei must be washed in a medium containing relatively low Mg^{++} concentration.

For the same type of tissue, TMg can vary according to the tissue conditions, e.g., liver nuclei behave in different ways depending on whether the liver is embryonic, newborn, adult, or regenerating.

III.   Dispersion of Extranucleolar Chromatin and Isolation of Nucleoli

A very fine dispersion of extranucleolar chromatin can be obtained by means of nuclear sonication in a medium containing a suitable concentration of Mg^{++} and in the presence of Ficoll, which favors dispersion and avoids the formation of thick threads of chromatin that are very difficult to separate from the nucleoli.

A.   Technical Examples

Solution A: Ficoll (Pharmacia Fine Chemicals, Inc., Uppsala, Sweden) 2.1% (w/v); tris-HCl 10 mM, pH 7.4; $MgCl_2$ 5.5 mM; mercaptoethanol 0.25 mM; polyvinylsulfate 10 μg/ml.

Solution B: identical except the $MgCl_2$ concentration is 0.5 mM.

The nuclear pellet obtained as described above is suspended for sonication in 3 ml of solution A in a conical 15-ml Pyrex tube and submitted to ultrasonic oscillations generated by an M.S.E. 100-watt ultrasonic disintegrator (22 kc/sec) fitted with a titanium probe of 0.3 cm in diameter inserted 0.5 cm deep in the solution. The tube is kept in an ice bath. Complete disruption of nuclei is obtained by 24 successive waves of sonication of 5 seconds each with 5-second rests in between. Then 3 ml of ice-cold solution B is added and 18 additional sonications (5 seconds with 5-second rests) are carried out. The sonicate is centrifuged at 400g (1600 rpm) in a Sorvall HB4 rotor for 5 minutes. The pellet contains only nucleoli. If its examination under the phase-contrast microscope proves that disruption is not satisfactory, a further purification is required. The nucleoli are resuspended in solution A and reexposed to 5 more waves of sonication and then directly centrifuged.

The Raytheon Sonic Oscillator model DF 101 (Raytheon Co., Waltham, Mass.) (10 kc, 200 W) can be used under slightly modified conditions. Nuclei are suspended in 6 ml of solution A in a 50-ml polypropylene tube (Sorval No. 218) refrigerated by circulating ethyl alcohol ($-15°C$). Disruption of nuclei is obtained with 10 successive sonication waves of 10 seconds each, separated by periods of 20 seconds; 6 ml of solution B is then added and 5 to 10 additional, similar waves are required to obtain purified nucleoli.

Any other generator of sonic or ultrasonic oscillations can be used, provided the proper conditions are determined. In particular the suspension temperature has to be maintained between 0 and 2°C.

B. Discussion

1. DISPERSION OF EXTRANUCLEOLAR CHROMATIN

A fine dispersion of the chromatin is essential to obtain a good nucleolar separation.

It is to be noted that in the example quoted the Mg^{++} concentration during the successive sonications is equal to or above 3 mM, while TMg is 1.5 mM.

In fact two extreme types of material can be distinguished: (1) Cells where the chromatin is easily dispersed in the presence of a relatively high Mg^{++} concentration. This is the most frequent case for ascites or cultured cells (e.g., Zajdela's hepatoma, HeLa cells, etc.) despite the fact that, for these concentrations, the chromatin seems fairly condensed under phase-contrast observation. Generally, a two-step operation, as described above, is best with, first, a relatively high Mg^{++} concentration that breaks the chromatin *homogeneously,* and, subsequently, a lower Mg^{++} concentration for finer dispersion of the chromatin. (2) Cells where the chromatin can only be dispersed finely at relatively low Mg^{++} concentration near the TMg (the most frequent case for solid tissues, e.g., the liver)— an Mg^{++} concentration at which the nuclei should appear empty but the nucleoli still compact and well contrasted. Direct sonication at the Mg^{++} concentration that is nearest the TMg is, in this case, frequently preferable. The preparation can be helped by the use of Mn^{++} instead of Mg^{++} or combined with it. Sometimes addition of ethylene-glycol-bis (aminoethyl ether) NN' tetraacetic acid (EGTA) at very low concentration is helpful.

The pH of the medium of sonication generally lies between 7.2 and 7.4. Nevertheless, it must be determined for each type of cell. Sometimes no buffer should be present in the medium. Ficoll is practically indispensable to a good chromatin dispersion, especially if the nuclear outer membrane has not been removed by nonionic detergent treatment. Its concentration must be experimentally determined, and sucrose can be used in combination with it.

2. ISOLATION OF NUCLEOLI

The dimensions and homogeneity of the nucleoli vary with the cellular type. The speed of centrifugation should vary according to the nucleolar size. The greater this is, the less should be the speed of centrifugation and, inversely, the speed should be increased when the nucleolar size is smaller.

The nucleolar separation from the suspension is aided by the presence of Ficoll, which reduces contamination of the nucleolar pellet with dispersed chromatin. In the example given, a Ficoll concentration of 2.1% was found experimentally to be best. This gives to the medium a viscosity equivalent to a 0.3 M sucrose solution. The Ficoll concentration can be raised or lowered before centrifugation. At this last step sucrose can also be combined with Ficoll at a concentration which must be experimentally determined.

IV. Conclusion

The principles applied here are sufficiently general to be applied to different cellular types.

The operations are simple, very rapid, and can be applied massively with high yield.

REFERENCES

Amalric, F., Simard, R., and Zalta, J-P. (1969). *Exp. Cell Res.* **55**, 370.

Anderson, N. G., and Wilbur, K. M. (1951). *J. Gen. Physiol.* **34**, 647.

Anderson, N. G., and Wilbur, K. M. (1952). *J. Gen. Physiol.* **35**, 781.

Birnstiel, M. L., Chipchase, M. I. H., and Bonner, J. (1961). *Biochem. Biophys. Res. Commun.* **6**, 161.

Busch, H., and Smetana, K. (1970). "The Nucleolus," p. 531. Academic Press, New York.

Desjardins, R., Smetana, K., Steele, W. J., and Busch, H. (1963). *Cancer Res.* **23**, 1819.

Maggio, R., Siekevitz, P., and Palade, G. E. (1963). *J. Cell Biol.* **18**, 267 and 293.

Monty, K. J., Litt, M., Kay, E. R., and Dounce, A. L. (1956). *J. Biophys. Biochem. Cytol.* **2**, 127.

Muramatsu, M., Smetana, K., and Busch, H. (1963). *Cancer Res.* **23**, 510.

Penman, S., Smith, T., and Holtzman, E. (1966). *Science* **154**, 786.

Philpot, J. St. L., and Stanier, J. E. (1956). *Biochem. J.* **63**, 214.

Poort, C. (1961). *Biochim. Biophys. Acta* **46**, 373.

Zajdela, F. (1963). "Colloque Franco-Soviétique." Gauthier-Villars, Paris.

Zajdela, F., and Morin, G. A. (1952). *Rev. Hematol.* **7**, 628.

Zalta, J., Zalta, J-P., and Simard, R. (1971). *J. Cell Biol.* **51**, 563.

Zalta, J-P., Rozencwajg, R., Carasso, M., and Favard, P. (1962). *C. R. Acad. Sci.* **255**, 412.

Chapter 11

A Technique for Studying Chemotaxis of Leukocytes in Well-Defined Chemotactic Fields

GARY J. GRIMES AND FRANK S. BARNES

Department of Electrical Engineering, University of Colorado, Boulder, Colorado

I. Introduction

Chemotaxis of human polymorphonuclear leukocytes (PMN) has most often been studied using the Boyden (1962) chamber technique, which reliably measures the relative degree of response to attractants, but gives no information about the behavior of individual cells in known chemotactic fields. Others have studied the chemotaxis of PMN on microscope slides using such techniques as long exposure photomicrography with dark field illumination (Harris, 1953; Lotz and Harris, 1956) and time lapse cinemicrography (Ramsey, 1972). Our technique, which we have called the stripe-source diffusion technique, makes it possible to track individual cells on microscope slides in well-defined fields of chemotactic agent concentration and concentration gradients as functions of temperature, pH, etc. The results of the study presented here indicate the existence of a threshold gradient and a saturation concentration for PMN chemotaxis using cyclic adenosine-3',5'-monophosphate (cyclic AMP) as the chemotactic agent (Gamow *et al.*, 1971; Leahy *et al.*, 1970), and we are able to present values of these parameters for normal body temperatures and pH. We have also taken this opportunity to compare the chemotactic responses of PMN to known chemotactic fields with the responses generated in our earlier work (Hu and Barnes, 1970) by laser-damaged erythrocytes.

II.   Methods and Materials

A.   Description of the Stripe-Source Diffusion Technique

This technique was developed in order to create known concentration distributions with appreciable gradients on microscope slides with good reproducibility with a minimum of equipment and time. The basic method consists of depositing a lateral stripe consisting of a known amount of solution of the chemotactic agent using a calibrated 20-microliter micro pipet (Dade, Division of American Hospital Supply, Miami, Florida) across the microscope slide (Fig. 1a). We have found it convenient to mark the location of the stripe with a diamond-tipped marker before the stripe is deposited. Once the stripe solution is thoroughly dry, a known amount of the sample solution (leukocytes and medium) is deposited on the slide, also from a calibrated micro pipet (Fig. 1b). After smearing the sample, if necessary, the cover slip is placed over it so that

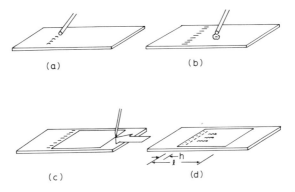

Fig. 1.   The sample is prepared for stripe-source diffusion by (a) depositing the stripe, (b) depositing the leukocytes, and (c) sliding the cover slip. The stripe width h and sample length l are shown in the completed sample (d).

the edge of the cover slip nearest the stripe is still slightly separated from the stripe of the chemotactic agent. After waiting a few seconds to allow any currents in the sample to cease, the cover slip should be gently pushed along the slide until the stripe is completely covered. This is best accomplished with a small instrument such as a hypodermic needle (Fig. 1c) in such a way that no forces are exerted on the cover slip perpendicular to the slide so that no currents are initiated in the sample. The sample is now ready to be sealed and placed in a constant temperature environment for incubation or observation while the chemotactic agent diffuses slowly across the slide (Fig. 1d). Figure 2 shows how the distance x is measured.

The accuracy of this technique depends on the assumption that the chemotactic agent will redissolve in a time which is shorter than that required for a significant amount of diffusion to take place and which is much longer than that required for any currents in the sample to cease. The inherent slowness of the liquid diffusion process will ensure the first condition in almost all cases, and the second can easily be checked by watching for currents in the vicinity of the dissolving chemotactic agent crystals using a microscope.

Whenever a high-power oil immersion objective is used, special care must be taken to ensure that the microscope objective does not push directly against the cover slip and initiate currents.

The validity of this technique has been confirmed in two ways: (1) photodensitometry has been conducted using a dye of known and distinct molecular weight (methylene blue) substituted for the chemotactic agent in the stripe; and (2) the biological assays are consistent, i.e., the directionality of leukocyte migration may be consistently predicted from the space and time coordinates of the sample.

There are several variations of this technique which may be used. It may be necessary to use a chemotactic agent such as casein (Baum et al., 1971) which is only slightly soluble in the normal range of plasma pH (7.40–7.45). In this case the solution is quickly frozen in the stripe configuration before evaporation can take place on a precooled microscope slide. Then, taking advantage of the low thermal conductivity of

FIG. 2. The distance x is measured from the edge of the cover slip on the same end with the stripe.

FIG. 3. For this geometry we may have two independent one-dimensional diffusion processes.

glass, the portion of the slide to one side of the stripe is heated to 5°C by immersing it in 5°C distilled water, without melting the stripe. Then a room temperature leukocyte sample and cover slip are placed on the slide. The remainder of the procedure is identical to the foregoing. The sample should quickly be warmed and kept at the desired temperature since the diffusion coefficient is temperature-dependent. In order to avoid solving the diffusion equations for the case of a time-varying diffusion coefficient, each sample should be kept at one constant temperature over the entire duration of incubation and observation.

Another variation of this technique may be realized by placing a stripe of a different chemical on the side of the slide as shown in Fig. 3. This allows the study of the chemotactic effects of one agent in the presence of varying known concentrations of another. Unless the two substances are involved in a chemical reaction involving one another as reactants or enzymes, the diffusion process of one is independent of the other so long as small enough concentrations of each are used to prevent the alteration of liquid properties which affect diffusion.

The stripe-source diffusion technique may also be modified by introducing a substance into the leukocyte sample before the smear is made. This is useful for observing chemotaxis in the presence of a constant concentration of a different agent, or, if the same substance that is contained in the stripe is added in this manner, the concentration of the chemotactic agent is increased uniformly at all points on the slide without affecting the gradient. This latter technique was used in this cyclic AMP study.

B.   Mathematical Techniques

The mathematical theory of diffusion is based on Fick's partial differential equations which are, for the one dimensional case in an isotropic medium,

$$J = -D \frac{\partial C}{\partial x} \tag{1a}$$

$$\frac{\partial C}{\partial t} = D \frac{\partial^2 C}{\partial x^2} \tag{1b}$$

The first equation specifies the rate of motion of diffusing particles through a unit cross-sectional area as a function of the diffusion coefficient and the concentration gradient, while the second specifies the change in the concentration at all points as a function of the diffusion coefficient and the second partial space derivative of the concentration.

Since complete analytical treatments of this problem may be found in common texts (Crank, 1956; Carslaw and Jaeger, 1959), only the results will be stated here. Although we solved this problem using numerical approximation techniques to evaluate the analytical solutions, iteration techniques for direct numerical solution exist for this set of equations (Richtmyer and Morton, 1962).

The general solution of this set of partial differential equations is available in two distinct forms, both of which are formally valid over the entire range of space and time. We are interested only in the solutions pertaining to the boundary conditions found in the one-dimensional stripe-source diffusion technique. Here we have a source of finite width diffusing into a medium with two distinct impenetrable boundaries, one immediately adjacent to the stripe and another a distance l away at the opposite end of the cover slip. The initial conditions may be stated mathematically as follows:

$$\text{At } t = 0 \text{ seconds, } C = \begin{cases} 1 & \text{for} \quad 0 \leqq x \leqq h \\ 0 & \text{for} \quad h < x \leqq l \end{cases} \tag{2}$$

This is shown graphically in Fig. 4.

The first type of solution consists of a series of error functions or related integrals and is most convenient for use in numerical evaluation for small values of time. The second form of solution is a Laplace transform derived trigonometric series and converges most rapidly for large values of time. The solutions considered here will be given for an initial stripe of concentration C_0, a sample length l, a stripe width h, and the distance x which is the distance from the edge of the cover slip on the same end with the stripe to the observation point as illustrated in Fig. 2.

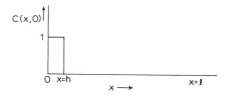

FIG. 4. The solutions of the diffusion equations used in this paper correspond to this particular initial distribution.

The solutions for the concentration, C, given in both of these forms are

$$C = \begin{cases} \dfrac{C_0}{2} \displaystyle\sum_{n=-\infty}^{\infty} \left[\mathrm{erf}\left(\dfrac{h + 2nl - x}{2\sqrt{Dt}}\right) + \mathrm{erf}\left(\dfrac{h - 2nl + x}{2\sqrt{Dt}}\right) \right] \\[2em] C_0 \left(\dfrac{h}{l} + \dfrac{2}{\pi} \displaystyle\sum_{n=1}^{\infty} \left[\dfrac{1}{n} \sin\left(\dfrac{n\pi h}{l}\right) \exp\left(\dfrac{-Dn^2\pi^2 t}{l^2}\right) \cos\left(\dfrac{n\pi x}{l}\right)\right]\right) \end{cases} \tag{3}$$

The solutions for the concentration gradient, grad C, are given in terms of the scalar component, $\mathrm{grad}_x C$, since we are only concerned with one dimension, x.

$$\mathrm{grad}_x C = \begin{cases} \dfrac{C_0}{2\sqrt{\pi Dt}} \displaystyle\sum_{n=-\infty}^{\infty} \left[-\exp\left\{-\left(\dfrac{h + 2nl - x}{2\sqrt{Dt}}\right)^2\right\} \right. \\[1.5em] \qquad\qquad \left. + \exp\left\{-\left(\dfrac{h - 2nl + x}{2\sqrt{Dt}}\right)^2\right\} \right] \\[1.5em] -\dfrac{2C_0}{l} \displaystyle\sum_{n=1}^{\infty} \sin\left(\dfrac{n\pi h}{l}\right) \exp\left(\dfrac{-Dn^2\pi^2 t}{l^2}\right) \sin\left(\dfrac{n\pi x}{l}\right) \end{cases} \tag{4}$$

A summary of the derivatives of these functions as well as numerical techniques for evaluating them may be found in the Appendix. Figure 5 shows some typical distributions of concentration, gradient, and their ratio for a particular value of time ($t = 2$ hours) for cyclic AMP in Eagle medium at 37°C. The concentration is given in arbitrary units with $C_0 = 1$, and the gradient component is given in the same units per centimeter, so that the ratio has units of centimeters. Here it can be seen that an enormous range of concentration and gradient values are normally generated on each sample, making it possible to observe cells in vastly different chemotactic fields on a single slide with all other variables held constant.

The diffusion coefficient may be calculated in $\mathrm{cm}^2\mathrm{sec}^{-1}$ using the empirical formula developed by Wilke and Chang (1955) for small concentrations of A in B:

$$D_{AB} = 7.4 \times 10^{-8} \left(\dfrac{\sqrt{\psi_B M_B} T}{\mu V_A^{0.6}}\right) \tag{5}$$

Here ψ_B is an "association parameter" for the solvent B (the recommended value for water is 2.6), M_B is the molecular weight of the solvent, T is the absolute temperature in degrees Kelvin, μ is the viscosity of the solution in centipoises, and V_A is the molar volume of the solute A in cm^3 gm-mole^{-1} as liquid at its normal boiling point.

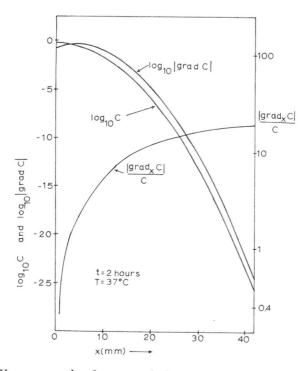

FIG. 5. Here we see that for a typical observation time ($t = 2$ hours) the concentration and the gradient each vary over many orders of magnitude, but their ratio is not nearly as rapidly varying. The diffusion constant is for cyclic AMP.

The viscosity of the medium is about 1.60 to 1.85 for plasma (Harkness, 1971). Between 15°C and 37°C, the viscosity changes about 2–3% per degree Centigrade (Harkness, 1971). Since 98–99% of the viscosity difference between plasma and water is owed to proteins while the small molecular components of plasma such as inorganic salts, urea, sugar, and cholesterol only account for 1–2%, we assumed that the viscosity of Eagle medium was about 1.05 that of water. The temperature dependence of the viscosity of water in the region 20°C to 100°C is given by (Borel, 1970):

$$\log_{10}\left(\frac{\mu_{H_2O}(T)}{\mu_0}\right) = \frac{1.3272(20 - T) - 0.001053(T - 20)^2}{T + 105} \tag{6}$$

where $\mu_0 = 1.002$ centipoises.

A Librascope L-3055 computer was used to generate extensive tables of the concentration and the concentration gradient and their derivatives with respect to time, temperature, and the diffusion coefficient as functions of time and space. The program, which consists of 317 executable FORTRAN statements plus documentation, generates tables of the afore-mentioned parameters for 15-minute time intervals and 1-mm space intervals. Our tables give values for the parameters assuming that the initial stripe concentration is one, since the values may be easily modified owing to the fact that all the solutions are linear in the initial stripe concentration C_0. It was decided that the tables should contain the derivatives as mentioned above so that if it is necessary to use a value of a parameter not used in the compilation of the table an approximate value for the concentration or gradient could be found by linear inter-polation using a truncated Taylor's series of the form

$$f(x_0,y_0 + \Delta y, z_0) = f(x_0,y_0,z_0) + \Delta y \left. \frac{\partial f}{\partial y} \right|_{x_0,y_0,z_0} (x,y,z) \tag{7}$$

It is also useful to have the time derivatives available so that a region of the sample may be found in which the concentration and the gradient are relatively nonvarying so that as many measurements can be taken as possible on a single leukocyte before the chemotactic field changes appreciably. If a computer is immediately available at the site of the research it might be desirable to make velocity measurements on the cells while recording their time and position coordinates, and find their indi-vidual chemotactic fields from these data at a later time.

If only estimates of the concentration and its gradient are required, or if the only points of interest are those involving the regions far from the stripe at short times, the expression for diffusion from an infinitesimally thin stripe into a semi-infinite body may be used which is

$$C(x,t) = \frac{Q}{\sqrt{\pi D t}} \exp\left[-\frac{x^2}{4Dt} \right] \tag{8}$$

where Q is the source strength in moles, atoms, or whatever, per square centimeter.

The solutions for a nonconstant diffusion coefficient are considerably more complex than these, and are usually expressed in terms of integrals which must be evaluated numerically. This is why we need to keep each sample at a constant temperature for its entire duration. This discussion has also assumed that none of the diffusing substance is being generated or destroyed. Care must be taken to account for these effects if they are present. For the most usual case in which the substance is being destroyed at a rate proportional to its concentration (which is found in an enzymatic reaction in which the enzyme concentration is constant and the reactant concentration is relatively small), we merely have to multiply the expected concentration by an exponential decay factor. The necessity for doing this may be evaluated in two ways, both of which involve a biological assay technique: (1) Cells may be observed at different times and places on the sample such that their responses should be identical if the total amount of chemotactic substance is constant, or (2) The sample medium may be varied to change the enzyme concentration to see if this has an appreciable effect.

One should note that when preparing the solution of the chemotactic agent for the stripe, it is convenient to calculate in advance what concentration will give the desired stripe concentration. For example, suppose that a stripe of width $h = 2$ mm is used in conjunction with a cover slip of length $l = 42$ mm. If the sample has a volume of 20 microliters (μliter), then the sample volume placed directly over the stripe is $2/42 \times 20$ μliter, or about 1 μliter. Then if a stripe of original unevaporated volume of 10 μliter is used, the stripe will be about 10 times as concentrated as the original solution. A 0.01 M cyclic AMP stripe was most often used in this study.

It might be desirable to use other geometries, such as a circular spot source, if larger gradients are desired for a given concentration. The mathematics of the linear stripe are quite tractable when we consider the actual case of a finite sized stripe diffusing into a finite medium. However, the mathematics of a finite circular source diffusing into a rectangular sample is more involved unless one is content to consider the simplified case of an infinite medium, in which case the solution will be valid for short times only.

C.   Preparation and Maintenance of Leukocyte Samples

The samples for this study were prepared in the following ways. All leukocyte samples were taken from the same adult male.

(1) Capillary blood was extracted by the finger-stick method in the absence of toxic agents, collected in nonoxalated hematocrit tubes (which had been previously cooled to 3°C) and centrifuged for 5 minutes at 250g at 3°C. After breaking the tube at the buffy coat-erythrocyte interface the leukocytes were suspended in Eagle medium (MEM) with Hank's salts but without L-glutamine or sodium bicarbonate (from Grand Island Biological Co., Grand Island, New York) and as small an amount of serum as possible.

(2) This preparation is the same as the foregoing except that heparinized tubes were used at room temperature to collect the blood and the samples were centrifuged as above at room temperature.

(3) A 5-ml venous sample was mixed with 0.25 ml of a 3.8% sodium citrate solution, allowed to settle for 10 minutes, and centrifuged for

3 minutes at 150g. The leukocytes were drawn from the buffy coat with a 20-μliter micro pipet for placement on the slide.

The samples were then placed on either freshly cleaned glass microscope slides or freshly cleaned plastic petri dishes (Falcon Plastics, Oxnard, California), the cover slips (ordinary glass) were manipulated into place as previously described and the samples were sealed with Vaseline. The samples were then maintained in a constant and uniform temperature environment either in an incubator or on a forced-air heated microscope stage of our own design. We monitored the temperature on the stage using copper-constantan thermocouples.

Eagle medium preparations are generally superior for the following reasons: (1) The leukocytes usually remain motile longer than if pure serum or plasma is used. (2) Eagle medium has been previously shown to exert no chemotactic effects (Borel, 1970). (This should make no difference since the concentration is uniform; however, we have not yet investigated whether chemotactic agents enhance random motion.) (3) The absence of protein in the medium drastically reduces the variability in viscosity from sample to sample and makes the viscosity far less temperature-sensitive. This is important since the viscosity must be known accurately in order to calculate a useful diffusion coefficient. An Eagle medium preparation would be especially helpful in pathological samples where increased immunoglobulins and other large proteins might drastically alter the viscosity.

D.   Methods of Making Measurements

In our first experiments using the stripe-source diffusion technique we measured the gross migration of granulocytes across a microscope slide for periods typically from 9 to 24 hours. At regular intervals the samples were removed from the incubator and the density of the cells as a function of x, the distance from the edge of the cover slip over the stripe to the observation point, was measured. The results were obtained simply by positioning the microscope stage at 1-mm intervals using the vernier scale before looking into the eyepieces so that human bias was minimized. The data collected in this way consist of the total number of cells at each sampling position calculated from the data of several nonoverlapping passes which are illustrated in Fig. 6.

A far more powerful technique may be readily used with this sample preparation technique by tracking individual granulocytes and measuring their velocities as a function of their time and space coordinates and thus their chemotactic field environment in terms of concentration and gradient. This was accomplished by tracking the cells using a Leitz

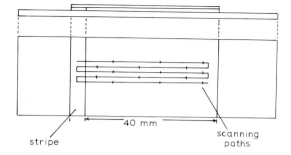

FIG. 6. These scanning paths indicated by the small arrows were the ones used in the gross migration studies.

Ortholux phase-contrast microscope in conjunction with a Sony ½-inch video tape recording system. The video screen was calibrated so that the change in the position of the part of the cell being observed could be measured with a precision of 1μ. The position of the leukocyte on the slide was known to 0.01 mm. The data collected represent the components of the PMN average velocity along the chemotactic agent concentration gradient, denoted by $\text{grad}_x C$, in microns per minute. In order to find the average velocity component at least 6 velocity measurements of 15 seconds duration each, spaced 1 minute apart, were made on each cell. The velocity measurements were not made in succession since the velocity of the cell is dependent on the immediately preceding orientation and velocity, and independent measurements were desired. It is desirable to space the measurements far apart in time to ensure their independence, but it is also desirable to make as many measurements as possible before the concentration and the gradient change significantly. Since a compromise is needed, the situation can be aided by choosing time and position coordinates on the slide so that the time rate of change of concentration and gradient is minimal. This is why we chose to include these time derivatives in our computer-generated table.

Only obviously healthy leukocytes which had adhered to the top of the slide and were actively migrating by ameboid motion were tracked. This essentially ensured that the data were taken only on neutrophils rather than on neutrophils, basophils, and eosonophils proportionate to their number, since it is known that eosonophils and basophils are less motile than neutrophils. However, the direction of motion of a leukocyte was not used as a criterion for choosing a cell for observation.

We first tracked cells using only the stripe as a means of adding chemotactic agent. Later, in order to separate the effects of the concentration and the gradient, we added fixed amounts of the chemotactic agent (cyclic AMP) directly to the samples before placing them on the slide. Then, velocity measurements were made using PMN in regions of the sample where the concentration due to the stripe was small compared with the uniform concentration of the original sample so that the concentration changed very little over a large range of gradients.

III.   Measurements of the Sensitivity of Polymorphonuclear Leukocytes to Cyclic Adenosine-3′,5′-monophosphate Concentrations and Gradients

A typical result from our initial gross migration studies is shown in Fig. 7 which represents the totals of five nonoverlapping passes of a relatively low density leukocyte preparation which had been incubated for 9 hours at 37°C in the presence of a 0.01 M cyclic AMP stripe. Although one can calculate average migration velocity components of about 5–7 microns per minute for 9- or 24-hour periods, this type of result has little to offer that cannot be accomplished as effectively with the Boyden chamber technique.

The results of tracking individual PMN are shown in Figs. 8 and 9. In Fig. 8 the concentration is allowed to vary as well as the gradient and in Fig. 9 the gradient varies as well as the concentration. This cannot be avoided since the concentration and the gradient are so closely linked as shown previously in Fig. 5. Since these graphs are not plane sections of

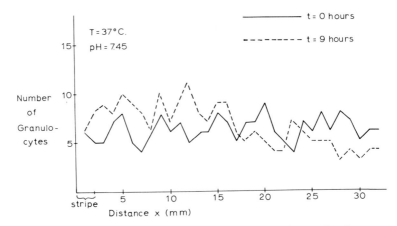

FIG. 7. This graph shows a typical shift in the granulocyte distribution in response to a 0.01 M cyclic AMP stripe on the left.

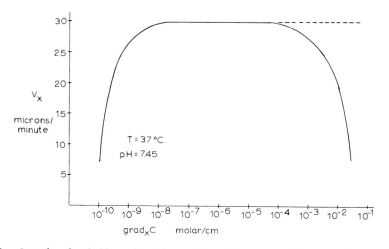

Fig. 8. The threshold gradient for cyclic AMP may be determined from this graph.

a three-dimensional graph of the form $V_x = V_x(C, \text{grad}_x C)$, we might expect them to contain artifacts.

In order to get a different functional relationship between V_x and the concentration and the gradient, three different background molarities of cyclic AMP were mixed with the Eagle medium. The results shown in Fig. 10 indicate that we have a threshold gradient and a saturation concentration. The threshold gradient is quite independent of concentration over the range of concentrations examined. It might also be said that the

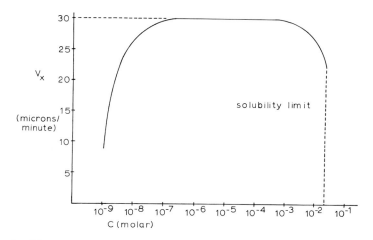

Fig. 9. The important feature of this curve is the partial reduction in granulocyte chemotaxis near the cyclic AMP solubility limit.

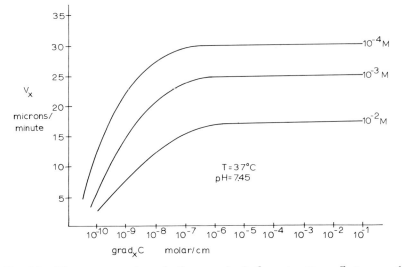

FIG. 10. These curves show further proof of the saturation effect near the solubility limit.

saturation is relatively independent of the gradient. Figures 8 and 9 might be improved by replacing the solid lines with the dashed ones as shown. It should be noted that no limited dynamic range for chemotaxis bounded on two sides can be specified in terms of either concentration of gradient as one might be tempted to do from the uncorrected results of Figs. 8 and 9.

Figure 11 shows that the average migration velocities for the various sample preparation techniques using different substrate materials and anticoagulants are quite similar. Since all of the means lie within the error bars placed at plus and minus one standard deviation, the data from the different types of samples were lumped into the foregoing results rather than being analyzed separately. At least 30 leukocytes and three different preparations were used to compile the data for this diagram. It is of interest that the leukocyte velocities were similar in Eagle medium and plasma, since the viscous forces on them would be quite different.

There are two justifications for the fact that no provision was made for a varying total amount of cyclic AMP: (1) Results were compared between regions of theoretically similar chemotactic fields which were separated considerably in time. Since the results obtained were statistically indistinguishable, it was concluded that the total amount of cyclic AMP did not change significantly over the time used to make measurements (typically several hours). (2) The results obtained using plasma

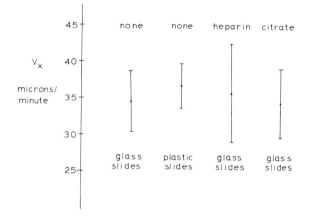

FIG. 11. These mean chemotactic velocities along with the corresponding error bars placed at plus and minus one standard deviation show that the anticoagulants and substrate materials tried have little effect on chemotaxis.

or serum were compared with those using Eagle medium and found not to differ. Since Eagle medium presumably does not contain phosphodiesterase, the principal enzyme responsible for the destruction of cyclic AMP, the reaction rate would be much higher for the pure serum or plasma samples as for the Eagle samples. This difference should have become evident if the rate constant for cyclic AMP destruction in plasma was significant. To be more precise, the cyclic AMP is not actually destroyed, but only converted to the 5′-monophosphate, a biologically inactive form, by splitting the ester bond which joins the phosphate to the 3′ carbon of the deoxyribose ring. It is hoped that soon we will be able to use the UV fluorescence technique of Secrist et al. (1972) to assay the cyclic AMP directly.

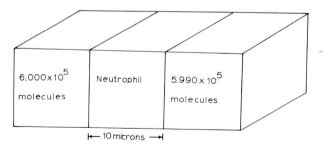

FIG. 12. This schematic of a neutrophil represented by a 10^3 μ^3 cube shows a typical molecular distribution of cyclic AMP during chemotaxis in imaginary compartments of equal volume on either side of the cell, one of which is "up" the concentration gradient and the other is "down."

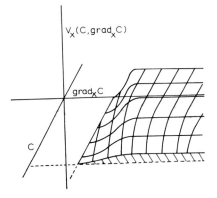

FIG. 13. This generalized representation of the chemotactic velocity as a function of the cyclic AMP concentration and gradient was postulated on the basis of the information in Figs. 8, 9, and 10.

We originally speculated that there might be a functional relationship between V_x and the ratio of the gradient to the concentration. However, on the basis of a plot of the data collected over the regions of active migration, no functional relationship was indicated.

It is of interest to note that a typical chemotactic environment illustrated in Fig. 12 has a gradient such that the concentration differs by only about 1% from the "head" of the leukocyte to the "tail."

The general shape of the $V_x(C, \mathrm{grad}_x C)$ surface for cyclic AMP is shown in Fig. 13. Note the discontinuity at the solubility limit.

IV. Discussion

It may be of interest to determine why the velocities measured in this study were generally higher than those measured by Ramsey (1972) in a similar recent study. This may be at least partially explained by the fact that we selected only obviously healthy and migrating cells for making measurements, while Ramsey made measurements on all the cells in his field. Ketchel and Favour (1955) noted that the ameboid motion velocities of granulocytes may vary over a factor of 10 from individual to individual. We have no way of commenting on this since all of our samples were derived from the same subject.

We found that the velocity variations of a single cell from time to time and those between cells are far greater than the averaged velocity variations between various samples.

Since we recorded only one component of the velocity in our initial study using this technique, we cannot accurately compare our results with the chemotactic factors reported by others (Ramsey, 1972; Dixon and McCutcheon, 1936). Since we found it common to observe scalar velocities of 40–45 microns per minute and the directed velocities, V_x, were typically 30 microns per minute, we might expect to calculate a chemotactic factor (the ratio of the component of the total path length in the direction of chemotaxis to the total path length) of about 0.6 to 0.7 which is closely comparable to the findings of Ramsey (1972) and Harris (1953).

It is also of interest that we have made a number of calculations showing that we can indeed simulate a microscopic diffusion field on a macroscopic microscope slide. Since both the concentration and the gradient remain linearly proportional to the initial stripe concentration as can easily be seen in Eqs. (1a,b) and (2) and since a large range of concentrations and gradients are generated across each sample slide, an enormous range of concentrations and gradients is possible even if only a small number of samples is used. However, the range of the gradient to concentration ratio (see Fig. 5) cannot be varied over such a wide range for this geometry. Fortunately, the range of the gradient to concentration ratio generated on the microscope slide by the stripe-source diffusion technique produces nearly the same ratio range on a microscopic scale that would be generated on a microscopic scale by a single or a small group of ruptured cells.

Further calculations have shown that the amount of cyclic AMP released by some typical mammalian cells when ruptured produces comparable concentration and gradient fields which have been shown to produce chemotaxis in this study. This is particularly interesting since a previous response-time modeling study by Hu and Barnes (1970), which involved thermally damaging a single erythrocyte with a 3-μ diameter argon ion laser beam, showed that molecules of molecular weight less than 10,000 must be responsible for the chemotactic response. It might be postulated that cyclic AMP is quite a universal chemotactic agent for leukocytes since most mammalian tissues examined for cyclic AMP have revealed (Robinson et al., 1971) high enough concentrations to produce chemotaxis. It is of interest that several studies have shown that body fluids such as plasma, cerebrospinal fluid, and gastric juices contain cyclic AMP in concentrations of about 2×10^{-8} M, which is considerably less than the saturation concentration discovered in this study.

The fact that the chemotactic response is diminished by about half by the time the plasma is saturated with cyclic AMP explains the "bunching" effect we have observed around laser-damaged erythrocytes as the granulocytes underwent positive necrotaxis (chemotaxis toward a necrotic source). This fact also ensures that the chemotactic ability of the PMN will not be totally impaired by too high cyclic AMP concentrations.

V.   Appendix: Techniques for Evaluating the Diffusion Equations

The purpose of this Appendix is to provide numerical techniques both for evaluating the solutions of the diffusion equations given in the mathematical development section and for finding the derivatives of these solutions with respect to various parameters.

For $Dt \leq 0.6$ the error function type solutions were used in our program, and for $Dt > 0.6$ the Laplace transform type trigonometric series solutions were used. For the evaluation of the error function for small arguments (we used for $|x| \leq 3$) we may use

$$\text{erf } x = \frac{2}{\sqrt{\pi}} \sum_{n=0}^{\infty} (-1)^n \frac{x^{2n+1}}{n!(2n+1)} \tag{A-1}$$

For large arguments (we used for $|x| > 3$) the error function may be evaluated by using the semiconvergent series

$$1 - \text{erf } x \equiv \text{erfc } x = \left(\frac{e^{-x^2}}{\sqrt{\pi}}\right) \sum_{n=1}^{\infty} (-1)^n \frac{1 \cdot 3 \cdots (2n+1)}{(2x^3)^n} \tag{A-2}$$

This series should be terminated when the desired accuracy is reached, but in no case should values of n be permitted such that $(2n+1) > (2x^2)$, since the series will begin to diverge.

Derivatives of the error function solutions may be found using (d/dx) erf $x = (2/\sqrt{\pi}) \exp(-x^2)$. For example,

$$\frac{\partial C(x,t)}{\partial t} = -\frac{C_0}{4\sqrt{\pi D}\, t^{3/2}} \sum_{n=-\infty}^{\infty} \left(\exp\left[-\left(\frac{h + 2nl - x}{2\sqrt{Dt}}\right)^2 \right] (h + 2nl - x) \right.$$
$$\left. + \exp\left[-\left(\frac{h - 2nl + x}{2\sqrt{Dt}}\right)^2 \right] (h - 2nl + x) \right) \tag{A-3}$$

$$\frac{\partial}{\partial t} \text{grad}_x C(x,t) = \frac{C_0}{4\sqrt{\pi D}\, t^{3/2}} \sum_{n=-\infty}^{+\infty} \left[-\exp\left\{ -\left(\frac{h+2nl-x}{2\sqrt{Dt}} \right)^2 \right\} \right.$$

$$\left. + \exp\left\{ -\left(\frac{h-2nl+x}{2\sqrt{Dt}} \right)^2 \right\} \right] + \frac{C_0}{8\sqrt{\pi}\, D^{3/2} t^{5/2}} \sum_{n=-\infty}^{\infty}$$

$$\left[-\exp\left\{ -\left(\frac{h+2nl-x}{2\sqrt{Dt}} \right)^2 \right\} (h+2nl-x)^2 \right.$$

$$\left. + \exp\left\{ -\left(\frac{h-2nl+x}{2\sqrt{Dt}} \right)^2 \right\} (h-2nl+x)^2 \right] \quad \text{(A-4)}$$

The corresponding Laplace transform trigonometric series time derivatives are:

$$\frac{\partial}{\partial t} C(x,t) = -C_0 \frac{2\pi D}{l^2} \sum_{n=1}^{\infty} n \sin\left(\frac{n\pi h}{l} \right) \exp\left[-\frac{Dtn^2\pi^2}{l^2} \right] \cos\left(\frac{n\pi x}{l} \right) \quad \text{(A-5)}$$

$$\frac{\partial}{\partial t} \text{grad}_x C(x,t) = \frac{2C_0 \pi^2 D}{l^3} \sum_{n=1}^{\infty} n^2 \sin\left(\frac{n\pi h}{l} \right) \exp\left[-\frac{Dtn^2\pi^2}{l^2} \right] \sin\left(\frac{n\pi x}{l} \right)$$

$$\text{(A-6)}$$

Using these time derivatives the derivatives with respect to the diffusion coefficient are available using

$$\frac{\partial}{\partial D} C(x,t) = \left(\frac{t}{D} \right) \frac{\partial}{\partial t} C(x,t) \quad \text{(A-7)}$$

$$\frac{\partial}{\partial D} \text{grad}_x C(x,t) = \left(\frac{t}{D} \right) \frac{\partial}{\partial t} \text{grad}_x C(x,t) \quad \text{(A-8)}$$

The derivatives with respect to the variables found in the expression for the diffusion coefficient (Eq. 5) may be easily found using the chain rule for differentiation in the form

$$\frac{\partial C}{\partial T} = \frac{\partial C}{\partial D} \frac{\partial D}{\partial T} = \left(\frac{t}{D} \right) \frac{\partial C}{\partial t} \frac{\partial D}{\partial T} \quad \text{(A-9)}$$

Further explanation of the use of the simplified solution (Eq. 7) as well as solutions for other geometries may be found in Crank (1956) and will not be presented here.

REFERENCES

Baum, J., Mowat, A. G., and Kirk, J. A. (1971). *J. Lab. Clin. Med.* **77**, 501.
Borel, J. F. (1970). *Int. Arch. Allergy Appl. Immunol.* **39**, 247.

Boyden, S. (1962). *J. Exp. Med.* **155**, 453.

Carslaw, H. S., and Jaeger, J. C. (1959). "Conduction of Heat in Solids." Oxford Univ. Press (Clarendon), London and New York.

Crank, J. (1956). "The Mathematics of Diffusion." Oxford Univ. Press (Clarendon), London and New York.

Dixon, H. M., and McCutcheon, M. (1936). *Proc. Soc. Exp. Biol. Med.* **34**, 173.

Gamow, R. I., Bottger, B., and Barnes, F. S. (1971). *Biophys. J.* **11**, 860.

Harkness, J. (1971). *Biorheology* **8**, 171.

Harris, H. (1953). *J. Pathol. Bacteriol.* **66**, 135.

Hu, C. L., and Barnes, F. S. (1970). *Biophys. J.* **10**, 958.

Ketchel, M. K., and Favour, C. B. (1955). *J. Exp. Med.* **101**, 647.

Leahy, D. R., McLean, E. R., and Bonner, J. T. (1970). *Blood* **36**, 52.

Lotz, M., and Harris, H. (1956). *Brit. J. Exp. Pathol.* **37**, 477.

Ramsey, W. S. (1972). *Exp. Cell Res.* **70**, 129.

Richtmyer, R. D., and Morton, K. W. (1962). "Difference Methods for Initial-Boundary Problems," 2nd ed. Wiley (Interscience), New York.

Robinson, G. A., Butcher, R. W., and Sutherland, E. W. (1971). "Cyclic AMP." Academic Press, New York.

Secrist, J. A., III, Barrio, J. R., Leonard, N. J., Villar-Palasi, C., and Gilman, A. G. (1972). *Science* **177**, 278.

Wilke, C. R., and Chang, P. (1955). *AIChE J.* **1**, 264; or see R. B. Bird, W. E. Stewart, and E. N. Lightfoot, "Notes on Transport Phenomena." Wiley, New York, 1958.

Chapter 12

New Staining Methods for Chromosomes

H. A. LUBS, W. H. McKENZIE, S. R. PATIL, AND S. MERRICK

Department of Pediatrics, University of Colorado Medical Center, Denver, Colorado

I. Introduction

A. Background: Two Key Experiments

Virtually every aspect of mammalian cytogenetics has been altered by the development of simple techniques for the differential staining of metaphase chromosomes. The ability to identify each chromosome, as well as small homolog differences, have brought cytogenetics to the brink of an exciting new phase. The difficulties encountered by many laboratories in obtaining consistent, satisfactory banding with these techniques have led to frustration and delay in the realization of much of this potential and have prompted this review.

Each of the subsequently developed techniques can be traced directly to two separate but complementary lines of investigation. Caspersson *et al.* (1967), in a collaborative project between the Karolinski Institute and Harvard Medical School, reported the production of clearly demarcated fluorescence bands in metaphase chromosomes of *Vicia faba,* *Trillium erectum,* and *Cricetulus griseus* (Chinese hamster) which were stained with quinacrine mustard dihydrochloride (QM). This compound was found after a systematic search for agents that would stain adjacent regions of chromosomes differentially. Three years later, Caspersson *et al.* (1970) reported the identification of each human chromosome by quinacrine mustard staining and also presented clear quantitative evidence of the consistency of the major banding patterns using a microdensitometric system. In this same year, a second major line of work was reported from the Department of Biology, Yale University, by Pardue and Gall (1970). In a remarkable experiment, they demonstrated *in vitro* hybridization between mouse centric heterochromatin and labeled mouse satellite DNA as well as its derivative RNA which had been transcribed *in vitro.* The methodology included denaturation of metaphase DNA with 0.07 N NaOH, incubation with labeled RNA or labeled denatured DNA at 66°C for 10 hours, removal of RNA that was not hybridized by a subsequent RNase treatment, and autoradiography. They also observed that the same regions in which hybridization had been effected also stained more darkly with Giemsa. This stimulated careful study of each step in their procedure by many laboratories in an effort to understand the basis of the differential staining and to develop better routine staining techniques. Differential staining along the arms as well as the centromeres was noted in these later experiments. Knowledge of the quinacrine banding patterns led to quick recognition and acceptance of the similar banding patterns that were observed with Giemsa after a variety of treatments.

B. Description of the Major Techniques

Four types of banding patterns have been recognized. In man, these were termed Q, G, C, and R bands at the IVth International Conference on Standardization in Human Cytogenetics in Paris (National Foundation—March of Dimes, 1972) (Table I and Figs. 1–4). The bands obtained by the use of quinacrine dihydrochloride or quinacrine mustard are called "Q" bands. Bands obtained with Giemsa after various heat and alkali treatments are termed Giemsa or "G" bands. The majority of the bands which fluoresce brightly with quinacrine, stain intensely with Giemsa and vice versa. A variety of more rigorous treatments resulted in differential staining of centromeres (as observed by Pardue and Gall), the distal end of the Y chromosome in man, and other areas containing "constitutive heterochromatin"; these darkly staining regions are called "C" bands. After heating at 87°C for a short time, bands which are essentially the reverse of G bands are obtained; these are termed "R" bands. It is likely that these general terms will be utilized in describing comparable bands in other species.

C. An Approach to Banding

The production of excellent banding requires greater care and precision at almost every step in the preparation of metaphases than is often employed. Good banding rarely results from the simple application of a "new stain" to available material without regard to the method of harvesting and slide preparation. For this reason, we began our use of the new techniques by restudying the classical preparative techniques, repeating the various banding methods as originally described, and adjusting the harvesting and staining procedures to obtain optimal banding. We have also attempted to determine how certain steps in the banding techniques affect chromosomes and have reconsidered the postulated mechanisms of banding in the light of this work. The same course will be followed in this review in an attempt to give the reader a broad view of a still developing field. At least one technique for each type of band is currently in routine use in the laboratory in a variety of studies involving large numbers of subjects, and our suggestions are based on this experience. Those who would first like a view of the probable mechanisms of banding, may wish to read Section V first. The orientation of our work is almost exclusively toward work with human cells, but an appreciation of the principles involved should make application to other species considerably easier.

TABLE I

COMPARISON OF G, R, AND C TECHNIQUES

Author	Hypotonic	Fixation	Slide preparation	HCl	Enzyme	NaOH	Buffer		Stain
							Concentration	Temperature and time	
G BANDING									
Sumner, Evans, and Buckland (ASG)	0.075 M KCl	Methanol/acetic acid, 3:1	Air dried	—	—	—	2XSSCᵃ	60°C for 1 hour	Giemsa, pH 6.8
Patil, Merrick, and Lubs (Giemsa 9)	0.075 M KCl	Methanol/acetic acid, 3:1	Blowing and heat dried on hot plate	—	—	—	—	—	Giemsa, pH 9.0 (5% Na phosphate buffer)
Schnedl	Gey's solution, 1:3	Methanol/acetic acid, 3:1	Air dried	—	—	0.07 N NaOH 90 seconds 20°C	Sorenson buffer (pH 6.8)	59°C for 24 hours	Giemsa, pH 70
Seabright (trypsin)	0.075 M KCl	Methanol/acetic acid, 3:1	Air dried by blowing	—	Trypsin 0.25%, 10–15 seconds, 25°C	—	—	—	Leishman, pH 6.8
Wang and Fedoroff (trypsin)	0.075 M KCl	Methanol/acetic acid, 3:1	Flame dried	—	Trypsin 0.025–0.05%, 10–15 minutes, 25°C	—	—	—	Giemsa, pH 7.0, or Wright's and Giemsa

R BANDING

Dutrillaux and Lejeune	Dilute horse serum, MgCl$_2$ hyaluronidase	Carnoy's	—	—	—	—	Phosphate buffer 20 mM pH 6.5	86–87°C for 10 minutes	Giemsa, pH 6.7

C BANDING

Arrighi and Hsu	Varied	50% Acetic and Carnoys	Flame dried or squash	0.2 N HCl, 30 minutes, 25°C	RNase, 60 minutes, 37°C	0.07 N NaOH, 2 minutes, 25°C	2XSSC	65°C for 12–24 hours	Giemsa, pH 6.8
Craig-Holmes and Shaw	0.075 M KCl	Methanol/acetic acid, 3:1	Flame dried	0.2 N HCl, 15 minutes, 25°C	—	0.07 N NaOH with 0.112 M NaCl, 60 seconds, 25°C	2XSSC, pH 7.0	65°C for 24 hours	Giemsa, pH 6.8
McKenzie and Lubs	0.075 M KCl	Methanol/acetic acid, 3:1	Blowing and heat dried on hot plate	0.2 N HCl, 15 minutes, 25°C	—	—	2XSSC, pH 7.0	65°C for 16–24 hours	Giemsa, pH 6.8

[a] 0.3 M sodium chloride and 0.03 M Trisodium citrate.

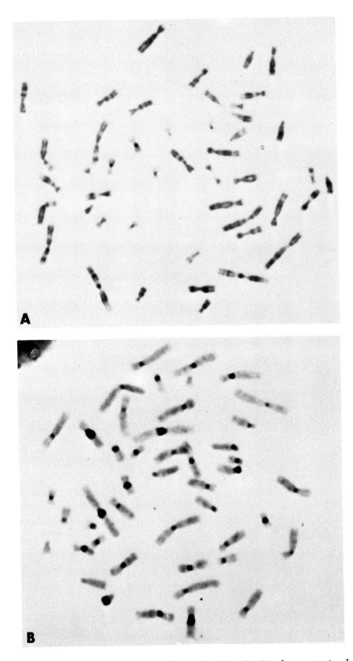

FIG. 1a. Metaphase cell from a male with Down's Syndrome stained by the Giemsa 9 method.

FIG. 1b. Metaphase cell from a normal male illustrating the darkly stained C regions of each chromosome.

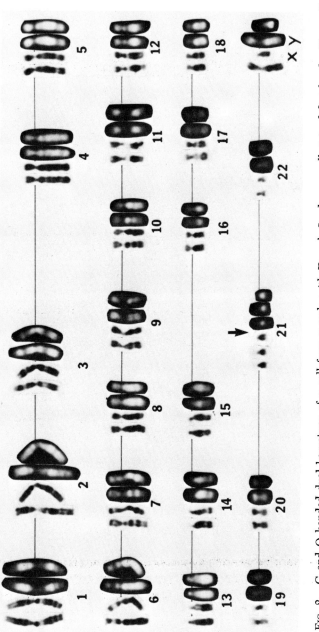

Fig. 2. G and Q banded dual karyotype of a cell from a male with Down's Syndrome; cell stained first by the Giemsa 9 method, destained, and restained with quinacrine dihydrochloride.

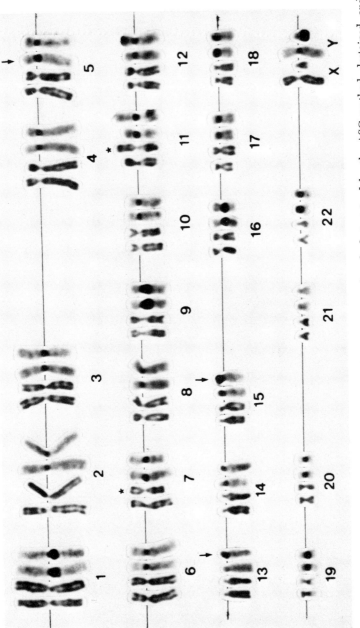

Fig. 3. G and C banded dual karyotype of a cell from a normal male first stained by the ASG method, destained, and restained after HCl and 2XSSC treatment to produce C bands. Note the translocation involving chromosome arms 7q and 11p, best observed in the G-banded preparation. Also note the homolog size disparity in the C regions of chromosomes 5, 13, and 15.

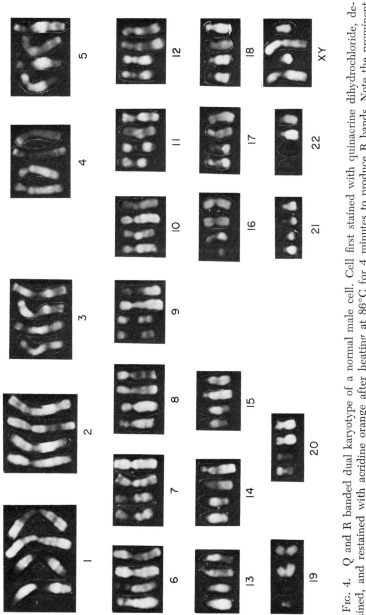

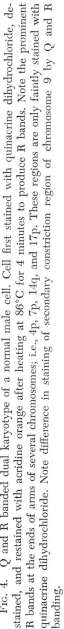

FIG. 4. Q and R banded dual karyotype of a normal male cell. Cell first stained with quinacrine dihydrochloride, destained, and restained with acridine orange after heating at 86°C for 4 minutes to produce R bands. Note the prominent R bands at the ends of arms of several chromosomes; i.e., 4p, 7p, 14q, and 17p. These regions are only faintly stained with quinacrine dihydrochloride. Note difference in staining of secondary constriction region of chromosome 9 by Q and R banding.

II.   Preparative Techniques: A Reevaluation

A.   General Comments

A number of related needs have stimulated our interest in the methodology of cytogenetics. Prior to the introduction of these new techniques we had already observed marked differences between laboratories in many parameters which subsequently proved related to banding, such as mitotic rate, degree of chromosome compaction, demonstration of known heterochromatic variants, and the clarity or fuzziness of whole chromosomes and centromeres (Patil *et al.*, 1970). In addition, the quality of metaphases has proven to be a major limiting factor in the development of automated chromosome analysis, with both conventional and new staining techniques. More recently, our interest in determining the precise differences and similarities between the several types of bands by observing them sequentially in the same cell has necessitated the development of a simple preparative technique which would work for Q, G, C, and R banding. The use of customary local procedures in harvesting cells without regard to their possible effects on banding and morphology must be questioned regularly.

B.   Initial Preparation of Metaphases

1.   MEDIA AND CULTURE CONDITIONS

No specific type of media has been shown to affect banding. It is quite clear, however, that the increased technical demands of the various banding procedures require excellent growth. Only in cultures with a high mitotic index can cells be selected from earlier stages in metaphase with few or no overlaps, clear centromeres and bands, and even spreading, with a reasonable expenditure of time. A poor culture is better repeated. Most phytohemagglutinin is now consistently mitogenic, but should not be used if in solution for more than a week. For both lymphocyte and long-term cultures, fetal calf serum or protein source, of course, remains critical and we would recommend testing several batches to obtain one which produces the maximal mitotic index. A culture time of 64–66 hours for human lymphocytes is optimal. Many of the failures to obtain satisfactory banding result from an accumulation of small errors or less than optimal procedures.

2. MITOTIC ARREST

Both Colcemid and colchicine are effective. The smallest amount consistent with an adequate accumulation of metaphases should be used. Similarly, the duration of Colcemid treatment should be adjusted to the shortest time producing satisfactory numbers of metaphases.

3. HYPOTONIC TREATMENT

The most widely employed hypotonic treatment now is $0.075\,M$ KCl, but both KCl and 1% Na citrate permit excellent G, Q, and C banding (we have not evaluated R banding in this respect). Dilute Hank's or other salt solutions, both in our hands and from observations of slides sent from other laboratories using Hank's, have resulted in consistently poor banding (especially Q banding), because of the fuzziness of the chromosome morphology. This was also observed by Crossen (1972) in a similar evaluation of banding methods, as were good results with KCl or Na citrate.

4. FIXATION

Three to one methanol–acetic acid is the most generally used fixative in human cytogenetics, because of the clear morphology and excellent spreading that can be obtained with it. It has proven satisfactory for all banding techniques. We have found that utilization of high quality, fresh reagents is essential and that large containers of either agent tend to deteriorate and accumulate water. When formaldehyde is employed in the fixative or other solutions to prevent denaturation it is extremely difficult to obtain adequate spreading of metaphases.

5. SLIDE PREPARATION

Slide preparation is the most varied step in metaphase preparation. It is still too little appreciated how critical this step is for subsequent banding. Often the preparative technique employed is either inadequately described or omitted altogether in literature reports. "Standard techniques" include direct air drying at room temperature; blowing on the slide with or without heat; heating on a hot plate, over a Bunsen burner, with a hair dryer, or in an incubator; and ignition. It is often not clear whether "flame" drying was ignition or heating over a flame or both. Slides are used dry or wet, at room or refrigerator temperature, without regard to possible significance.

Unless trypsin is employed, heating at 60–70°C at some stage of processing is essential for the production of G banding and heating at 87°C is critical for R banding. (Heat is not necessary for Q banding but procedures which enhance G banding generally also enhance Q banding, hence, the discussion is also pertinent to Q banding.) Therefore, any procedure which does not control heat satisfactorily, produces highly variable banding. Ignition drying, for example, produces highly variable results with Q and G banding. None of the spreading procedures employing heat are entirely satisfactory, but we have found that as long as overheating (which produces refractile, irregularly staining chromosomes) is avoided, the overall chances of producing good banding are enhanced by heat drying. (See below for details.)

6. PREPARATIVE TECHNIQUE ROUTINELY USED

Unless one component of the preparative technique was under study, a standard preparative technique was used throughout our evaluation of the new techniques. Because certain features are unusual, and generally good Q, G, C, and R banding has been produced with it, the procedure will be described in detail. It should be pointed out, however, that in spite of our efforts at standardization, some differences are observed in the quality of results obtained by two people in the same laboratory following the same protocol.

The procedure we use is as follows: Lymphocytes are grown for 64–66 hours in GIBCO McCoys 5A medium with 15% fetal calf and phytohemagglutinin. A final concentration of Colcemid of 0.1 μg/ml of media is used for 2 hours. Hypotonic treatment is carried out for a total of 16–18 minutes (including 6 minutes centrifugation) with room temperature 0.075 M KCl. Following two changes in fresh 3:1 methyl alcohol-acetic acid (total 25 minutes) and careful resuspension of the cells, the cell suspension is dropped onto a glass slide held at a 45° angle on which there is an even film of cold water. This is maintained by storing the slides (after cleaning with 7X) in a staining dish filled with cold water. Spreading of metaphases is carried out first by blowing vigorously at the cell suspension at a right angle to the slide and then by placing the slide directly on a hot plate for 1½–2 minutes. The hot plate should be set to maintain a beaker of water at 65°C. The slide should be hot to touch but not so hot as to burn. Using a thermister, the hot plate surface temperature at a constant setting was found to vary from time to time and place to place (from 70 to 120°C). This method of "heat drying" is critical for obtaining banding with the Giemsa 9 technique and enhances banding with other methods. The combined use of wet slides and blowing provides optimal spreading, with minimal overlaps, i.e., maximal numbers of usable metaphases. The subsequent sequence of procedures is shown in Fig. 5.

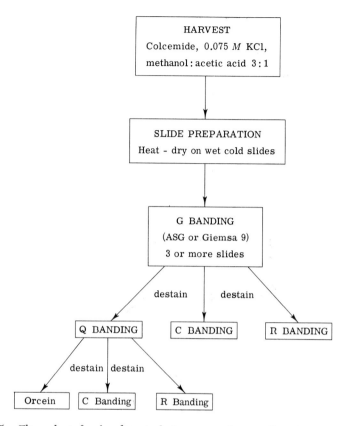

FIG. 5. Flow chart for banding techniques now in use. Trypsin treatment can be tried on slides which do not show banding with other banding techniques.

III. Principal Methods:
Original and Recommended Procedures

A. Conventional Staining

The combined use of a conventional stain and a banding technique has a number of advantages. Delineation of centromeres and ends of arms is often unclear if the banding is faint at either of these critical regions. Moreover, definition of many small conventional details such as the short arms, secondary constrictions, and satellites of human acrocentric chromosomes are often inconsistent and poorly defined with the new techniques. Occasionally, as shown in Fig. 6, the size of an arm may vary

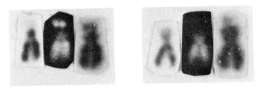

21

FIG. 6. Triple karyotype of chromosome number 21 from the same individual sequentially stained with conventional Giemsa, quinacrine dihydrochloride, and orcein. Certain chromosomal features are more easily detected by some stains than by others. The chromosome on the left has prominent satellites visible with quinacrine dihydrochloride and orcein stains. The chromosome on the right has a long short arm visible only with orcein staining. A small portion of the distal end of the short arm does not fluoresce with quinacrine dihydrochloride.

significantly with different stains. Orcein is significantly superior to Giemsa in respect to defining morphological details. Since we have not been able to develop a satisfactory destaining procedure for orcein, which permits subsequent Q banding, it can only be employed as the last step in the series of stains. The ease with which Giemsa can be destained, however, permits a variety of sequences to be followed. The type and sequence of staining depend upon the question being asked. If one of the many other available stains is to be used, we would suggest a careful comparison with the stains discussed above, in order to define its relative strength and weakness in demonstrating details of morphology rather than assuming comparability.

B. Q Banding

1. METHOD OF CASPERSSON, LOMAKKA, AND ZECH (1971a) FOR QUINACRINE MUSTARD DIHYDROCHLORIDE (QM)

Following treatment of a peripheral blood culture with Colcemid (final concentration 0.04 μg/ml) and hypotonic treatment with 0.95% Na citrate for 20 minutes, chromosomes were prepared by air drying. QM treatment was performed as follows: The slides were transferred from absolute ethanol through an alcohol series, to buffer (MacIlvaine's disodium phosphate/citric acid buffer, pH 7.0), and into the staining solution. An aqueous solution of QM was added to the buffer to give a final concentration of 50 μg/ml of the fluorochrome in the staining solution. Slides were stained for 20 minutes at 20°C, washed three times in buffer, and sealed with a cover slip in buffer.

2. METHOD OF LIN, UCHIDA, AND BYRNES (1971) FOR QUINACRINE DIHYDROCHLORIDE

Air-dried slides were stained in 0.5% aqueous quinacrine dihydrochloride solution at pH 4.5 for 15 minutes. Slides were washed in distilled water at pH 4.5 for a total of 10 minutes, air dried, and mounted in 0.1 M phosphate buffer or distilled water at pH 4.5.

3. COMMENTS AND SUGGESTIONS

As with all of the techniques, good banding resolution will only be obtained in generally excellent cultures and in cells which are relatively early in metaphase. Short, stubby, or poorly defined chromosomes will yield only a minimum of Q banding. Of all banding techniques, Q banding appears to be the least sensitive to slide preparation and heating but there is a direct correlation between good Giemsa banding and good Q banding. Therefore, the heat-drying procedure designed to produce G banding may enhance Q banding. Both quinacrine dihydrochloride and QM have yielded excellent results in our experience as have both epi illumination and transmitted dark field illumination. Poor resolution of the bands may result from using hypotonic Hank's or distilled water for hypotonic treatment (see Section II,B,3), a generally poor culture, or inadequate fixation. There is also variation between batches of quinacrine mustard in the clarity of banding. An ethanol rinse before staining prevents swelling of the chromosomes and results in better resolution of bands. The layer of buffer should be made as thin as possible by blotting the cover slip gently to prevent light scattering. The pH of the buffer may vary from 4.5 to 7.0 without adverse effect. QM can be stored for 2–4 weeks in a refrigerator without deterioration, and quinacrine dihydrochloride for months. Slides can subsequently be restained by other procedures (see Section IV,D: Destaining). It is not clear, however, whether QM or quinacrine dihydrochloride is in fact removed, but restaining with other stains after the destaining procedure is effective. An alcohol rinse is helpful in removing debris present after QM or quinacrine dihydrochloride staining.

C. Giemsa Banding Techniques

1. ASG METHOD OF SUMNER, EVANS, AND BUCKLAND (1971)

The essential step is an incubation in 2XSSC for 1 hour at 60°C.
Following hypotonic treatment with 0.075 M KCl, lymphocytes were fixed in 3:1 methanol–acetic acid, air dried and incubated for 1 hour at

60°C in 2XSSC (2XSSC = NaCl 0.3 M and 0.03 M Na citrate), rinsed briefly with deionized water, stained with Giemsa (Gurr's R66). Stain was prepared as a 1:50 dilution with Gurr's buffer at pH 6.8. Staining time was 1½ hours. After staining, slides were rinsed in water, blotted dry, soaked in xylene, and mounted.

2. GIEMSA 9 METHOD OF PATIL, MERRICK, AND LUBS (1971)

The treatment consists of staining with Giemsa at pH 9.0 after heat drying.

Cultures were treated with 0.075 M KCl for 16 minutes, fixed in 3:1 methanol–acetic acid with two changes, suspended, and dropped on a wet slide which was then placed on a hot plate (set to maintain a beaker of water at 60–65°C) for 1½ to 2 minutes (see Section II,B,6 for details). Slides were stained with Giemsa adjusted to exactly pH 9.0 in 0.003 M sodium phosphate buffer. Since the pH may change rapidly when acid slides are introduced into the stain, only a few slides should be stained at one time. Ignition drying produced poor results.

3. METHOD OF SCHNEDL (1971a,b)

Both brief sodium hydroxide treatment and incubation in buffer are employed.

Fibroblasts and peripheral cultures were treated with Colcemid (3 hours), 1:3 hypotonic Gey's solution for 10 minutes at 37°C, centrifuged for 8 minutes, fixed in two changes of 3:1 methanol–acetic acid, suspended, and air dried by placing drops on a clean slide which had been dipped in ice cold, distilled water. Slides were placed on a horizontal board and allowed to air dry. The critical treatment was 0.07 N NaOH at 20°C for 90 seconds, followed by washing in 70% ethanol, 95% ethanol, absolute ethanol, and drying. Slides were incubated for 24 hours in Sorenson buffer (pH 6.8) at 59°C and stained with buffered Giemsa solution at pH 7.0 for 20 minutes.

4. TRYPSIN TECHNIQUES

a. Method of Seabright (1971). A brief period of trypsin treatment is used prior to staining.

Following hypotonic KCl treatment and fixation, slides were air dried. Flame drying was found to affect the quality of the banding adversely. Slides were flooded with 0.25% BACTO trypsin (DIFCO) solution for 10–15 seconds and rinsed twice in isotonic saline. Preparations were examined by phase microscopy to check for the presence of swelling and banding of chromosomes. Longer treatment was employed if these were not observed. Slides were immediately stained with Leishman's stain diluted 1:4 with buffer (pH 6.8), treated for 5 minutes, washed in buffer, dried, rinsed in xylol, and mounted.

b. *Method of Wang and Fedoroff (1972; Wang et al., 1972)*. This independently reported technique has been used with considerable success by many laboratories. It differs principally in using a lower concentration for a longer time.

Colcemid (0.06 μg/ml) for ½–2 hours and 0.075 M KCl for 10–15 minutes were employed. Flame drying was used successfully. Slides were treated with 0.025–0.05% trypsin (DIFCO or GIBCO) in calcium- and magnesium-free balanced salt solution or trypsin-Versene (1 part 0.025–0.05% trypsin and 1 part 0.02% EDTA at pH 7.0) for 10–15 minutes at 25–30°C. Slides were rinsed in two changes each of 70% and 100% ethanol, air dried, and stained for 1–2 minutes with Giemsa, rinsed in distilled water, air dried, and mounted. Giemsa staining was initially used but better results were later reported (Wang *et al.*, 1972) with Wright's solution or a combination of Wright's and Giemsa.

5. COMMENTS ON G BANDING

A number of other similar techniques have been reported (Drets and Shaw, 1971; Ridler, 1971; Bhasin and Foerster, 1972; Chaudhuri *et al.*, 1971; Crossen, 1972) but not discussed here in detail because they represent only minor changes from the spectrum described above. Various enzymes such as α-chymotrypsin (Finaz and de Grouchy, 1971), Pronase (Dutrillaux *et al.*, 1971), and Pancreatin (Müller and Rosenkranz, 1972) have been used for the production of G bands. Methods involving pretreatments with urea (Shiroishi and Yosida, 1972) and balanced salt solution with or without Mg^{++} and Ca^{++} (Dev *et al.*, 1972) have also been reported. With adequate slide preparation both orcein and conventional Giemsa staining may result in some G banding, however. Hence some caution is in order in interpreting band-inducing treatments.

We have not evaluated many of these procedures because they have been reported quite recently or represent only trivial changes from previously reported techniques. Moreover, judging from the published photographs in these reports none appear superior to the methods described in detail. A detailed study of the effects of pretreatment with a wide variety of solutions including inorganic salts, buffers, acid and alkali, and detergents prior to Giemsa staining of Chinese hamster cells has recently been published by Kato and Moriwaki (1972). One of their principal conclusions was that an alkaline pH was more important than the chemical composition of the solution and, in most instances, Na^+ was also required. In a few instances, high concentrations of salts were ineffective although low concentrations of the same salt were potent band inducers.

This was attributed to a salting-out effect which rendered chromosomal protein insoluble. They also found that flame drying inhibited the production of G banding.

Each of the various methodologies reported in detail in this review has significant advantages. Treatment with trypsin, perhaps, is more versatile since new and old slides can be used in obtaining Giemsa banding and this has been reproduced easily in many laboratories. The concentration and duration of trypsin treatment must be adjusted for each lot and source of trypsin. We prefer a lower concentration and longer time, which permits easier control of the desired degree of treatment. Too little treatment produces poor banding; too much treatment produces chromosome destruction or a combination of G and C banding (Merrick *et al.*, 1972, and Fig. 7a). Conflicting reports of the effect of slide preparation on subsequent trypsin banding have been presented above; in our experience ignition drying has not produced consistently good results. Either heat or air drying is recommended. Kato and Moriwaki (1972) suggested that the long SCC incubation in the Drets and Shaw technique (1971) was necessary to overcome the effects of flame drying. Trypsin can also be used with some success on slides where Giemsa banding was not initially obtained with other techniques.

The ASG technique has also found wide acceptance but does not produce banding in all cultures or all metaphases. Longer periods of SSC incubation may enhance banding. Lomholt and Mohr (1971) have reported a similar technique in which heating was employed at 69–70°C for 3 hours. We have not evaluated this technique. This temperature was chosen because of the knowledge that AT-rich regions uncoil at 69°C and GC-rich regions uncoil at 110°C (Marmur and Doty, 1959). With all Giemsa techniques it is essential to adapt the staining conditions to the desired results (see Section IV on trouble shooting).

The chief advantage of the Giemsa 9 technique lies in its simplicity and minimal destructive effect on chromosomes. Comings *et al.* (1972) found only a 0.7% loss of DNA. No extra work is involved except the adjustment of the pH of the Giemsa stain to 9.0, provided slides have been prepared by the heat-drying technique. Most laboratories, however, have had difficulty in reproducing banding consistently with this technique. This may be due in part to the lack of emphasis, in the original report, on the necessity of using heat drying of metaphases or to difficulty in exactly reproducing the same heat-drying conditions. The technique of Schnedl (1971a,b) is effective but destructive, as shown by chromosomal appearance and biochemical studies (Comings *et al.*, 1972). The trypsin technique is more reliable, simpler, and less destructive, provided the incubation time is short.

For virtually all types of cytogenetic work, diagnostic or experimental, there is every reason to use a Giemsa banding technique routinely as a first step. Little extra work is involved and much information may be gained. Cells are found more quickly and easily than with quinacrine fluorescence, and the same cells can be restudied with fluorescence and C banding with minimal expenditure of time and effort.

Mounting of Giemsa banded slides is not necessary and the removal of mounting agents with acetone or other solvents often has an adverse effect on chromosome morphology. Therefore, mounting should be avoided if a second banding technique is to be used. Staining time with Giemsa is proportional to the molarity of buffer in the staining solutions (Patil et al., 1971). We suggest the following staining solution for conventional Giemsa staining and for the ASG technique or trypsin technique: 1 ml Giemsa, 1 ml pH 6.8 MacIlvaine's buffer, 48 ml H_2O. Staining time is between 3 and 5 minutes. The long staining time originally employed in the ASG technique was likely due to a high concentration of buffer. Many authors have not described the details of their Giemsa staining and it is, therefore, difficult to know whether longer staining times in different laboratories represent a difference in staining solutions or relative degrees of chromosome destruction, since longer staining is required if chromatin material has been lost.

D. C Banding Techniques

1. METHOD OF ARRIGHI AND HSU (1971)

This technique was directly derived from the Pardue and Gall hybridization experiments. Subsequently reported C banding techniques generally represent simplifications of this basic methodology by omitting one or more steps.

The original procedure was as follows: Colcemid, hypotonic treatment (several were used), a fixation in 50% acetic acid or Carnoy's, and flame drying were used for lymphocyte and fibroblast cultures. (For squash preparation, slides were subbed by coating alcohol-cleaned slides with a solution of 0.1% gelatin and 0.01% chrome alum. Cells were squashed without staining and the coverslip removed by the dry ice method). Slides were treated as follows: 0.2 N HCl at room temperature for 30 minutes, rinsed in water, pancreatic RNase at 37°C in a moist chamber for 60 minutes, rinsed twice in 2XSSC followed by 70% and 95% ethanol, and air dried. Slides were then treated with 0.07 N NaOH for 2 minutes at room temperature, rinsed in 70% ethanol and several changes of 95% ethanol, and subsequently incubated overnight in 2XSSC at 65°C.

Finally, slides were rinsed in 70% ethanol and 95% ethanol and stained with Giemsa for 15–30 minutes (pH 6.8).

An essentially identical technique was reported concurrently by Chen and Ruddle (1971).

2. METHOD OF CRAIG-HOLMES AND SHAW (1971)

The principal differences are omission of RNase treatment and a shorter incubation with HCl and NaOH.

The essential treatment, following a 3:1 methyl alcohol fixation and flame drying, consisted of 15 minutes 0.2 HCl, 1 minute 0.07 N NaOH in 0.112 M NaCl, pH 12.0 (Craig-Holmes *et al.*, 1972), and incubation in 2XSSC for 24 hours. Because of the faint staining, cells must be photographed with phase microscopy but very clear demarcation of C bands is obtained. Incubation in 2XSSC was carried out by placing each slide horizontally in a petri dish containing 2XSSC, but elevated above the level of 2XSSC. A layer of 2XSSC and a cover slip were placed on the slide.

A similar technique was reported by Gagné *et al.* (1971) which differed in using a higher concentration of NaOH and SSC in conjunction with a shorter NaOH treatment time.

3. METHOD OF McKENZIE AND LUBS (1972)

An additional step, NaOH treatment, is omitted in this technique. Cells must be heat dried.

Cells were heat dried following hypotonic 0.075 M KCl treatment and fixation in 3:1 methyl alcohol–acetic acid. Ignition drying did not result in satisfactory results. Treatment consisted of 0.2 N HCl for 15 minutes, and 16–24 hours of incubation in 2XSSC in a coplin jar. It was essential to maintain the pH close to 7.0 during this incubation (see below). Photomicroscopy can be carried out either by bright field or phase microscopy.

4. GENERAL COMMENTS ON C BANDING

Treatment with NaOH alone or with NaOH and SSC is very destructive (Comings *et al.*, 1972). HCl and SSC incubation are significantly less so. More destructive treatment is required when lower slide temperatures are employed in slide preparation. If flame drying or air drying, which involve relatively low slide temperatures, is employed, the sequence of HCl, NaOH, and SSC treatment is recommended. If heat drying is employed only HCl and SSC incubation are necessary (McKenzie and Lubs, 1972). NaOH treatment must be monitored carefully since overtreatment produces swelling and ghostlike chromosomes. During an SSC incubation of

16–24 hours in a coplin jar the pH becomes progressively more alkaline, rising from 7.0 to 7.5 when one slide is incubated in 50 ml of 2XSSC. When four or more slides are incubated in 50 ml of SSC the pH may increase to 8.5. We recommend one slide per 50-ml coplin jar and transferring slides to fresh 2XSSC at pH 7.0 for the last 4–6 hours of incubation. McKenzie and Lubs (1972) found that if the pH was adjusted to 6.0 during the overnight incubation, G banding was obtained; with pH 7.0, good C banding was obtained; with pH 8.0, C banding was obtained but the chromosomes were ghostlike and difficult to stain. Presumably destruction was increased. This effect has been little appreciated and probably is responsible for some of the variable results with SSC incubation.

The quality of C band preparations is somewhat impaired by previous observation of quinacrine fluorescence. Only those cells previously photographed for quinacrine fluorescence have diminished C band clarity. Adjacent cells on the same slide (presumably only minimally exposed to UV wavelengths) have excellent C bands. G banding techniques have no effect on subsequent C banding results. Therefore, G band, rather than Q band, homolog identification should precede most C band investigations.

E.   R Banding

1.   Method of Dutrillaux and Lejeune (1971)

The essence of this method is a short incubation at 87°C.

A complex hypotonic solution (15 volumes of horse serum diluted 1:6, 1 volume of 3.39% $MgCl_2$ in water, and 2.5 International Units of hyaluronidase), fixation in Carnoy's, and spreading on cold, wet slides with air drying were employed. Slides were then placed in a phosphate buffer at pH 6.5 for 10 minutes at 86–87°C, rinsed in tap water, and stained with Giemsa (Giemsa commercial solution 4%, phosphate buffer 4%, pH 6.7, double distilled water 92%; Dutrillaux, 1972). R banding was not obtained if a longer heating time was employed.

2.   Comments and Recommendations

This technique is particularly important when used in conjunction with G or Q banding techniques, since it provides a much clearer definition of the ends of many of the chromosomes arms. For the same reason, it is especially useful in defining the exact location of translocations which occur near the ends of arms.

Dutrillaux and LeJeune (1971) stated that banding cannot occur if the chromosomes have been previously stained. In our experience, however, Giemsa and quinacrine staining can both be employed prior to R banding. The destaining procedure for these stains and the preparation of triple karyotypes is described below. We have not obtained as good definition of R banding with Giemsa stain as with other banding, however. The dual karyotype in Fig. 4 represents the best combination we have been able to evolve. Initially, slides were stained with quinacrine, destained as described, and stained with acridine orange. Definition of the ends of arms and the bands with acridine orange (AO) is excellent. See below (acridine orange) for details of staining. It is of particular importance in respect to the mechanism of R banding that the positive R regions show a green fluorescence with acridine orange and the negative R regions are orange-red (see Sections III,F, and V).

The incubation solution should be between 85° and 87°C at the time the slides are immersed. If incubation is begun with the buffer at room temperature, and then placed in the incubator for the described period of time, G banding will be obtained. No banding or G banding generally indicates too short an incubation or too low a temperature, and C banding or ghostlike chromosomes results from too long an incubation or too high a temperature. We have obtained R banding after 4 minutes incubation with heat-dried slides.

F. Acridine Orange (AO) Staining and Evaluation of Strandedness

1. METHOD OF DE LA CHAPELLE, SCHRODER, AND SELANDER (1971)

Using *Microtus agrestis* fibroblasts and bone marrow cells and human lymphocytes, cells were treated with hypotonic Na citrate or KCl, fixed and stained with acridine orange in a 0.5% aqueous solution for 20 minutes and sealed in MacIlvaine's buffer at pH 7.0. The absorption maximum of AO is at 500 nm and the emission maximum at 532 nm. Barrier filter 50, which transmits above 500 nm is satisfactory.

2. COMMENTS AND SUGGESTIONS

This useful technique was reviewed extensively by Riegler (1966), and makes use of the green fluorescence obtained from double-stranded DNA with AO and red fluorescence obtained from single-stranded DNA. It has been particularly helpful in studying the mechanisms of banding and in analyzing individual steps in the many procedures described above. We have also found it useful for the demonstration of both R and C banding.

Mounting the slides in buffer at pH 5.0 results in better differentiation between green and red fluorescence than at pH 7.0. Combinations of colors between green and red may be obtained, presumably indicating varying proportions of strandedness. Stockert and Lisanti (1972) have demonstrated that renaturation occurs in a matter of seconds in repetitive DNA and in a matter of minutes in nonrepetitive DNA. Formaldehyde binds to single-stranded DNA and prevents renaturation. Therefore, it is critical to add formaldehyde to the incubating solutions or at the termination of incubation if strandedness is to be determined precisely without the complicating effects of cooling.

G.   Dual and Triple Karyotypes of the Same Cell

As discussed in Section III, sequential study of the same cell is often helpful. A clearer definition of many features of chromosome morphology can be obtained when two or more techniques are employed. Several combinations have proven particularly helpful: conventional Giemsa staining and Q banding or Q banding followed by orcein, G banding followed by C banding, G and R banding, or Q and R banding (with AO). The sequence of G, Q, and C can be carried out, but the chances of loosing a cell or having imperfect detail in the banding with one technique are significant. Therefore, it is a more efficient routine to use two or more slides, each having a different dual sequence, beginning with G banding (Fig. 6). G or Q banding should always precede C banding, since C banding alone does not permit the distinction between many similar chromosomes.

IV.   Special Procedures and Troubleshooting

Many of the procedures discussed here are common to several banding procedures and have been referred to previously. Even though certain of the comments are repetitive, a review of all the general effects of certain treatments should prove helpful in adjusting methods to specific needs.

A.   Buffers and Heat

The effect produced by incubation with SSC or other salt solutions depends on the time, temperature, and pH employed. Longer incubation in SSC produces increasing fuzziness of bands and chromosomal mor-

phology. Although the concentration of SSC has been increased to 6XSSC
we have not observed that the concentration was important. Kernell and
Ringertz (1972) have clearly demonstrated the increasing rapidity and
completeness of the denaturation of DNA in solution as temperatures are
increased from 65 to 100°C. Renaturation occurs rapidly upon cooling,
however, unless formaldehyde is added and the primary effect of heat
in producing banding is more subtle and apparently involves a DNA-
protein interaction (see Section V). The means by which incubation at
87°C produces R banding is obscure, but likely involves a different effect
on the same mechanism.

B.   pH, Acid, and Alkali Treatment

Treatment with acid solution alone produces no banding. Staining
with Giemsa adjusted to pH 9.0 (provided cells have been heated during
preparation) produces excellent G banding; and pH greater than 10 produces chromosome destruction. NaOH also causes complete denaturation
of DNA as shown by AO studies (de la Chapelle *et al.*, 1971; Comings
et al., 1972). More significantly, the combination of NaOH treatment and
SSC incubation results in considerable loss of DNA and protein as shown
by the biochemical studies of Comings *et al.* (1972). The interrelation-
ships of treatment at varying pH's and the mechanisms of banding are
discussed in Section V.

C.   Trypsin

The work of Comings *et al.* (1972) showed little loss of DNA and
protein during a brief period of trypsin treatment. Little difference in
banding is apparent when ASG and a short trypsin treatment are compared (Fig. 7a). With longer treatment, however, considerable destruction does occur. Both our observations and those of Comings suggest that
this treatment must be carefully controlled or a combination of G and C
banding will be obtained; the result is similar, therefore, to alkali treatment. Utilizing diluted trypsin (0.025%), in order to control the treatment
more easily, we have demonstrated the sequential loss of G banding and
appearance of C banding (Figs. 7a and b). In this work, even a brief
period of trypsin treatment produced a relative increase in density of
staining in certain C regions (Fig. 7b). Possible evidence of chromatin
loss was demonstrated by an increased staining time (from 4 to 10 minutes) during the sequential treatment.

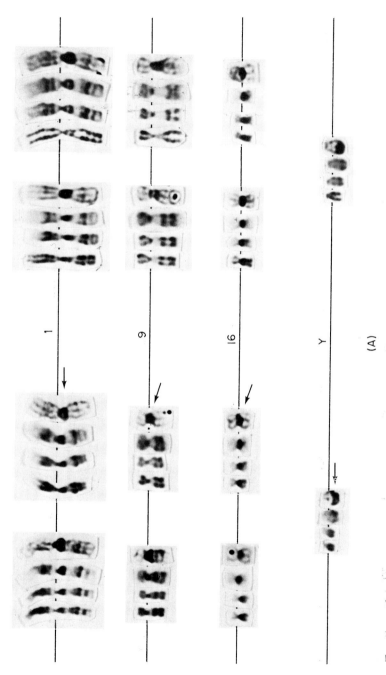

Fig. 7a. Chromosomes 1, 9, 16, and Y from two human metaphase cells showing loss of G and appearance of C banding during sequential trypsin treatment. From left to right, for each chromosome, banding after Giemsa 9 staining, and totals of 4, 12, and 24 minutes of 0.025% trypsin treatment. Cells were destained between trypsin treatments. In 1 and 16, the secondary constriction (h) regions and adjacent centromeres are stained darkly and the area stained does not change during the progression from G to C banding. In 9, the h region becomes darkly stained and the C band includes both the h region and the darkly stained centromere. In the Y, the distal two-thirds of the long arm becomes darkly stained after 24 minutes trypsin, but gradual darkening is evident.

(A)

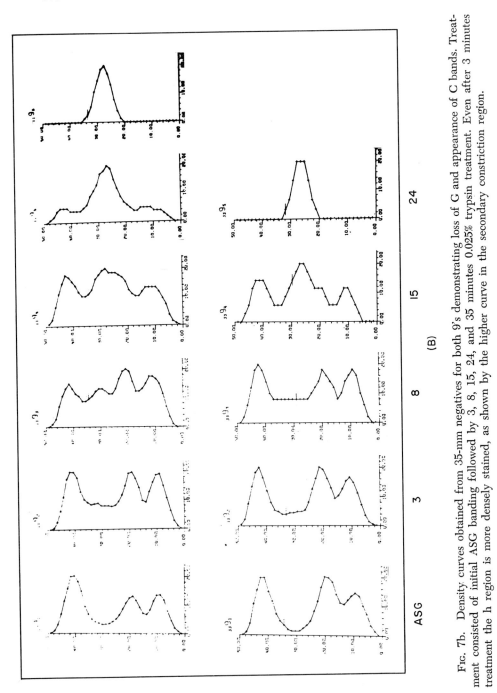

FIG. 7b. Density curves obtained from 35-mm negatives for both 9's demonstrating loss of G and appearance of C bands. Treatment consisted of initial ASG banding followed by 3, 8, 15, 24, and 35 minutes 0.025% trypsin treatment. Even after 3 minutes treatment the h region is more densely stained, as shown by the higher curve in the secondary constriction region.

D. Destaining

Giemsa stained preparations can be destained easily by the following procedure: xylene (2 minutes), equal parts xylene–100% ethanol (1 minute), 100% ethanol (2 minutes), 95% ethanol (2 minutes), and 70% ethanol with 1% HCl (½ minute), a final 70% ethanol rinse (2 minutes), and air drying. The HCl is not necessary but enhances destaining. The short treatment time does not affect banding. This destaining sequence has been carried out as many as six times without change to Giemsa banding, staining time, or morphology. (The first three steps are used only if slides are mounted or there is oil on the slides.) Destaining after Giemsa banding permits subsequent Q banding, C banding, or orcein staining. Although destaining after Q banding is apparently not complete, satisfactory G, C, and R banding can usually be obtained. To date, we have been unable to derive a means of destaining orcein (see below, Section IV,F) which will permit subsequent banding.

E. Microscopy and Photography

1. G Banding

With Giemsa staining bright-field optics should be used. Giemsa banding can best be seen with a combination of light staining and a green filter. Kodak high contrast copy film using an ASA of 6.3–12.5 produces the contrast needed for G banding. A lower ASA setting, i.e., longer exposure time, should be used for Giemsa banding than for conventional Giemsa in order to retain both the weakly and darkly staining bands. Printing should be done on a soft paper (#1 or #2) to retain the full range of banding.

2. Q Banding and Acridine Orange (AO) Fluorescence

We use a Zeiss photomicroscope with an epi illumination or transmitted dark-field illumination, excitation filters BG12 and BG38, and a 50–53 barrier filter. Kodak Tri-X, plus X, and Panatomic X films have been used successfully. Tri-X is a faster, but extremely grainy film. Exposure times may vary from 11 seconds with Tri-X to 2 minutes with Panatomic X. Because Q banding fades rapidly under UV light exposure, it is recommended that photographs of each cell be taken prior to more careful examination under the microscope. Printing should be done on a hard paper such as #4. Good color reproduction can be obtained with high speed Ektachrome, which is developed as for an ASA rating of 400. The same optical and photographic systems work well for acridine orange and quinacrine fluorescence.

3. C BANDING

Generally the same system as for G banding should be used but a slightly higher ASA setting may be needed to ensure resolution of the non-C regions. Printing should be done on a hard paper to emphasize the contrast between C and non-C regions.

4. R BANDING

Phase photography gives the best results with Giemsa R banding, however, the bands are visible using bright-field optics. The procedures for AO photography have been described above.

5. GENERAL COMMENTS

Photography plays a larger role in routine cytogenetics than heretofore. With Q banding, there is not sufficient time to identify and pair each chromosome by eye through the microscope because of fading. As a routine, we have found it easier to print a 5×7 enlargement of each cell to be analyzed, and to write the chromosome number adjacent to each chromosome. This permits almost as good a comparison of banding details as the construction of a single or dual karyotype and saves considerable time (a dual karyotype takes about 1–1½ hours to construct).

For G, Q, and R banding it is important to compare the print with the image from the projected negative, since the dimly staining or fluorescing regions are difficult to print without obscuring the banding, i.e., a slightly underexposed print will not show the ends of many arms, and a slightly overexposed print may show an apparently nonbanded chromosome. The margin for error is much lower than with conventional staining; often it is necessary to make two prints to bring out both types of details.

F. Summary of Troubleshooting

Most of the important points have been mentioned in relation to specific techniques but certain principles warrant reiteration. With Giemsa staining, ghostlike unstained chromosomes indicate either understaining for G banding or overheating or overtreatment during C banding. Uniform staining may result from overstaining, poor slide preparation, or an inadequate banding procedure. If chromosomes are highly refractile they may have been subjected to too much heat. If the clarity of the bands is particularly poor, either overtreatment with hypotonic solution or overtreatment with various agents which destroy chromosome material is likely (see also Q banding comments). Variable results indicate that a sensitive parameter, such as pH or temperature, is not being maintained constantly at some critical point in the methodology. Adjustment of the photographic procedures to the desired end is critical, as discussed in the section on photography.

Comings *et al.* (1972) observed that older slides showed a red, rather than the usual green fluorescence with AO staining and suggested that slow denaturation was occurring with time. We have confirmed this observation and found that denaturation in our preparations begins around 6 months, and clearly affects the ease of banding, morphology, and even the type of banding obtained under various conditions. For example, in some 1-year-old slides kept at room temperature, C banding was obtained under conditions that should have produced G banding. G banding with both ASG and trypsin treatment was either poor or not obtainable, in cells with red AO fluorescence. Q banding is somewhat more reliable in older slides but the quality tends to deteriorate with age. Thus good Q banding may also be difficult to obtain in older slides.

G. Adaptation of Methods to Other Tissues and Species

Difficulties in obtaining adequate banding in other species are more likely due to technical differences in slide preparation than to biological differences. The earlier discussion on slide preparation should prove helpful in this regard. For example, adequate spreading and clear morphology are more difficult to obtain with fibroblasts, fused cells, and malignant cells than with lymphocytes. Longer hypotonic treatment and additional changes of fixative may prove helpful.

For meiotic cells, sequential analysis with conventional staining, Q banding, and C banding is essential. The first is necessary to delineate chiasmata and other morphological details clearly, the second for identification of the homologs, and the third to clearly establish the centromere location. Caspersson *et al.* (1971b) have reported the use of the first two techniques, and Hulten (1972) has successfully added C banding, with considerable improvement in the reliability of the analysis. The sequence consisted of Q banding, orcein, and C banding by the Arrighi and Hsu technique without intervening destaining.

Both X and Y bodies in interphase cells can be identified in buccal mucosal and other tissues using quinacrine fluorescence (Pearson *et al.*, 1970; Greensher *et al.*, 1971). Similar work has proven to be a particularly effective means of identifying X and Y mosaicism and aneuploidy in amniotic membrane in man (Greensher *et al.*, 1971).

V.   Mechanisms of Banding

Achievement of uniformly excellent banding depends upon knowledgable manipulation of the various treatment variables. Complete understanding of the mechanisms of banding is essential, but not yet available. Enough is known, however, to provide certain guidelines.

Denaturation and renaturation of DNA were an inherent part of the original report by Pardue and Gall and were initially postulated as the basis for G and C banding. Studies with acridine orange have demonstrated uniformly double-stranded DNA throughout the length of G banded chromosomes (de la Chapelle *et al.*, 1971; Stockert and Lisanti, 1972; Comings *et al.*, 1972; Lubs, 1972) and either double- or single-stranded DNA in C bands (Comings *et al.*, 1972; Lubs, 1972) depending on the type of pretreatment. Therefore, differential renaturation is clearly not the mechanism by which these bands are produced, even though both denaturation and renaturation occur during various treatments.

C banding appears to result primarily from greater loss of chromatin material in nonrepetitious regions than in repetitious regions—i.e., the C regions stain more darkly because there is more chromatin. This idea is supported by direct morphological observation of apparent destruction, a concomitant marked decrease in staining with a variety of stains in the non-C regions in man (Lubs, 1972), and most conclusively by the densitometric and biochemical studies of Comings *et al.* (1972), which showed a differential loss of labeled nucleic acids and protein after NaOH and SCC treatment in these regions in Chinese hamster and mouse cells. Because of the relatively lower density ratio of C to non-C regions with Giemsa than Feulgen in untreated chromosomes, and greater ratio in treated chromosomes, Comings *et al.* (1972) postulated that a DNA-protein interaction was also important in C banding.

The ease with which G banding can be produced by trypsin treatment is strong evidence that protein is intimately involved in the banding mechanisms. There is little loss of protein or DNA either during 1-hour SSC incubation, during staining at pH 9, or during a short trypsin treatment (Comings *et al.*, 1972). Therefore, a protein–DNA interaction, which is both sensitive to pH and temperature, very likely is involved. Kato and Moriwaki (1972) have postulated only solubilization or loss of some chromosomal proteins as the likely mechanism. Perhaps heating at 65°C or staining at pH 9 prevents easy access of Giemsa (or causes a decrease in affinity for Giemsa) at the weakly staining G regions by affecting certain classes of proteins. This differential affinity, however, can be easily overcome; increasing the staining time at pH 9 from 3 to 8 minutes, for example, produces uniform staining rather than banding.

Since Q banding is affected by these variables to a much lesser degree and can be obtained without heat or alkaline treatment, it is difficult to devise a simple hypothesis which satisfactorily explains the similarity of G and Q bands. Enhancement of fluorescence by AT-rich DNA and quenching of fluorescence in GC-rich regions have been clearly demonstrated (Ellison and Barr, 1972; Weisblum and de Haseth, 1972), and appears to be the best explanation for Q banding.

The suggestion that positive Q bands are AT rich and negative Q bands are GC-rich has now been confirmed by Miller (1972), who demonstrated that Q banding was obtained with fluorescent anti-adenosine antibody. Moreover, Bram (1971) has shown a difference in the secondary structure of AT and GC rich DNAs. Quite possibly such differences affect the composition of overlying proteins. Given a varying response to heat and alkaline treatment by different classes of proteins, a close interrelationship between G, Q, and R banding mechanisms can then be envisioned. Thus, the protein and DNA base ratio hypotheses are complementary not contradictory.

It has now also been shown by Miller (1972) that R positive regions are GC rich, since R banding was produced with fluorescent anticytosine antibody. Moreover, if acridine orange fluorescence is employed after R band treatment (heating at 87°C) the R positive regions have a green fluorescence and the R negative regions have an orange-red fluorescence (Lubs, 1972). Differential fluorescence is not observed, however, if formaldehyde is added to the R banding solution at the end of treatment. Apparently, permanent denaturation ordinarily occurs only in the AT-rich regions and the GC-rich regions renature following the R banding treatment. These results with metaphase cells are similar to, but not identical to, those obtained by Marmur and Doty (1959) with GC- and AT-rich DNA in solution. Whether R banding with Giemsa is due primarily to the difference in base composition shown by Miller, the differential renaturation that occurs, or a specific change in DNA–protein interaction at the GC- or AT-rich regions is not currently clear.

In summary, because of their marked similarity, related but complex mechanisms are very likely involved in Q, G, and R banding. Q banding can be explained satisfactorily by positive quinacrine fluorescence at AT-rich regions. Conversely, R positive regions appear to be GC-rich. Differential renaturation or a DNA–protein interaction, however, also may play a role in R banding. G banding clearly involves both proteins and DNA and the most likely explanation appears to be that proteins with different responses to temperature, alkali, and trypsin treatment overly AT- and GC-rich regions. C banding is largely due to differential loss of chromatin in non-C regions.

VI. Significance and Applications of the New Techniques

Population studies in man using the new techniques have shown a high degree of polymorphism (National Foundation—March of Dimes, 1972). McKenzie (1972), in a series of 50 newborns, has found an average of six chromosomal polymorphisms per infant using a combination of Q, G, and C banding techniques. Variants were observed in nearly every chromosome. Examples of Q polymorphisms are shown in Fig. 8 and C polymorphisms in Fig. 9.

The use of banding techniques will make possible important contributions to the study of population differences, in evolution and speciation, and in gene localization (by family studies and cell hybridization studies). The new techniques have also provided a ready means of studying certain aspects of chromosome structure. In addition, they have given a greater impetus to development of completely automated techniques both because of resurgence of interest in cytogenetics and because it is now possible to identify each chromosome precisely (Fig. 10). These automated procedures can be adapted to any of the banding techniques in any species, provided excellent slides or photographs are presented to

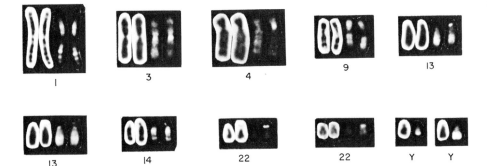

Fig. 8. A sample of Q polymorphisms found in a series of 50 normal newborn babies. Each homolog pair is from a different child. Cells were first stained by the ASG technique, destained, and subsequently stained with quinacrine dihydrochloride. The polymorphic chromosome is on the right in each case. Note the increased length of the h region in chromosomes 1 and 9, the brightly fluorescing region of chromosomes 3, 4, 13, 14, and 22 and the length difference in the two Y chromosomes.

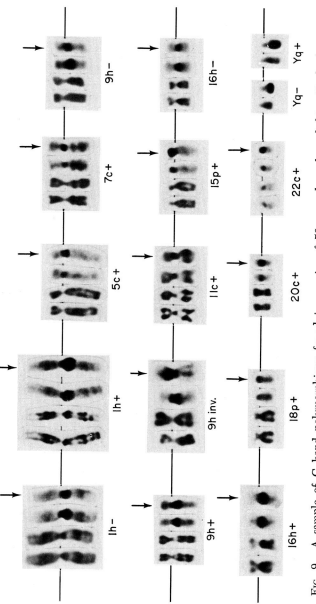

Fig. 9. A sample of C band polymorphisms found in a series of 50 normal newborn babies. Each chromosome shown was first stained by the ASG technique, destained, and subsequently stained with Giemsa after HCl and 2XSSC exposure to produce C bands. The variant chromosome, in each case, is indicated by an arrow. (+ = Increased C band length with or without a concomitant increase in chromosome length; − = decreased C band length with or without a concomitant decrease in chromosome length.)

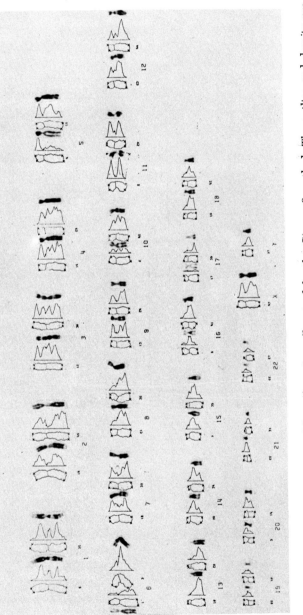

FIG. 10. Printout of an automated karyotype of a cell stained by the Giemsa 9 method. The outline and density profile for each chromosome were displayed automatically. A photograph of each chromosome from which these were derived was placed next to the plots for comparison.

the scanning and analysis system. This approach should provide still further quantification to cytogenetic investigation.

The relative merits of the many techniques have been discussed in Section III. The general orientation of this review has been to provide the basis for investigators to adapt specific techniques to their particular needs, not to recommend specific techniques. It is apparent that a sufficiently wide variety of methods have now been described to meet almost any need and it is likely that more biochemically defined techniques will continue to be developed very rapidly. Demonstration of differences in base ratios in the DNA of the C regions of 1, 9, and 16 in man has already been accomplished by Jones and Corneo (1971) and Saunders et al. (1972), and possible concomitant differences in reaction to vigorous alkaline treatment between these regions have been shown by Bobrow et al. (1972). These and related studies should provide still further stimulation to the growth of cytogenetic investigation.

REFERENCES

Arrighi, F. E., and Hsu, T. C. (1971). *Cytogenetics* (*Basel*) **10**, 18.

Bhasin, M. K., and Foerster, W. (1972). *Humangenetik* **14**, 247.

Bobrow, M., Madan, K., and Pearson, P. L. (1972). *Nature* (*London*), *New Biol.* **238**, 122.

Bram, S. (1971). *Nature* (*London*), *New Biol.* **232**, 174.

Caspersson, T., Farber, S., Foley, E., Kydynowski, J., Modest, E., Simonsson, E., Wagh, U., and Zech, L. (1967). *Exp. Cell Res.* **49**, 219.

Caspersson, T., Zech, L., Johansson, C., and Modest, E. (1970). *Chromosoma* **30**, 215.

Caspersson, T., Lomakka, G., and Zech, L. (1971a). *Hereditas* **67**, 89.

Caspersson, T., Hulten, M., Lindsten, J., and Zech, L. (1971b). *Hereditas* **67**, 147.

Chaudhuri, J. P., Vogel, W., Voiculescu, I., and Wolf, U. (1971). *Humangenetik* **14**, 83.

Chen, R. R., and Ruddle, F. H. (1971). *Chromosoma* **34**, 51.

Comings, D. E., Avelino, E., Okada, T. A., and Wyandt, H. E. (1972). *Exp. Cell Res.* (in press).

Craig-Holmes, A. P., and Shaw, M. W. (1971). *Science* **174**, 702.

Craig-Holmes, A. P., Shaw, M. W., and Moore, F. B. (1972) (in press).

Crossen, P. E. (1972). *Clin. Genet.* **3**, 169.

de la Chapelle, A., Schroder, J., and Selander, R. (1971). *Hereditas* **69**, 149.

Dev, V. G., Warburton, D., and Miller, O. J. (1972). *Lancet* **1**, 1285.

Drets, M. E., and Shaw, M. W. (1971). *Proc. Nat. Acad. Sci. U. S.* **68**, 2073.

Dutrillaux, B. (1972). Personal communication.

Dutrillaux, B, and Lejeune, J. (1971). *C. R. Acad. Sci.* **272**, 2638.

Dutrillaux, B., de Grouchy, J., Finaz, C., and Lejeune, J. (1971). *C. R. Acad. Sci.* **273**, 587.

Ellison, J. R., and Barr, H. J. (1972). *Chromosoma* (in press).

Finaz, C., and de Grouchy, J. (1971). *Ann. Genet.* **14**, 309.

Gagné, R., Tanguay, R., and Laberge, C. (1971). *Nature* (*London*), *New Biol.* **232**, 29.

Greensher, A., Gersh, R., Peakman, D., and Robinson, A. (1971). *Lancet* 1, 920.

Hulten, M. (1972). Personal communication.

Jones, K. W., and Corneo, G. (1971). *Nature (London)*, *New Biol.* 233, 268.

Kato, H., and Moriwaki, K. (1972). *Chromosoma* 38, 105.

Kernell, A. M., and Ringertz, N. R. (1972). *Exp. Cell Res.* 72, 240.

Lin, C. C., Uchida, I. A., and Byrnes, E. (1971). *Can. J. Genet. Cytol.* 13, 361.

Lomholt, B., and Mohr, J. (1971). *Nature (London)*, *New Biol.* 234, 109.

Lubs, H. A. (1972). *In* "Nobel Symposium XXIII" (T. Caspersson, ed.). Stockholm, Sweden (in press).

McKenzie, W. H. (1972). Personal communication.

McKenzie, W. H., and Lubs, H. A. (1972). Chromosoma (submitted for publication).

Marmur, J., and Doty, P. (1959). *Nature (London)* 183, 1427.

Merrick, S., Ledley, R. S., and Lubs, H. A. (1972). *Pediat. Res.* (submitted for publication).

Miller, O. J. (1972). *In* "Nobel Symposium XXIII" (T. Caspersson, ed.). Stockholm, Sweden (in press).

Müller, W., and Rosenkranz, W. (1972). *Lancet* 1, 898.

National Foundation—March of Dimes. (1972). "IVth International Conference on Standardization in Human Cytogenetics" (*Basel*). *Cytogenetics* 11, 313.

Pardue, M. L., and Gall, J. G. (1970). *Science* 168, 1356.

Patil, S. R., Lubs, H. A., Summitt, R., Moorhead, P. F., Gerald, P., Cohen, M., and Hecht, F. (1970). Presented at the American Society of Human Genetics Meeting, Indianapolis, Indiana.

Patil, S. R., Merrick, S., and Lubs, H. A. (1971). *Science* 173, 821.

Pearson, P. L., Bobrow, M., and Vosa, C. G. (1970). *Nature (London)* 226, 78.

Ridler, M. A. C. (1971). *Lancet* 2, 354.

Riegler, R. (1966). *Acta Physiol. Scand.* 67, Suppl. 267, 1–121.

Saunders, G. F., Hsu, T. C., Getz, M. J., Simes, E. L., and Arrighi, F. E. (1972). *Nature (London)*, *New Biol.* 236, 244.

Schnedl, W. (1971a). *Nature (London)*, *New Biol.* 233, 93.

Schnedl, W. (1971b). *Chromosoma* 34, 448.

Seabright, M. (1971). *Lancet* 2, 971.

Shiroishi, Y., and Yosida, T. H. (1972). *Chromosoma* 37, 75.

Stockert, J. C., and Lisanti, J. A. (1972). *Chromosoma* 37, 117.

Sumner, A. T., Evans, H. J., and Buckland, R. A. (1971). *Nature (London)*, *New Biol.* 232, 31.

Wang, H. C., and Fedoroff, S. (1972). *Nature (London)*, *New Biol.* 235, 52.

Wang, H. C., Fedoroff, S., and Dicksinson, C. (1972). *Cytobios* (in press).

Weisblum, B., and de Haseth, P. (1972). *Proc. Nat. Acad. Sci. U. S.* 69, 629.

AUTHOR INDEX

Numbers in italics refer to the pages on which the complete references are listed.

SUBJECT INDEX

A

Acanthamoeba, 190
Aceto-orcein, 199
Acholeplasma, 149
Acholeplasma laidlawii, 147, 161
Acholeplasma sp., 147
Acridine orange, 85, 366, 368, 371–374
Acriflavine, 84–85, 94
Actinomycin D, 214
 resistance to, 231
Adenosine, prototrophy, 268
Adenosine triphosphatase, 241–243
Adenosine triphosphate, 151
Adenovirus, 154, 164–167, 169, 172, 176
 bovine, 168
 avian, 168
Adenylate kinase, 298, 303
Aerosolization, 196
Aerosols, 161, 186
African green monkey, 166
Air filter, particulate, 186
Alanine, 6
Alcaligens fecalis, 180
Alcohol dehydrogenase, 303
Alginic acid, 2
α-amanitin, 214
 cells resistant to, 231
 resistance to, 275
Amebae, 188–193
Ameboid motion, 340
American Type Culture Collection, 202
Amethopterin, 44, 214, 233
Amethopterin B, 157
Aminobenzoic acid, 8
α-Aminobutyric acid, 6
Aminopterin, 214, 233–234
p-Aminosalicyclic acid, 309
Amphotericin B, 192
Aneuploidy, 218
Animal Cell Culture Collection, 194
Antibiotics, 162–164, 179, 185–186
Antifolate drugs, 214
Antifungal medium, 181
Antiglobulin, 175
Antiviral serum, 175

1-β-4-Arabinofuranosylcytosine, 214, 246–247, 257
Arboviruses, 174
Arginine, 6, 71, 151–152, 154, 156–157
Arginine dihydrolase, 151
Argon ion laser beam, 341
Ascorbic acid, 7
Asparagine, 6
Aspartic acid, 6
Auxotrophic lines, 211, 214, 255
 cross feeding in, 254
Auxotrophic markers, 212, 217, 227, 250
 drug-resistance, 263
Auxotrophic mutants, 220, 224–225, 247, 250, 276
 isolation of, 246
Auxotrophic mutations, 216, 249
Auxotrophic phenotypes, 250
Auxotrophs, 214, 252
 induction of, 268
 isolation of, 250
 nutritional, 214, 218–219, 249–250
Auxotrophy, 251, 268, 270, 273
 glycine, 252
 nutritional, 267
Avian leukosis viruses, 174, 176
Avian sarcoma, 168
8-Azaguanine, 213, 228, 231–232, 235–236, 238, 244
 cells resistant to, 231–232, 235, 237
 cells sensitive to, 237
 resistance to, 214, 216–217, 226, 230–231, 234, 270, 273, 275
 resistant clonal lines, 233, 234, 237
 resistant mutants, 231, 235, 240, 271
 resistant phenotypes, 271–272
8-Azaguanosine, 231–233
8-Azahypoxanthine, 231–232, 235
Azahypoxanthine resistance, 267

B

Bacillus, 180
Bacterial L forms, 161
Basic fuchsin, 83
Basophils, 335
BHK cell line, 78, 261